BRILL'S STUDIES IN INTELLECTUAL HISTORY

VOLUME 78

John Murdoch in 1977

TEXTS AND CONTEXTS IN ANCIENT AND MEDIEVAL SCIENCE

Studies on the Occasion of John E. Murdoch's Seventieth Birthday

EDITED BY

EDITH SYLLA AND MICHAEL McVAUGH

BRILL
LEIDEN · NEW YORK · KÖLN
1997

This book is printed on acid-free paper.

Library of Congress Cataloging-in-Publication Data

Texts and contexts in ancient and medieval science : studies on the occasion of John E. Murdoch's seventieth birthday / edited by Edith Sylla and Michael McVaugh.
p. cm. — (Brill's studies in intellectual history, ISSN 0920-8607 ; v. 78)
Includes index.
ISBN 9004108238 (alk. paper)
1. Science, Ancient. 2. Science, Medieval. 3. Murdoch, John Emery, 1927– . I. Murdoch, John Emery, 1927– . II. Sylla, Edith Dudley. III. McVaugh, M. R. (Michael Rogers), 1938– . IV. Series.
Q124.95.T49 1997
509.3—dc21 97–18497
CIP

Die Deutsche Bibliothek - CIP-Einheitsaufnahme

Texts and contexts in ancient and medieval science : studies on the occasion of John E. Murdoch's seventieth birthday / ed. by Edith Sylla and Michael McVaugh. – Leiden ; New York ; Köln : Brill, 1997
(Brill's studies in intellectual history ; Vol. 78)
ISBN 90–04–10823–8

ISSN 0920-8607
ISBN 90 04 10823 8

PRINTED IN THE NETHERLANDS

PREFACE

This book had its origins at a session of the 1992 History of Science Society meetings in Washington, D.C., planned on the occasion of John Murdoch's sixty-fifth year, at which talks were given by Michael McVaugh, William R. Newman, and Michael Shank, with Edith Sylla as chair. Of the papers contained here, that by Michael Shank is a revision of the talk he gave in 1992. The other papers were newly written for this volume.

All the authors in the volume, and especially the editors, owe a vote of profuse thanks to our copyeditor, Julie McVaugh, for her expertise, keen eye, balanced judgment, and memory for detail—even when our lesser powers of attention failed us in carrying out her directives consistently. We would also like to thank Keith Jewett for his assistance in producing the camera-ready copy and index—while simultaneously learning how to use a new computer and word-processing system and taking care of a new baby.

We are sad to report that Wilbur Knorr died on 18 March 1997, just as he was completing the proofreading of the camera-ready copy of his paper. We are glad that his paper was in essentially final form at the time of his death, but accept responsibility for any aspects of the paper that may lack final polish. Which of his "forthcoming" articles (referred to as such in the notes) will eventually appear is uncertain at this time.

CONTENTS

PLATES

CONTRIBUTORS

Joan Cadden is Professor of History at the University of California, Davis. She is the author of *Meanings of Sex Difference in the Middle Ages: Medicine, Science, and Culture* (Cambridge, 1993), which received the History of Science Society's Pfizer Prize. She is currently researching Pietro d'Abano's explanations for homosexual desire and the rhetorical strategies of medieval natural philosophers.

Jean De Groot is Assistant Professor in the School of Philosophy at the Catholic University of America in Washington, D.C. Her publications include *Aristotle and Philoponus on Light*, "Form and Succession in Aristotle's *Physics*," and "Teresa of Avila and the Meaning of Mystical Theology." She lives in Arlington, Va., with her husband and three daughters.

Edward Grant is Distinguished Professor Emeritus of History and Philosophy of Science at Indiana University, Bloomington. He is the author of numerous articles and books on medieval science and natural philosophy. His latest books are *Planets, Stars, and Orbs: The Medieval Cosmos: 1200–1687* (Cambridge, 1994) and *The Foundations of Modern Science in the Middle Ages* (Cambridge, 1996).

Wilbur Knorr, at the time of his death, was Professor of History and Philosophy of Science at Stanford University and author of four books and fifty articles on ancient and medieval mathematical sciences.

Michael R. McVaugh is William Smith Wells Professor of History at the University of North Carolina. His research centers on the transmission of Greek and Arabic scientific knowledge to the Latin Middle Ages and on the development of medicine and surgery in the thirteenth and fourteenth centuries.

George Molland, after taking his first degree in mathematics at the University of Cambridge, switched to history and philosophy and gained his Ph.D. in 1967. He has worked extensively on medieval and early modern science and mathematics, and has recently been paying attention to the achievement of Roger Bacon.

William R. Newman received his Ph.D. in the History of Science from Harvard University in 1986. He has published *The "Summa perfectionis" of pseudo-Geber: A Critical Edition, Translation and Study* (Brill, 1991) and *Gehennical Fire: The Lives of George Starkey, An American Alchemist in the Scientific Revolution* (Harvard University Press, 1994).

Katharine Park teaches history at Wellesley College. She has published numerous articles on the history of medicine and the life sciences in late medieval and Renaissance Europe, with an emphasis on Italy. Her most recent book, *Wonders and the Order of Nature*, co-authored with Lorraine Daston, is forthcoming shortly from Zone Books.

Michael H. Shank is Associate Professor and Chair of the History of Science Department at the University of Wisconsin–Madison. His research interests include the history of natural philosophy and astronomy in the late fourteenth and fifteenth centuries, with special attention to developments associated with the University of Vienna.

Edith Sylla is Professor of History at North Carolina State University. Her research interests include fourteenth-century natural philosophy, mathematics, logic, and theology, and the transition from medieval to modern science in the seventeenth century. She is currently working on Jacob Bernoulli, Margaret Cavendish, and Gottfried Wilhelm Leibniz.

Johannes M. M. H. Thijssen holds degrees in philosophy, history, and medieval studies. He is Associate Professor of ancient and medieval philosophy and director of the research program "From artes to science" at the Catholic University of Nijmegen, the Netherlands. He is founder and editor of the journal *Early Science and Medicine*. His current projects include a study of John Buridan's philosophy of nature.

William A. Wallace is a Dominican priest and professor emeritus of philosophy and history at the Catholic University of America. Currently, he teaches at the University of Maryland, College Park. He has published extensively in the history of medieval and early modern science, especially on Galileo and his Latin notebooks.

INTRODUCTION

EDITH D. SYLLA AND MICHAEL R. MCVAUGH

At the end of the thirteenth century Arnald of Villanova unleashed a storm of controversy by prophesying the imminent end of the world and the second coming of Christ. A few years later Henry of Harclay opposed Arnald in a question entitled, "Whether astrologers or any sort of calculators may prove the second coming of Christ."[1] This was no mere academic matter, but one of great public interest. In the same years, Harclay composed scholastic questions on the possible past and future eternity of the world, in the process exploring and developing ideas of infinity and continuity. Why did Harclay direct his attention to the possible future eternity of the world? Was he merely fulfilling his academic responsibility to lecture on Peter Lombard's *Sentences*, where this issue came up? Were his interests actually in the mathematical issues he raised, for instance concerning the properties of infinity? Or was he perhaps trying to quiet a sneaking fear that Arnald might have been right, and that the end of the world was indeed at hand? These possibilities, none of which excludes the others, force us to ask how much of the setting, social and cultural as well as academic and intellectual, must we try to recapture in order to understand the meaning of this and other scholastic texts.

In 1962, the importance of the social context of science was forcibly brought to the attention of historians of science by the publication of Thomas Kuhn's *Structure of Scientific Revolutions*. That year John Murdoch, a 1957 University of Wisconsin Ph.D., was teaching at Princeton University. Under the direction of Marshall Clagett, Murdoch had edited Thomas Bradwardine's *De continuo*, but under the direction of Julius Weinberg and William Hay in the Wisconsin philosophy department—and with the advice of Alexandre Koyré of the École Pratique des Hautes Études—he had accompanied his edition with an extensive philosophical analysis (the critical text of *De continuo* does not begin until page 339!). Murdoch's dissertation, although not infrequently used on microfilm, never has been published, perhaps in part because, shortly after he completed it, ideas about how medieval science ought to be approached came under the influence of

[1] "Utrum astrologi vel quicumque calculatores possint probare secundum adventum Christi," in Franz Pelster, "Die Quaestio Heinrichs von Harclay über die zweite Ankunft Christi und die Erwartung des baldigen Weltendes zu Anfang des XIV. Jahrhunderts," *Archivio Italiano per la Storia della Pietà* 1 (1951): 25–82.

Kuhn's *Structure*. In his dissertation, Murdoch treated *De continuo* "not only as a document of fourteenth-century philosophy of science, but also as one way of speaking, and possibly speaking sensibly, about continuity itself."[2] As he began to think about turning his dissertation into a book, rather than analyzing Bradwardine's *De continuo* from the point of view of modern philosophy and mathematics, in terms of Dedekind or Cantor, he began to look at it in terms of the fourteenth-century "atomists" against whom Bradwardine was arguing—for instance, Henry of Harclay, Walter Chatton, Gerard of Odo, and Nicholas Bonetus. In 1962, however, his philosophical orientation was still strong. One of us remembers vividly a question John posed that very year on graduate prelims: "How might a medieval Hilbert have axiomatized Aristotelian physics?"

In retrospect, that question can be seen, not only as reflecting a philosophical approach, but also as embodying the assumption that Aristotelian natural philosophy consisted of a coherent set of beliefs and assumptions about the natural world, one that medieval scholastics shared—in other words, that Aristotelian physics had some of the properties of a modern scientific paradigm in the Kuhnian sense. But Murdoch's approach was also shaped by the work of Anneliese Maier, which, in the 1960s, was setting the model for how to embed the history of medieval science within the wider context of medieval philosophy and theology. Whereas Pierre Duhem had presented the ideas of the scholastics only in French translation, Anneliese Maier presented a seamless web of German and Latin, somehow enabling readers not perfectly fluent in either German or medieval Latin to begin to comprehend scholastic natural philosophy in its own terms. Whereas Duhem linked medieval scientific ideas to modern ones, Maier saw them more clearly in their own particularity.

After his move from Princeton to Harvard in 1963, Murdoch gave a series of talks directed toward presenting fourteenth-century natural philosophy not as the "before" picture of the Scientific Revolution, but, as Maier had done, in its own terms. Utilizing a distinction first made by Ernest Moody, Murdoch described fourteenth-century natural philosophy as "analytic," as compared to the more "synthetic" approach of the thirteenth century; fleshing out this distinction with concrete examples, he described the "analytical languages" that embodied the fourteenth-century analytic approach, ranging from Thomas Bradwardine's function for describing the relations of forces, resistances, and velocities in motion, to supposition theory, the analysis of first and last instants, maxima and minima, and so forth.

---

[2] John Murdoch, "Geometry and the Continuum in the Fourteenth Century: A Philosophical Analysis of Thomas Bradwardine's *Tractatus de Continuo*" (Ph.D. dissertation, University of Wisconsin, 1957), p. ii.

While Pierre Duhem, as a scientist himself, had approached medieval science with interests shaped by modern science, and while Marshall Clagett had followed him in looking for the medieval background to Galileo, John Murdoch, always with a stronger bond to philosophy than to science narrowly conceived, repeatedly argued for the importance of logic within fourteenth-century science. If mathematics provides the instrument of choice in modern physics, it was often logic that best served the medieval natural philosopher, as is clear, for instance, from Curtis Wilson's work on William Heytesbury, or from the work of Norman and Barbara Kretzmann on Richard Kilvington's sophismata. As work in the history of medieval logic has flourished in the last decades, the quality and interest of that logic has become more and more apparent, and Murdoch has played a leading role in the effort to convince historians of medieval science that they ignore medieval logic at their peril.

As the influence of *The Structure of Scientific Revolutions* spread through the history of science community in the 1960s and early 1970s, pushing more and more scholars to reflect on the intellectual and social forces conditioning the development of scientific ideas, Murdoch organized the First International Colloquium on Philosophy, Science, and Theology in the Middle Ages, held in North Andover, Massachusetts, in September 1973. This colloquium was designed in part to respond to these issues, not, however, in a programmatic way but on the basis of firmly grounded reading of texts. If others were arguing that in general the external social setting penetrates scientific thinking and changes it, the participants in the colloquium were expected to make a close examination of case studies of medieval philosophy, science, and theology, to see what in fact was going on.

Moreover, it was not obvious that there was anything in medieval learning that corresponded to modern normal science in the Kuhnian sense, which would demand that there be, not only a shared paradigm, but also a well-defined community dedicated to working according to the dictates of that paradigm. In his own paper for the colloquium, John Murdoch argued for the "unitary character of late medieval learning," saying that fourteenth-century science *was* philosophy and that shared techniques of analysis created a looser union of medieval philosophy and theology as well. He based this insight into the nature of fourteenth-century natural philosophy on a broad and deep familiarity with the texts of medieval science, from which he gave selected examples.

In a sense, Murdoch's characterization of medieval science was a reiteration of the long-held view that natural philosophy was what Edward Grant has recently called "the mother of all sciences," saying that, aside from mathematical sciences such as astronomy and optics, "almost all the other sciences—physics, chemistry, biology, geology, meteorology, and psychology, and all their subdivisions and branches—emerged as independent disciplines from within the matrix of natural philosophy during the seventeenth to nineteenth centuries."[3] But, post-Kuhn, to point out the unitary character of late medieval science said something more than simply to assert that the other sciences emerged, only during the seventeenth century and later, one-by-one from natural philosophy: if Murdoch's characterization is true, then medieval natural philosophy, not only by its method or content but by its institutional or social structure as well, had features that differentiate it from modern normal science according to Kuhn's model.

If Murdoch's insight into the character of late medieval learning has yet to be fully absorbed and made use of by historians of medieval science, a variety of conferences over the last twenty-five years has helped to make the outlines of the intellectual context of medieval science somewhat clearer, and Murdoch has had an organizing hand in many of them. He founded the informal Medieval Science Colloquia that were held at Harvard in 1969 and 1971, moved to North Carolina in 1973, 1975, and 1981, and later still to Collegeville, Minnesota, in 1984 and 1987. An active member of the Société Internationale pour l'Étude de la Philosophie Médiévale, Murdoch was an organizer for the Louvain-la-neuve Congress in 1982 and president at the Helsinki Congress in 1987. He also organized international conferences on the history and philosophy of science at Jyväskylä, Finland, in 1973, at Pisa in 1978, and at Montreal in 1980. More than once he led summer institutes in the history of medieval science, for instance a Southeastern Institute of Medieval and Renaissance Studies seminar on problems of the eternity of the world in North Carolina in 1975 and one on the edition of medieval texts at Harvard in 1977. And most recently John Murdoch, with William Newman, organized an enthusiastically received summer workshop-conference on late medieval and early modern corpuscular matter theory in St. Andrews, 6–15 August 1996.

In 1973, when the North Andover conference was held, and for the next few years, it still seemed somewhat unrealistic if not unprofitable to hope to go beyond the shared world of medieval scholastics so as to understand the role of the wider, nonacademic context in shaping their ideas—such as the possible role of Arnald of Villanova's apocalypticism in shaping Henry of Harclay's views on the future eternity of the world. While historians of recent science were paying increasing attention to the social and institutional contexts within which science is practiced, historians of ancient and medieval science apparently had much less material to work with if they chose to explore the social context of premodern science—although in some cases historians of medieval medicine had more luck in this regard.

[3] Edward Grant, *The Foundations of Modern Science in the Middle Ages: Their Religious, Institutional, and Intellectual Contexts* (Cambridge: Cambridge University Press, 1996), pp. 192–94.

And yet, as many have been coming to realize, there is more opportunity to see medieval science "in the round"—including not only the remaining texts, but also the setting within which they were produced—than might at first be realized. To give only one example, of which John's remarkable *Album of Science* volume (1984) has made us aware: the images, the pictures that accompany some medieval scientific texts, can often, when studied for themselves, give us a new and different perspective on the meaning of those texts and on their authors' intentions. If, in the end, our knowledge of the context of medieval science most often comes from other texts, both those produced within universities and those whose origin was outside, nevertheless using a broader range of texts or using the same texts more inventively can help to reconstruct context, by which we can understand the original texts better.

The history of medieval science today is at its best a history not only of texts but of texts within their contexts, not only of writings but of the intellectual, institutional, and social settings that called them forth. Its exponents have taken up and extended the program implicit in the title given to the papers resulting from the North Andover conference, *The Cultural Context of Medieval Learning* (1975). In the past, historians who studied medieval scientific texts usually had interests that extended beyond the content of medieval science in its own terms. Frequently, medieval science has been forced to play the role of the ancien régime, the mistaken approach to nature that had to be overthrown by Galileo, Descartes, Newton, and the other heroes of the Scientific Revolution. In other cases, interest in medieval natural philosophy has been the handmaiden to a greater interest in the history of the medieval church—for instance to an interest in Thomism or in Franciscan authors, and so forth. The present volume is meant to suggest the kinds of approaches to medieval science being explored today in which the primary interest is to understand medieval science cum philosophy in its own terms, as an approach to understanding the world that may or may not have similarities to modern alternative approaches.

Taking our cue, then, from John Murdoch's approach to the history of later medieval natural philosophy, we have sought in the papers included in this volume always to begin from and remain grounded in texts, but to make plain how looking at the particular contexts of the ideas we examine may enable us to see in a different light and to understand better what is really going on in the texts. Rather than floating grand new theories, most authors here adopt the role that Guy Beaujouan at the 1973 North Andover conference called that of the "modest worker of history," or, as George Molland calls it here, that of historical "underlaborer." Nevertheless, many authors show what wider implications may be drawn from small details, carefully interpreted. Not a few of us share the sorts of philosophical interests reflected in Murdoch's dissertation—we find the intellectual issues considered by our authors to be of abiding and not only historical interest.

Two of the papers that follow include new editions, a possibility that we offered in inviting contributions. In his paper on "Armengaud Blaise as a Translator of Galen," Michael McVaugh edits Armengaud's "translation" of Galen's "On an awareness of one's own defects and vices" (*De cognitione propriorum defectuum et viciorum*), showing that it was not exactly a translation, since it omitted whole sections and reorganized what remained in order to emphasize the desirability of accepting criticism and admitting one's mistakes. McVaugh asks why Armengaud might have chosen to translate this particular work, which was not medical but moral. In 1289, Armengaud had received his degree in medicine from the local bishop over the violent objections of the Montpellier medical faculty. About 1300 he dedicated his translation of *De cognitione* to Guillaume de Mazet, master of medicine at Montpellier, and he promised to dedicate future translations to other men, "as long as there shall be an awareness of their own vices, otherwise not. For what is given to someone who does not appreciate it is rightly considered lost; and one regrets having given what is seen to have been lost." Historians have tended to assume either that translators do their work altruistically, simply to enable others to read important texts, or for personal reward, for instance the hope of personal advancement or patronage. McVaugh speculates that in this case there may have been a more complicated and personal motivation arising from the relevance of the content of *De cognitione* to Armengaud's personal history of rejection by the Montpellier medical faculty.

The other text edition included here is made by William A. Wallace in his paper, "Domingo de Soto's 'Laws' of Motion: Text and Context." Almost thirty years ago, Wallace pointed out that Domingo de Soto had applied the expression "uniformiter difformis" not to some imagined motion or quality, but to the motion of naturally falling bodies. Here Wallace for the first time edits and translates the relevant texts from Domingo de Soto's commentary on Aristotle's *Physics*, including also Soto's treatment of Aristotle's rules for the relations of forces and resistances and velocities in motion. Soto, Wallace shows, was clearly familiar with and accepted Bradwardine's use of ratios of ratios, while recommending that when Aristotle speaks of "half the resistance," he be interpreted to mean that resistance to which the power has double the ratio it has to the entire resistance, and so forth. Later Jesuit authors at the Collegio Romano and at Coimbra, while referring to Domingo de Soto's teachings, seem to have rejected Bradwardine's and Soto's interpretations of Aristotle's "law." As in previous cases, Wallace here significantly advances our knowledge of sixteenth-century progressive Aristotelianism by making readily available texts not previously utilized by historians of science. And, as in his previous studies of the Collegio Romano, Wallace shows not only what individual authors had to say on a particular physical topic, but also how their works related to each other. Sixteenth-century scholastic authors, like medieval scholastics, did not, in general, claim originality or credit for new ideas: if a particular idea struck them as correct they were likely to say it was what Aristotle intended if the texts could plausibly be so interpreted, rather than claiming it for themselves. This makes it difficult for historians looking for intellectual priority to determine who first had a new idea, but it is also a trait of medieval science that helped to give it its particular characteristics as an intellectual endeavor.

Whereas McVaugh and Wallace edit new texts in this volume, other authors work closely with textual details to reach conclusions that might not be apparent from a less careful reading. Most prominent in this regard is Wilbur Knorr's paper on "The Latin Sources of *Quadrans vetus*: What They Imply for Its Authorship and Date." As he has done in earlier papers, Knorr compares the detailed wording of parallel texts from different works to reach conclusions about the identity or nonidentity of their authors. Here he argues that the author of *Quadrans vetus* was almost certainly not the Robertus Anglicus who wrote a commentary on the *De spera* of Sacrobosco in Montpellier in 1271: whereas *Quadrans vetus* is concise and direct, the commentary is prolix and formalistic; whereas, for instance, the commentary frequently uses gerundives like "dicendum" and "notandum," *Quadrans vetus* does not, and there are many other linguistic differences. On the other hand the commentary on *De spera* and *Quadrans vetus* have closely parallel passages describing methods of finding one's latitude from the equinoctial meridianal altitude of the sun or from the altitude of the celestial pole. By detailed comparisons, Knorr argues that in these latter cases both texts derive ultimately from the still earlier *Artis cuiuslibet consummatio*, but that the commentary on *De spera* has only those parts of *Artis* also found in *Quadrans vetus*, so that the commentary must be later than, and dependent upon, *Quadrans vetus*. This means that *Quadrans vetus* was earlier than 1271—and Knorr has advanced our knowledge by establishing that point.

Plunging, then, into another set of detailed textual comparisons, this time of *Quadrans vetus* to his own working edition of Campanus's *Practica quadrantis*, Knorr argues that *Quadrans vetus* uses Campanus's *Practica quadrantis*, and hence must have been written after it. But then the compositional method of the author of *Quadrans vetus*—most probably to be named "Johannes Anglicus in Montepessulano," as Knorr argues—becomes clear: the work is a compilation drawn systematically from a number of earlier works, as Knorr shows in a table of its contents as compared to earlier works. If it was, then, so derivative a work, why was it so popular, as evidenced by the number of existing manuscripts? Its frequency is to be explained, Knorr argues, by its adoption as a textbook at the University of Paris. An examination of existing manuscripts—of which four examples are shown in plates—shows that many of them were written in the same atelier, probably by the same scribes, and for use at the University of Paris. Again we have an example of a medieval scientific work being valued, not for its originality, but, in this case, for its qualities as a textbook. But then, if *Quadrans vetus* was a popular textbook, perhaps we must revise our ideas about the theoretical and impractical nature of medieval university science, for the work embodies a solid knowledge of how to make an instrument constructed on the basis of more than elementary geometrical and astronomical knowledge, as shown, for instance, in the complex division of the cursor. Not all scholars have accepted previous efforts to reveal the secrets of *Quadrans vetus*: thus Emmanuel Poulle rejected Nan Hahn's previous conclusion, with which Knorr here agrees, that the author of *Quadrans vetus* was most probably named "Johannes," and not "Robertus." It remains to be seen whether readers will accept all of Knorr's conclusions as proved beyond a reasonable doubt, but if one can reach firm conclusions on the basis of minute textual clues, Knorr has gone a long way toward doing it here.

George Molland, in his paper, "Roger Bacon's *De laudibus mathematicae*: A Preliminary Study," takes an approach not unlike Knorr's, but he writes, as his title indicates, with a tone of inquiry, rather than of demonstration. Quoting John Leland, who famously said that it would be easier to collect the leaves of the Sibyl than the names of the books that Bacon wrote, Molland describes his work here as "underlaboring"—the sort of unglamorous spade work that still remains in part to be done to sort out the chaos of Roger Bacon's manuscript legacy. *De laudibus mathematicae* is a work, or part of a work, praising mathematics, which has strong parallels to Bacon's *Opus majus*, Part IV. On the basis of the two known manuscript copies of the work (too long to be edited here), Molland inquires into the differences between *De laudibus mathematicae* and the parallel sections of the *Opus maius*, and speculates about their significance. The sections of *De laudibus* independent of the *Opus maius* have mostly to do with astrology, in which Bacon firmly believed. In both *De laudibus* and the *Opus maius*, Bacon is clear that the stars do not override human free will, but his advocacy of the work of astrology in *De laudibus* is freer and more naturalistic than in the *Opus maius*, where he asserts that astrologers know that God can change all things, and so end their predictions with the expression, "so will it be if God should wish it." On the basis of this difference in nuance, Molland concludes that *De laudibus* was likely intended for a "scientific" recipient—Bacon calls it a letter—and not, as was the *Opus maius*, for the pope. Nevertheless, as he makes clear, the status of astrology in the Middle Ages was not a simple one. We should not assume that a given individual was either for astrology or against it, but should rather expect a more complex negotiation, one in which relations between experts and nonexperts were an issue, as well as more direct concerns about the ways in which the heavenly bodies may or may not influence the terrestrial realm and the human beings therein. As will be seen, this issue arises again in several of the other papers.

Up to this point, we have described the four papers that relate most directly to given texts, indicating what can be learned from their detailed reading. Closely related to this approach, but displaying even more ingenuity, is the paper of Jean De Groot, "Eudoxan Astronomy and Aristotelian Holism in the *Physics*," which uses what can be inferred from reports about Eudoxus's no-longer-extant *On Speeds* to propose a modified interpretation of the nature of Aristotelian physics. It has previously been proposed that Eudoxan geometry had an influence on Aristotle's physics, for instance through ideas of the structure of the continuum. De Groot here suggests that it was not so much Eudoxan geometry, with a focus on topological issues, that influenced Aristotle's physics, but Eudoxan astronomy. As De Groot points out, it was an inevitable feature of Greek astronomy that the speed of revolution of one star or planet was always measured in terms of another, the rotation of the heavens themselves serving as the most reliable available clock. But then one never has absolute measures, only relative ones—the motion of a planet, for instance, may be said to be "slower" because when the heavens have returned to a given position, the planet has yet to complete its circuit. De Groot calls this a situation of "internal measurement." Furthermore, a kind of holism results because the units of measure are based on completed motions, for instance a day, as the time in which the stars return to their starting position, or a year, as the time in which the sun returns to a given star. One motion is thus taken as an "indexical whole," and other motions are related to it. It would, then, be Eudoxus's conception of ratios in astronomy—including ratios of velocities understood as magnitudes linked to direction and destination, not abstracted from them—that influenced Aristotle's thinking about ratios in the *Physics*. In sum, De Groot argues that to understand Aristotle's *Physics* we need to understand the context of questions addressed and answers expected in fourth-century B.C. astronomy, which was at that time the most advanced existing discipline studying motion. By understanding how speed was handled in astronomy we gain insight into what Aristotle expected from the measurement of sublunar motions—not, that is, the abstract measurement of geometrical magnitude, but one that retained more of the properties of physical things.

If Eudoxus's lost *On Speeds* provides a context necessary for understanding Aristotle's measurement of motion in the *Physics*, according to De Groot, it is the hidden theological context that provides the necessary background for understanding Henry of Harclay's ideas of the relations of indivisibles to continua, as Edith Sylla argues in "Thomas Bradwardine's *De continuo* and the Structure of Fourteenth-Century Learning." A long tradition of arguments going back at least to John Philoponus was carried on by medieval authors to argue that the world could not always have existed, because then paradoxes of infinites would result. Like several other authors before him, Harclay rejected these arguments and embraced the conclusion that infinites can be unequal—the infinite revolutions of the moon in an eternal time could, in this view, be both infinite and twelve times as numerous as the infinite revolutions of the sun. But then Harclay also adopted an idea put forth by Robert Grosseteste—proposed to overcome what De Groot calls the necessity of "internal measurement" or the measurement by "indexical wholes," given the apparent lack of absolute standards of measurement—that God now knows all the infinitely numerous indivisibles in continua and can determine, for instance, that a two-foot line is twice as great as a one-foot line on the grounds that it contains twice as many indivisibles; or that two hours are twice as long as one hour because two hours contain twice as many indivisible instants, one infinite multitude being twice as numerous as another.

Propelled by his conviction of God's omniscience, Harclay then went on to argue that the indivisibles in continua must be immediate to each other—if God sees all the points in a line, Harclay reasoned, he must see the point next to the end point, for to assume otherwise contradicts the assumption that he sees all the points. Thus Harclay and other fourteenth-century thinkers, by reasoning from the premise of God's omniscience or omnipotence, were led to contradict well-established Aristotelian and Euclidean conceptions of the continuum. Although in his *De continuo* Bradwardine excluded references to God, we cannot understand the dynamic of fourteenth-century discussions of the continuum unless we take account of the theological context from which their problems arose. Incidentally, it might be noted that in Harclay's views of the relations of indivisibles to continua and in Bradwardine's arguments against Harclay, enduring issues of infinity and continuity are raised, for the understanding of which the philosophical analysis of Murdoch's Ph.D. dissertation is still highly relevant; this may be an instance in which the swing from text (and scientific or philosophical content) to context went too far and too fast.

If in Sylla's paper the argument is made that a wider university context beyond the disciplines of the arts faculty must be examined, the remaining papers argue for the relevance of a still wider context, extending beyond the university walls. Among the papers in this category, Hans Thijssen's "What Really Happened on 7 March 1277? Bishop Tempier's Condemnation and Its Institutional Context" proposes to treat the Parisian condemnation of 1277 not as a unique event but as one of a series of condemnations, the nature of which can shed light on what was or was not special about the condemnation of 1277. Previous scholars have pointed out that to have one's views condemned in the medieval university was not a career-destroying event: in more than a few cases, when a scholar's opinions were labeled as heretical, the scholar made a public recantation and went on to pursue his career with no permanent disability. The cases that appear in the record, Thijssen argues, are those where this more-or-less expected pattern did not occur—when the scholar refused to recant, or when he recanted but then went on to espouse his heretical views again. The condemnation of 1277 is unusual in that it does not name those who advocated the theses listed, and Thijssen suggests that this might be explained if Siger of Brabant, for example, had already been investigated for holding views listed in 1277 and had been acquitted (or perhaps had recanted): then an established tradition prohibiting double jeopardy would have prevented Siger from being cited again by name for holding the same theses, and Bishop Tempier could only condemn the theses anonymously—or as espoused by Siger's followers—or not at all. Thijssen finds evidence to support the view that Siger had already been tried for his views in a letter preserved in a manual for inquisitors giving examples of how to compose various types of documents, which happens to include a letter of 23 November 1276 summoning Siger to appear before the court of the inquisitor in Saint Quentin.

The papers of Edward Grant, Joan Cadden, and Michael Shank also address an extra-university context. In "Nicole Oresme, Aristotle's *On the Heavens*, and the Court of Charles V," Edward Grant compares Oresme's Latin commentary on *De caelo* with his French translation and commentary on the same work, *Le livre du ciel et du monde*—taking advantage of a unique chance to see how an outstanding thinker like Oresme wrote differently for an intra- and an extra-university audience. Whereas the Latin questions on *On the Heavens* belonged within the university arts faculty and therefore exclude theological considerations for the most part, Grant shows that *Le livre* brought in theology much more often, discussing such issues as how the bodies of the blessed will be situated in paradise or how an angel can be at the same time in the heavens and in more than a hundred thousand distant places on earth without occupying more space than the equivalent of one foot. In agreement with most medieval scholars, Oresme insisted in *Le livre* that there is only one truth, that what is true in philosophy must also be true in theology, but he nevertheless expressed disagreement with Aristotle in numerous places—in Oresme's own table of notable materials, Grant counts 35 cases of opposition to Aristotle and 4 to Averroes. Among the themes most emphasized in *Le livre* were the eternity of the world and the status of the earth, its shape, and whether it rests or moves. In his Latin questions on *De caelo*, Oresme concluded that the hypothesis that the earth rotates could not save the phenomena as well as the assumption that the heavens rotate, but in *Le livre* he emphasized that when all the arguments have been taken into account, one could still believe that the earth rotates and not the heavens—even though he himself thinks that the heavens rotate and not the earth—because no experience demonstrates that the earth does not move. Here Oresme seems to want to give his lay readers ammunition against those who impugn the articles of faith on the grounds that they are against natural reason—what is against natural reason, he implies, may well be true.

In Grant's view, Oresme's motivations for including considerations in *Le livre* that had not appeared in his questions on *De caelo*, or for excluding issues that he had previously included, are not easy to discern. Perhaps he tended to exclude issues that were more dry and technical and to include subjects that would be more stimulating to lay readers, yet he included fairly difficult arguments about the incommensurability of the heavenly motions in order to refute Aristotle's argument that something could not begin to be after an eternity of nonexistence, or cease existing after an eternity of existence. Referring to Joan Cadden's suggestion in this volume that King Charles V was interested in promoting a certain level of debate among his courtiers, Grant points to a passage at the end of Oresme's commentary where he says that one of his purposes in writing was to encourage the young men of the court to argue against him and to correct him because of their love for the truth. If this was his purpose, in Grant's view, it might explain why Oresme highlighted his disagreements with Aristotle and why he emphasized that the answers to some questions of natural philosophy, such as the rotation of the earth, were undecidable by natural reason. But why, in the end, did Charles V commission Oresme to translate *De caelo* in particular? If we may suppose that Aristotle's *Ethics*, *Politics*, and *Economics* (of which he also requested translations) had obvious relevance to courtiers, why *On the Heavens*? Did Charles suppose that knowledge of *On the Heavens* would help with various kinds of government planning and forecasting, including the use of astrology? Grant finds implausible Albert Menut's assertion that Charles's interest in astrology was behind his request for a French translation of *De caelo*, supposing that he more likely valued it as a work of general education about the cosmos.

Here it may be worthwhile to compare Joan Cadden's paper for the light it sheds on the status of astrology at Charles V's court and the relation this may have had to Oresme's antiastrological writings. In "Charles V, Nicole Oresme, and Christine de Pizan: Unities and Uses of Knowledge in Fourteenth-Century France," Joan Cadden enters into the question of Charles V's motivations in greater detail. As we have known for a long time, Charles V frequently consulted astrologers, and yet Oresme, often a beneficiary of Charles's patronage, wrote extensive treatises against astrology on more than one occasion. How are we to understand this situation? Was Oresme simply a failure in the attempt to discredit astrology in Charles's eyes? Here

Cadden has the clever idea of using the work of Christine de Pizan, daughter of Charles V's one-time court astrologer, to illuminate the contemporary pro-astrology forces at Charles's court. As was the case in Roger Bacon's *De laudibus*, it becomes clear that the issue was not simply to decide for or against the resort to astrology, but rather to distinguish between responsible and irresponsible uses. And in light of Cadden's picture of Charles V's court, we should perhaps revise our views of the relation of the university-trained clergy and laity in the Middle Ages; there was not simply a confrontation between those with intellectual authority and those lacking it, but rather a more complex interaction and negotiation.

Similar themes are to be found in Michael Shank's "Intellectual Consulting in Fifteenth-Century Vienna: The Case of Astrology," one of the points of which is to argue against the stereotype of the medieval university as ivory tower. University expertise, Shank finds, was indeed valued beyond the confines of the lecture hall, as for instance when the city council of Vienna solicited advice on matters of taxation, interest, and property law in the early 1390s. If this might be expected in matters of law or medicine, it was also true for astrology. When Henry of Langenstein was still a master at the University of Paris, King Charles V asked him to discuss the significance of the comet of 1368: Langenstein replied that it was futile to read signs about princes and plagues in comets, and in his writings, Langenstein criticized astrology on epistemological grounds. Overlapping Langenstein's tenure in Vienna, however, Frederick of Drosendorf, canon of St. Stephen's cathedral, was known as a "great astrologer," and determined questions about the comet of 1402. Most astrology in Vienna seems to have been done by physicians, but in the mid-1400s masters of arts also seem to have been involved. John of Gmunden, who lectured at Vienna on the astrolabe and the albion, at his death willed to the faculty of arts a significant collection of astrological books—to be kept in a locked cabinet! Although Gmunden is known for his attack on the astrological prediction of a conjunction of all the planets made by James of Clusa for 1432, he has more to say about Clusa's incompetence than about astrology per se. A manuscript of Gmunden's works copied by one of his students and concluding with a table of planetary aspects in relation to the moon was evidently later used by Frederick III for astrological purposes. Somewhat later, Regiomontanus cast a horoscope for the son of Frederick III and the Empress Eleonor of Portugal. Thus, over a number of years, even if masters at the University of Vienna were skeptical about the value of astrology, they provided astrological consulting to the Habsburg court when requested to do so. Demands for academic consulting drew masters into more practically-oriented activities.

In "Art, Nature, and Experiment among Some Aristotelian Alchemists," William Newman continues Shank's inquiries into the more active or applied aspects of medieval and early modern scientific activity. Responding to the common assertion that ancient and medieval science were largely devoid of experiment, Newman finds that alchemists do not at all fit this stereotypical view. Building upon a well-known view of Shmuel Sambursky, Sarah Waterlow Broadie has argued that Aristotelians did not believe in experiment because they thought that natural substances exhibit their substance-typifying characteristics only in natural circumstances—by their artificiality, experiments could not lead to knowledge about natures. Peter Dear and Antonio Pérez-Ramos have argued along similar lines to explain why widespread experimentation does not appear before the end of the sixteenth century. As counterevidence to these views, Newman focuses on alchemy from the thirteenth century up through Daniel Sennert (1572–1637). Avicenna had famously argued that alchemists know they cannot transmute the species of metals. Against Avicenna, many alchemists responded that they could indeed make new metallic products, by assisting in natural processes. This heritage of alchemical ideas lies behind Sennert's claim that alchemical processes are not artificial but natural, using natural causes such as fire and heat, and that the products of chemistry, like natural things, have innate principles of change. On these grounds Sennert argued for laboratory experiments in careful isolation from the environment. In sum, Newman asserts that previous scholars have given a false view of medieval Aristotelianism at least in part because they have neglected the rich experimental literature of medieval and early modern alchemy.

If the evidence Newman uses to overturn stereotypes concerning Aristotelianism comes, most strongly, from the early seventeenth century, Katharine Park, in "The Meanings of Natural Diversity: Marco Polo on the 'Division' of the World," finds evidence from the thirteenth and fourteenth centuries for the avidity of at least certain medieval authors for new knowledge and the experience of diversity. Park's central text is the *Description of the World* of Marco Polo, representing an intersection between courtly and mercantile cultures. Diversity, in the *Description of the World*, was a highly positive and attractive quality, not least because it provided the basis for profitable trade between regions of the world. In the mid-thirteenth century, John Pian di Carpini and William of Rubruck had traveled to Asia without celebrating its diversity, as Polo was to do, but rather reporting on its inferiority or barbarism. Polo was more open to appreciate the Orient because he was not a missionary or ambassador, as were John and William, but rather a merchant in search of commodities to trade. But his views resonated with encyclopedic works such as the *De proprietatibus rerum* of Bartholomaeus Anglicus, the *De natura rerum* of Thomas of Cantimpré, and the *Speculum naturale* of Vincent of Beauvais.

In the works of scholastics like Albertus Magnus and Roger Bacon, the differences between physical places could be explained by the varying influences of the heavenly bodies, elaborated by Albertus in his *De natura loci.* As Park points out, both Albertus Magnus and Roger Bacon in his *Opus maius* emphasize how natural diversity results from what we might call astrological influence: the cones of rays coming from the heavens to a given point on the earth. Neither Marco Polo nor his co-author Rustichello of Pisa was in any position to read Albertus Magnus or Roger Bacon, but they had many views in common. Park speculates that Marco Polo and Rustichello might have imbibed their views from an intermediary work such as Gossouin's smaller vernacular *Image du monde*. But there is also the complementary possibility that scholastics like Albertus Magnus were influenced by the wider world around them. Although, as is well known, Aristotle had condemned commerce beyond the household as contrary to nature, later scholastic thinkers found loopholes justifying commerce and trade. Profit could be justified by the danger and risk incurred by the merchant. More relevant here, trade could be justified on the grounds that it brings things from places where they are abundant to places where they are scarce—even Aristotle had admitted that trade in this case is not contrary to nature. In this context, Polo's *Description of the World* can be seen, in Park's terms, as a sermon naturalizing and justifying trade. Thus Park's paper beautifully ties together the themes of several of the other papers—the interrelation of academic and nonacademic ideas, the influence of "astrological" ways of thinking, and even the possible influence of the outside real-world context on scholastic thinkers, for instance in their discussions of legal and economic issues.

Let us return to the episode we described at the outset, involving Henry of Harclay and his *questio* on the second coming of Christ. Harclay divided his answer to the question into three parts: first, he inquires whether the second coming of Christ can be proved by theology; second, whether it can be proved by poetry or songs; and third, whether it can be proved by philosophy or the sayings of astrologers. Harclay's first, theological response depends upon the Bible and the authority of the saints. That the antichrist will come is sure, but can it be known from the scriptures or saints *when* he will come? Harclay cites various predictions according to which the antichrist should already have come, and he concludes that it is clear that they are false, by "the experiment of time."[4] Arnald of Villanova's prediction was based on interpreting the days referred to in the prophecies of Daniel as years, which according to Harclay has no evidence or probability.[5] Thomas Aquinas cites Augustine to the effect that we cannot know the day or hour

[4] Pelster, "Quaestio" (n. 1), p. 56: "Sed iam constat istam opinionem falsam esse experimento temporis." Later (p. 57) he refers to the "experience of time" ("experientia temporum").

[5] Ibid., p. 60: "nullam habet evidenciam nec probabilitatem."

of Christ's second coming, but perhaps we might know the century or millennium—Aquinas does not say where Augustine says this, reports Harclay, and no wonder, because in truth Augustine never said it.[6] The *magister historiarum* [Petrus Comestor] predicted the end of the world, Harclay reports, but if he were alive at the time for which he predicted the end of the world, he would propose a new calculation so that he would not be caught lying.[7]

As to Harclay's second article, whether poems or songs can prove the end of the world, his main sources are Joachim of Flore, Hildegard of Bingen, and pseudo-Ovid—"thus probably, by the authority of both sexes."[8] Hildegard, he says, was a Teutonic nun, who wrote many books and letters, some to St. Bernard of Clairvaux and others to the Cistercians and many other persons and communities. In a letter that she wrote to the clergy of Cologne she prophesied about what would happen in the coming century—which prophecies have been fulfilled. In the third book of her *Scivias*, she wrote that she was now in the seventh millennium since the beginning of the world, and that there would not be another millennium, so that it seems the world will end in the seventh millennium.[9] Joachim and Hildegard,

---

[6] Ibid., p. 65: "Et notandum quod istam auctoritatem Augustini allegat Thomas de Alquino in Glosa Mathei 24. Sed non dicit ubi Augustinus dicit hoc, quia in rei veritate nusquam dixit hoc."

[7] Ibid., p. 68: "Ego credo quod, si iste magister viveret in anno predicto 1356 ab incarnacione et videret quod istud non esset impletum, ipse adhuc fingeret novam calculacionem annorum, ut ulterius differret hoc negocium, ne posset deprehendi de mendacio. Mirares quod sine omni probabilitate contra omnem exposicionem sanctorum et catholicorum doctorum aliquis catholicus vanas opiniones vel sequitur vel fingit. Non est opus improbare, sed tantum recitare. Sed cum recitatur, ex se reprobatur."

[8] Ibid., p. 73: "ita probabiliter per auctoritatem utriusque sexus."

[9] Ibid., p. 77–78: "Item Hildegardis monialis teutonica, que multos libros et epistolas conscripsit, aliquas ad sanctum Bernardum abbatem Claravalensem et alias ad Cistercienses et multis aliis personis et communitatibus. Unde propheta fuit et in epistola, quam scripsit ad clerum Coloniensem, prophetavit manifeste de hiis que post tempus suum contigerunt et per centum annos, que nos vidimus impleta. Inter alios librum scripsit unum magnum propheticum et doctrinalem continentem 3[es] libros vel 3 partes, quem librum, ut dicit, Bernardus iussit reponi in archivis Clarevallensibus, quod non esset, nisi esset magne utilitatis. Intitulat autem illum librum Liber *Si vias* (!), id est simplicis videntis et simpliciter prophete.... Dicit autem in principio istius libri quod est que <dixit et scripsit, non secundum adinvencionem cordis sui aut ullius hominis, sed ut est in celestibus, vidi et audivi et percepi et perfecta misteria Dei, et tum audivi vocem de celo michi dicentem: Clama ergo et scribe sic.> Hec ista in principio libri. Modo ad propositum dicit Hildegardis ... quod tempore suo mundus fuit in 7. millenario ab origine mundi et quod alius millenarius non esset futurus. Ergo videtur quod in isto septimo millenario finietur mundus. Videtur michi quod moderni racionatores multo magis increpandi sunt quam vel Ioachim vel Hildegardis. Non invenio istos calculatores aliquid determinasse, sed semper in dubio locutus fuisse.... ecce auctoritas Hildegardis de verbo ad verbum: < ... Sex enim dies operis sunt, 7[ma] est requiei: numerus dierum alius non est. Quid superest tibi, o homo, sciendum non est, sed in secreto Patris est. Sed vos, o homines de tempore hoc, tempus temporum in cursu vestro habetur, antequam homicida est veniat, qui fidem catholicam pervertere cupiet.> Hec ista.

Harclay comments, are much less to be blamed than are the presumptuous modern calculators who claim to know more exactly the end of the world.

In the third article, Harclay asks whether the second coming of Christ can be known by astronomy, and he quotes Albumasar, Ptolemy, and Albategni—the latter supposing that the world will end when the fixed stars complete their course, which will take 36,000 years according to Ptolemy. But perhaps, as the world comes to an end, the rate of rotation will accelerate, so that 365 revolutions will occur in a single day. The rapid motion will be the cause of the conflagration that ends the world. These opinions, Harclay says, are so vain that they require no disproof. It was found after the time of Ptolemy, he says, that the fixed stars do not move regularly, but sometimes move forward and sometimes backward—this is called the motion of accession and recession of the eighth sphere. This motion was found by Thabit the Christian, whose method is followed by all modern astronomers ["astrologi"], and all tables are based on this.[10]

The underlying argument of our collection of papers has been that the efforts of medieval thinkers—natural philosophers, theologians, physicians, astrologers—to make sense out of the natural world were shaped by a nested series of enveloping contexts. Our contributors have had to tease these contexts out, but Henry of Harclay's *questio* remarkably made many of them explicit: the academic community and its debates, the church and its writings, the scientific world and its technical reasoning—Harclay knew them all and could reply to them all, to each in its own terms. Indeed, he makes vividly clear how much more interchange there was between university scholars and people outside the university than has traditionally been assumed. We hope that the papers we present here will go some way toward suggesting how this interchange can be explored, by showing how studies of texts and of their various contexts may fruitfully interact.

[10] Ibid., p. 82: "Ista sunt vanissima; non indigemus inprobacione. Verum est quod posicio Tholemei et aliorum supra nominatorum de motu stellarum fixarum corrigitur per moderniores. Nam inventum est post tempus Tholomei quod stelle fixe non moventur semper uniformiter secundum successionem signorum ab ariete in taurum, sed aliquando procedendo, aliquando retrocedendo secundum tempus determinatum per spacium 9 vel 10 graduum. Et motus accessionis et recessionis 8. spere vocatur. Quem motum adinvenit Thebit christianus, quem modum secuntur omnes moderni astrologi, et fundantur tabule super hoc. Et tunc primus modus ponendi est vanus sicud omnes modi ponendi vani sunt."

PUBLICATIONS OF JOHN E. MURDOCH

"Thomas Bradwardine: Geometry and the Continuum in the Fourteenth Century," *Actes du IXe Congrès International d'Histoire des Sciences*, Barcelona-Madrid, 1960.

Rationes Mathematice: Un aspect du rapport des mathématiques et de la philosophie au Moyen Age (Paris: Université de Paris, 1962).

"The Medieval Language of Proportions: Elements of the Interaction with Greek Foundations and the Development of New Mathematical Techniques," *Scientific Change*, ed. A. C. Crombie (London: Heinemann, 1963), pp. 237–71, 334–43.

"Nicole Oresme's *Quaestiones super geometriam Euclidis*," *Scripta Mathematica* 37 (1964): 67–91.

"Superposition, Congruence and Continuity in the Middle Ages," *Mélanges Alexandre Koyré*, vol. 1, *L'aventure de la science* (Paris: Hermann, 1964), pp. 416–41.

"The Equality of Infinites in the Middle Ages," *Actes du XIe Congrès International d'Histoire des Sciences* (Warsaw/Cracow/Wroclaw, 1965), vol. 3, pp. 171–74.

"Two Questions on the Continuum: Walter Chatton (?), O.F.M. and Adam Wodeham, O.F.M.," *Franciscan Studies* 26 (1966): 212–88 (with Edward Synan).

"*Euclides Graeco-Latinus*: A Hitherto Unknown Medieval Latin Translation of the *Elements* Made Directly from the Greek," *Harvard Studies in Classical Philology* 71 (1966): 249–302.

"The Medieval Euclid: Salient Aspects of the Translations of the *Elements* by Adelard of Bath and Campanus of Novara," *Revue de Synthèse*, 3rd ser., nos. 49–52 (1968): 67–94.

"*Mathesis in philosophiam scholasticam introducta*: The Rise and Development of the Application of Mathematics in Fourteenth-Century Philosophy and Theology," in *Arts Libéraux et Philosophie au Moyen Age*, *Actes du Quatrième Congrès International de Philosophie Médiévale*, Montreal: Institut d'études médiévales/Paris: Vrin, 1969, pp. 215–54.

"Thomas Bradwardine," in *Dictionary of Scientific Biography*, vol. 2 (New York: Charles Scribner's Sons, 1970), pp. 390–97.

"Walter Burley," in *Dictionary of Scientific Biography*, vol. 2 (New York: Charles Scribner's Sons, 1970), pp. 608–12 (with Edith Sylla).

"Euclid: Transmission of the *Elements*," *Dictionary of Scientific Biography*, vol. 4 (New York: Charles Scribner's Sons, 1971), pp. 437–59.

"Naissance et développement de l'atomisme au bas moyen âge latin," *Cahiers d'études médiévales*. II. *La science de la nature, théories et pratiques* (Montreal: Bellarmin/Paris: Vrin, 1974), pp. 1–30.

"Philosophy and the Enterprise of Science in the Later Middle Ages," in *The Interaction between Science and Philosophy*, ed. Y. Elkana (Atlantic Highlands, N. J.:Humanities Press, 1974), pp. 51–113.

The Cultural Context of Medieval Learning, ed. and introduction (Dordrecht: Reidel, 1975) (with Edith Sylla).

"From Social into Intellectual Factors: An Aspect of the Unitary Character of Late Medieval Learning," in *The Cultural Context of Medieval Learning* (Dordrecht: Reidel, 1975), pp. 271–339.

"A Central Method of Analysis in Fourteenth-Century Science," *Proceedings of XIVth International Congress of the History of Science*, no. 2 (Tokyo-Kyoto, 1975), pp. 68–71.

"Richard Swineshead," *Dictionary of Scientific Biography*, vol. 13 (New York: Charles Scribner's Sons, 1976), pp. 184–213 (with Edith Sylla).

"Transmission and Figuration: An Aspect of the Islamic Contribution to Mathematics, Science, and Natural Philosophy in the Latin West," *Proceedings of International Al-Biruni Congress* (Teheran, September 1973), pp. 407–37, and in *Proceedings of the First International Symposium for the History of Arabic Science, Aleppo, April 5–12, 1976*, vol. 2, pp. 108–22.

"Music and Natural Philosophy: Hitherto Unnoticed Questiones of Blasius of Parma(?)," Pearl Kibre Festschrift, *Manuscripta* 20 (1976): 119–36.

"The Development of a Critical Temper: New Approaches and Modes of Analysis in Fourteenth-Century Philosophy, Science, and Theology," in *Proceedings of the Southeastern Institute of Medieval and Renaissance Studies*, Summer 1975, ed. S. Wenzel, *Medieval and Renaissance Studies* 7 (1978): 51–79.

"*Subtilitates Anglicanae* in Fourteenth-Century Paris: John Mirecourt and Peter Ceffons," in *Machaut's World: Science and Art in the Fourteenth Century, Annals of the New York Academy of Sciences* 413 (1978): 51–86.

"The Science of Motion," in *Science in the Middle Ages*, ed. David Lindberg (Chicago: University of Chicago Press, 1978), pp. 206–64 (with Edith Sylla).

"Propositional Analysis in Fourteenth-Century Natural Philosophy: A Case Study," *Synthèse* 40 (1979): 117–46.

"Utility versus Truth: At Least One Reflection on the Importance of the Philosophy of Science for the History of Science," in *Probabilistic Thinking, Thermodynamics, and the Interaction of the History and Philosophy of Science*, Proceedings of the 1978 Pisa Conference on the History and Philosophy of Science, ed. J. Hintikka et al, vol. 2 (Dordrecht/Boston: Reidel, 1981), pp. 311–19.

"*Scientia mediantibus vocibus*: Metalinguistic Analysis in Late Medieval Natural Philosophy," in *Sprache und Erkenntnis im Mittelalter*, Acts of the Sixth International Congress of Medieval Philosophy, Bonn, 29 August–3 September 1977), *Miscellanea Mediaevalia*, vol. 13 (Berlin/New York: De Gruyter, 1981), pp. 73–106.

"George Sarton and the Formation of the History of Science," in *Belgium and Europe, Proceedings of the International Francqui-Colloquium, Brussels-Ghent, 12–14 November 1980* (Brussels: Koninklijke Academie voor Wetenschappen, Letteren en Schone Kunsten van Belgie, 1981), pp. 123–38.

"Mathematics and Infinity in the Later Middle Ages," *Proceedings of the American Catholic Philosophical Association* 55 (1981): 40–58.

"Henry of Harclay and the Infinite," in *Studi sul XIV secolo in memoria di Anneliese Maier*, ed. A. Maierù and A. Paravicini-Bagliani (Rome: Edizioni di Storia e Letteratura, 1982), pp. 219–61.

"William of Ockham and the Logic of Infinity and Continuity," in *Infinity and Continuity in Ancient and Medieval Thought*, ed. Norman Kretzmann (Ithaca: Cornell University Press, 1982), pp. 165–206.

"Infinity and Continuity," in *The Cambridge History of Later Medieval Philosophy*, ed. N. Kretzmann, A. Kenny, J. Pinborg (Cambridge: Cambridge University Press, 1982), pp. 564–91.

"Mathematics and Sophisms in Late Medieval Natural Philosophy," in *Les genres littéraires dans les sources théologiques et philosophiques médiévales*, Actes du Colloque International de Louvain-la-Neuve, 25–27 Mai 1981 (Louvain-la-Neuve: Institut d'Études Médiévales de l'Université Catholique de Louvain, 1982), pp. 85–100.

"The Analytic Character of Late Medieval Learning: Natural Philosophy without Nature," in *Approaches to Nature in the Middle Ages*, Papers of the Tenth Annual Conference of the Centre for Medieval and Early Renaissance Studies, ed. Lawrence Roberts (Binghamton, N.Y.: State University of New York Press, 1982), pp. 171–213.

Album of Science: Antiquity and the Middle Ages (New York: Charles Scribners, 1984), xii + 403 pp.

"Atomism and Motion in the Fourteenth Century," in *Transformation and Tradition in the Sciences*, Essays in Honor of I. Bernard Cohen, ed. E. Mendelsohn (New York: Cambridge University Press, 1974), pp. 45–66.

"Thomas Bradwardine: Mathematics and Continuity in the Fourteenth Century," in *Mathematics and its Applications to Science and Natural Philosophy in the Middle Ages*, Marshall Clagett Festschrift, ed. E. Grant and J. E. Murdoch (New York: Cambridge University Press, 1987), pp. 103–37.

"Alexandre Koyré and the History of Science in America: Some Doctrinal and Personal Reflections," *History and Technology* 4 (1987): 71–79.

"From the Medieval to the Renaissance Aristotle," in *New Perspectives on Renaissance Thought: Essays in the History of Science, Education, and Philosophy in Memory of Charles B. Schmitt*, ed. John Henry and Sarah Hutton (London: Duckworth and Istituto Italiano per gli Studi Filosofici, 1990), pp. 163–76.

"Pierre Duhem and the History of Late Medieval Science and Philosopohy in the Latin West," in *Gli studi di filosofia medievale fra otto e novecento*, ed. R. Imbach and A. Maierù (Rome: Edizioni di Storia e Letteratura, 1991), pp. 253–302.

"Aristotle on Democritus's Argument Against Infinite Divisibility in *De generatione et corruptione*, Book I, Ch 2," forthcoming.

EUDOXAN ASTRONOMY AND ARISTOTELIAN HOLISM IN THE *PHYSICS*

JEAN DE GROOT*

It is generally acknowledged that Eudoxan mathematics exerts a strong influence in the motion books (IV–VIII) of Aristotle's *Physics*. This presents something of a paradox for interpretation, however. In Aristotle's classification of the sciences in *Physics* II.2 and *Metaphysics* E.1, physics is not a mathematical science; it is not even a natural science that incorporates the mathematical, as do the "mixed" sciences—astronomy, harmonics, and optics. This means that Aristotle's remarks about mathematical demonstration within a science of nature are not relevant to the interpretation of the mathematical element in his own *Physics*. Without programmatic guidance from Aristotle, scholars have been free to interpret the Eudoxan influence as they wish. In general, they have localized it to Aristotle's notion of magnitude (μέγεθος).They have understood μέγεθος as the idea of an undifferentiated geometrical continuum whose structure insures the infinite divisibility of all its parts. Based on this interpretation, Aristotle appears to have produced a mathematical physics of correspondence based on the structure of the continuum.[1] I propose that this picture is misleading in a number of respects.

Our best information is that the mathematical formulation of the continuum was first given a precise form by Eudoxus; and it is true that the continuum plays a central role—if not *the* central role—in the mathematical reasoning in Aristotle's treatment of motion. From these facts it does not follow, however, that the Eudoxan influence was such that Aristotle would conceive the continuum of the *Physics* abstractly, undifferentiated as to place and divisible throughout.[2] Eudoxus's conception of the continuum

* I am grateful to Kurt Pritzl, Robert Sokolowski, and Edith Sylla for advice on and criticism of the earlier version of this paper.

[1] The textual basis for this approach is Aristotle's statement that the continuity of motion and time follow upon the continuity of magnitude (219a10–14). The compositional structure of magnitude given in *Physics* V.3 and VI.1 may be mapped onto motion and time. For a general treatment of the mapping of structures, see Edward Hussey, *Aristotle's Physics: Books III and IV* (Oxford: Clarendon Press, 1983), pp. 141–46.

[2] The modern interest in topology associated with the theory of irrationals and the consequent modern understanding of fourth-century B.C. proportionality theory in terms of the theory of irrationals have led to an interpretation of Aristotle's mathematical treatment of motion almost exclusively in terms of the structure of the continuum. See G. E. L. Owen's seminal articles "Zeno and the Mathematicians" and "*Tithenai ta phainomena,*" in the collection of his articles edited by Martha Nussbaum, *Logic, Science, and Dialectic: Collected Papers in Greek Philosophy* (Ithaca, N.Y.: Cornell University Press, 1986), pp. 45–64, 239–51. Of more recent scholarship, see Michael J. White, *The Continuous and the Discrete*

depends on his concept of ratio, and his ratio-concept played an important role not only in geometry but also in Eudoxan astronomy.

The comparison of velocities, distances, and times for which Aristotle evokes the continuum principle was pioneered by Eudoxus in his lost treatise of astronomy, *On Speeds* (Περὶ Ταχῶν). We possess enough knowledge of *On Speeds* to know that it is one source of the cosmos of homocentric spheres of the *De caelo* and also that it compared velocities in terms of times, as Aristotle does in *Physics* IV.8 and VI.2.[3] If *On Speeds* is Aristotle's model in the *Physics*, it may indicate that Aristotle meant for the directionality and place that figure so prominently in astronomy to be included in the conception of magnitude appropriate to physics.[4] I would like to offer some evidence that this was the case—that, in fact, Aristotle thought it quite natural to look at the magnitude connected with motion in terms of direction and destination—because of the example of astronomy. The assumption is often made that the abstraction involved in the geometrical conception of magnitude is a necessary first step to achieving mathematical science, and that Eudoxus and Aristotle first took this step.[5] Yet, the

---

(Oxford: Clarendon Press, 1992); and David Bostock, "Aristotle on Continuity in *Physics* VI," in *Aristotle's Physics: A Collection of Critical Essays*, ed. Lindsay Judson (Oxford: Clarendon Press, 1991), pp. 179–212. Wolfgang Wieland criticizes the interpretation of Aristotle's continuum in mathematical terms in *Die aristotelische Physik* (Göttingen: Vandenhoeck & Ruprecht, 1962), pp. 278–82.

[3] The only direct reference to *On Speeds* from antiquity is in Simplicius's commentary on *De caelo* II.10, where the commentator treats in some detail the spheres responsible for the speeds of the different heavenly bodies: *Commentaria in de caelo*, ed. J. L. Heiberg (Berlin: G. Reimer, 1894), 7: 492.31–497.8. Fragments from Eudoxus's astronomical works, including Simplicius's account of *On Speeds* and other indirect references, are collected in François Lasserre, *Die Fragmente des Eudoxos von Knidos* (Berlin: Walter de Gruyter, 1966). Simplicius says nothing about axioms or definitions, or about the ratios that determined speeds. Reconstructions have been made, the most celebrated being Giovanni V. Schiaparelli's reconstruction of the Eudoxan hippopede that is meant to account for the retrograde motion of the planets ("Le sfere omocentriche di Eudosso, di Calippo e di Aristotele," *Pubblicazioni del R. Osservatorio di Brera in Milano*, 1875 [9]; reprinted in idem, *Scritti sulla storia della astronomia antica* [Bologna, 1926], 2: 3–112). For treatments of this reconstruction and information about Eudoxan astronomy in general, see Thomas L. Heath, *Aristarchos of Samos: The Ancient Copernicus* (Oxford: Clarendon Press, 1913), pp. 190–224; and D. R. Dicks, *Early Greek Astronomy to Aristotle* (Ithaca, N.Y.: Cornell University Press, 1970), pp. 151–89. A new reconstruction has been made by Erkka Maula, *Studies in Eudoxus' Homocentric Spheres, Commentationes Humanarum Litterarum*, vol. 50 (Helsinki: 1974). Lasserre's *Die Fragmente* includes useful commentaries by Lasserre concerning Eudoxan astronomy. See also François Lasserre, *La naissance des mathématiques à l'époque de Platon* (Fribourg: Éditions Universitaires, 1990), pp. 179–236.

[4] This would shed light on one of the puzzling passages of Book IV, Aristotle's contention that the before and after (τὸ πρότερον καὶ ὕστερον) are first in place (219a15). On this passage, see Jacques Marcel Dubois, *Le temps et l'instant selon Aristote* (Paris: Desclée de Brouwer, 1967), pp. 344–51.

[5] This view is expressed by Hans-Joachim Waschkies at the beginning of his article "Mathematical Continuum and Continuity of Movement," in *La Physique d'Aristote et les conditions d'une science de la nature*, ed. François de Gandt and Pierre Souffrin (Paris: J. Vrin, 1991), pp. 151–79.

treatment of speeds in astronomy is an example of an ancient mathematical science that relied on the continuum concept and yet did not depend on abstraction from directionality or place. For understanding ancient astronomy, it may be useful to distinguish between the mathematics of the continuum and the ancient mathematics of *continua*. Following the example of Eudoxan astronomy, Aristotle's concern in the *Physics* is not so much with the continuum, understood as a mathematical notion, as it is with different continua manifest in nature.[6]

To take account of the influence of Eudoxan astronomy on the *Physics* can correct some of the limitations of the geometrical interpretation of μέγεθος. The geometrical interpretation lends itself to a reading of *Physics* VI in which the technical aspects of Book VI are unrelated to the many references in that book to priority and posteriority in motion and to the involvement of parts of a motion with the end of motion. These are expressions of Aristotle's holism—his predilection for including the end of motion in the explanation of the motion itself—and their intimate association with the continuity doctrine of Book VI requires some accounting.[7] These references might be dismissed as an overlay of teleology unrelated to the technical treatment of the continuum. Yet, if the mathematics of motion in *Physics* IV–VI are mainly matters of topology and correspondence, as the geometrical interpretation of μέγεθος implies, then it is difficult to see why these aspects of modern mathematical science were separated from Aristotle's teleology and cosmology only in the Scientific Revolution and under the influence of a rival philosophy of nature, atomism.[8] It is appropriate to

[6] My position has a number of points of contact with the point of view of Wolfgang Wieland, particularly his view that the focus of Aristotle's continuum theory is motion, not the structure of magnitude, and his insistence that Aristotle's concern is objects that are continuous, not the mathematics of the continuum (*Die aristotelische Physik* [n. 2], pp. 279, 281). It seems, though, that Wieland would minimize a Eudoxan influence, because he denies the mathematical import of Aristotle's continuum. I maintain, in contrast, that the Eudoxan influence on the *Physics* was never purely mathematical but was always in the realm of emerging scientific theory.

[7] See in particular *Physics* VI.1, 232a2–6; 4, 234b10–20; 5, 235b16–30; 6, 236b32–28, 237b9–22. For the holistic implications of *Physics* VI.6, see below. The terms πρῶτον, πρότερον, and ὕστερον in the earlier chapters are usually taken as time references. I submit that Aristotle means for them to refer also to the order of stages of change. He believes that, within a change, the stages are related to one another by conditional necessity—or, as he puts it, as πρότερον to ὕστερον. On this subject, see Jean De Groot, "Form and Succession in Aristotle's *Physics*," in *Proceedings of the Boston Area Colloquium in Ancient Philosophy*, vol.10 (1994), ed. John Cleary and William Wians (Lanham, Md.: University Press of America, 1996), pp. 1–23.

[8] On the difference between late medieval and early modern mathematical treatments of motion, see A. G. Molland, "The Atomisation of Motion: A Facet of the Scientific Revolution," *Studies in History and Philosophy of Science*, 13/1 (1982): 31–54; idem, "Continuity and Measure in Medieval Natural Philosophy," in *Mensura, Mass, Zahl, Zahlensymbolik in Mittelalter*, ed. A. Zimmermann (Berlin: Walter De Gruyter, 1983), pp. 132–44, and idem, "Aristotelian Holism and Medieval Mathematical Physics," in *Studies in Medieval Natural Philosophy*, ed. Stefano Caroti (Florence: Leo S. Olschki, 1989), pp. 227–35.

separate the two issues of Aristotle's teleology and his holism. Teleology is understood to implicate the causes of motion, while holism is a way of conceptualizing motion. It seems likely that Aristotle's mathematical treatment of motion is integrally related to notions of progressivity and finality, quite apart from the issue of teleology. If so, his handling of the continuum must be a central feature of this mathematical understanding of progressivity.

Thus, to investigate the significance of astronomy for the *Physics* involves understanding the role of ratios and the continuum in astronomy. I shall begin with the evidence for incompleteness of abstraction in the mathematical approach typical of ancient astronomy, presenting the basic problematic of astronomy and how ratios function in that problematic. Then I shall show how that mathematical conception produces, in application to sublunary motion, a conception of motion that is holistic and implicates progressivity.

In support of the distinction betwen the mathematics of the continuum and the mathematics of continua, it is worth noting that neither stereometry nor the science of solids in motion had yet achieved the separation from experience that plane geometry had achieved. Plato complains about this in *Republic* VII. In the course of assembling the curriculum of studies for the guardians, he seems at first to say that it is appropriate to proceed from arithmetic to geometry to astronomy. Then, chiding Glaucon for stressing the utility of astronomy, he says that the more appropriate study to be next after plane geometry is a mathematics of solids taken purely as three-dimensional. The drawback of astronomy is that it treats the solid as already in motion (ἐν περιφορᾷ ὂν ἤδη στερεὸν λαβόντες) rather than treating it simply as a solid (528a9–b3). Glaucon sensibly replies that the principles and theorems needed to treat the solid solely as three-dimensional have not yet been discovered. Plato agrees, but insists that some progress has been made toward the development of this discipline.[9] This passage shows that it was the norm in fourth-century Greek science for the mathematical treatment of a body to include its being in motion. Plato believes this way of looking at solids to be very different from an approach without motion.

[9] He would have in mind the duplication of the cube and possibly theorems connected with spherics. On the duplication of the cube and the references to stereometry in this passage, see James Adam, *The Republic of Plato*, 2 vols. (Cambridge: Cambridge University Press, 1902) 2: 120–26. For approaches to the cube duplication problem in the Academy, see also Thomas L. Heath, *The Thirteen Books of The Elements*, 3 vols., 2nd ed. (New York: Dover, 1956), 2: 244–46, 255–58; and Wilbur Knorr, *The Ancient Tradition of Geometric Problems* (New York: Dover, 1986), pp. 50–76. On theorems of spherics that may have been known to Plato, see Ian Mueller, "Ascending to Problems: Astronomy and Harmonics in *Republic* VII," in *Science and the Sciences in Plato*, ed. John P. Anton (New York: Eidos, 1980), pp. 103–22; and J. L. Berggren, "The Relation of Greek Spherics to Early Greek Astronomy," in *Science and Philosophy in Classical Greece*, ed. Alan C. Bowen (New York: Garland, 1991), pp. 227–48.

The inclusion of motion is not the only difference between astronomy and a mathematics of solids: there is also the fact that astronomy treats objects of experience. Later in *Republic* VII, Plato addresses this feature of astronomy as a limitation (529c7–d5). Insofar as the motions of the heavenly bodies are traced in something visible (ἐν ὁρατῷ πεποίκιλται), they fall short of true motions, "those movements in which the really fast and the really slow [τὸ ὂν τάχος καὶ ἡ οὖσα βραδυτής]—in true number and in all the true figures (σχήμασι)—are moved with respect to one another and in their turn move what is contained in them."[10] The true number and true figures are the mathematical objects by which a more perfect astronomy would express the paths of moving bodies. It is notoriously unclear, however, whether Plato is referring to an ideal astronomy of the existing celestial bodies, or is recommending a general geometrical kinematics of circular motion. As an example of a general kinematics, Alexander Mourelatos cites the principle of *Laws* X that different circles contained within a rotating disk move at speeds that are in proportion to their distances from the center of the disk.[11] Some version of this principle still plays a part in the Eudoxan astronomy of *On Speeds* as Aristotle, in *De caelo* I.8 and 10, attests. Yet astronomy remained stubbornly linked to celestial phenomena, the omnipresent and spectacular exemplar of circular kinematics.[12]

Aristotle wrestled with the problems of astronomy: its inclusion of motion, and its dependence on existing things. In *Posterior Analytics* I.13, 78b35–39, he says that τὰ μηχανικά is related to στερεομετρία in the same way that optics is related to geometry, harmonics to arithmetic, and τὰ φαινόμενα to astronomy:

$$\frac{\text{optics}}{\text{geometry}} \propto \frac{\text{mechanics}}{\text{stereometry}} \propto \frac{\text{harmonics}}{\text{arithmetic}} \propto \frac{\textit{ta phainomena}}{\text{astronomy}}$$

Τὰ μηχανικά, the applied version of stereometry, is a τέχνη, a practical craft that uses natural principles to attain some end contrary to nature (*Mechanics* 847a10–19). The first thing to notice about this array of the sciences is its omission of a mathematical science of the principles of nature as such. This is most likely explained by the fact that Aristotle is correlating

[10] *Republic* VII 529c8–d4. The translation is by Allan Bloom, *The Republic of Plato*, 2nd ed. (New York: Basic Books, 1991). On this passage, see two articles by Alexander P. D. Mourelatos: "Plato's 'Real Astronomy': *Republic* 527d–531d," in Anton, *Science and the Sciences in Plato* (n. 9), pp. 33–73; and "Astronomy and Kinematics in Plato's Project of Rationalist Explanation," *Studies in History and Philosophy of Science* 12/1 (1981): 1–32.

[11] Mourelatos, "Astronomy and Kinematics," p. 3. The passage is *Laws* X, 893c3–d5.

[12] For evidence of the persistence of this link, see Mueller, "Ascending to Problems" (n. 9), pp. 105–8, and Berggren, "Greek Spherics" (n. 9), pp. 234–41. On the difficulty of reconciling this link with the idea of an abstract rational kinematics, see Mourelatos, "Astronomy and Kinematics" (n. 10), pp. 13–18.

sciences that already have some tradition or are in practice. Yet, why not make stereometry *also* the pure mathematics for the applied science of sublunary nature? Undoubtedly because the science of sublunary nature is centered on the phenomenon of motion (*Physics* II.1, 192b13–23). For this reason, its principles would have to come from a mathematics that incorporates motion.

The second thing to note is Aristotle's last comparison, relating ἡ ἀστρολογική to empirical astronomy. This suggests that he regards astronomy as a genuine mathematical discipline comparable to geometry or arithmetic, in spite of its being a science of motion. The comparability of astronomy to geometry and arithmetic suggested here must be read in the light of other passages where Aristotle makes astronomy a mixed science comparable to optics and harmonics.[13] This passage does not contradict those others if we recognize that the comparisons he is making in *Posterior Analytics* I.13 are not entirely parallel: of the four pairs related, only the relation of optics to geometry and of harmonics to arithmetic seem to match. Mechanics is a craft while optics, harmonics, and astronomy are not. The assemblage of the data of astronomy represented by τὰ φαινόμενα is different again, being closer to Aristotle's description of ἐμπειρία than to sciences like optics and harmonics.[14] What does appear from these comparisons, however, is that (1) there is no pure mathematical discipline to match up to astronomy, when astronomy is taken as the applied science; and (2) astronomy is mathematical enough to fill the role of the missing stereometry of motion. In *Metaphysics* M.3, Aristotle suggests this when he says that a science (ἐπιστήμη) is most precise if it is "without motion," but if it includes motion, it is most precise if it concerns the primary motion (circular motion) and primary motion that is also uniform (1078a11–13). He does not call this science of circular motion ἀστρολογία, but he evidently has astronomy in mind since he goes on to compare optics and harmonics to it.

These passages from Plato and Aristotle show the special status of astronomy among the mathematical sciences, due to the fact that it had not attained the degree of separation from its sensible substratum that geometry had attained. This was not wholly a weakness, however: it also reflected the strength of astronomy as including motion in its mathematical treatment. As Plato and Aristotle present it, the inclusion of motion in the problematic of astronomy was no theoretical impediment to a full mathematical treatment.

---

[13] On mixed sciences, see *Physics* II.2, 193b22–194a12, which treats optics, harmonics, and astronomy together; *Metaphysics* B.2. 997b34–998a6, which treats astronomy; and *Posterior Analytics* I.6, 75b7–17, where optics is the example.

[14] On the concept of experience and its relation to knowledge and craft, see *Metaphysics* A.1, 980b25–981a30. On the role of experience in astronomy, see *Prior Analytics* I.30, 46a18–24.

This is remarkable, given that, in practice, the inclusion of motion meant a comparison of speeds. Differences in speeds is a problem presented by experience. In fact, incompleteness of abstraction was entrenched in the mathematical treatment of astronomy. Let me begin with one central aspect of its incomplete abstraction.

In the mathematical treatment of astronomy as practiced, distance was understood in terms of elongations, separation from the fixed stars, or positions relative to segments of a great circle that is itself in motion. Distances were never absolute. This is because, although the heavens could be an eternal timekeeper for sublunary occurrences, no standard of time superior to the heavens is available. Accordingly, for the measurement of speeds in the heavens, a time is the finite period in which some indexical occurrence takes place or is repeated.[15] The most accessible such occurrences are risings and settings. Thus, the speed of some heavenly body is measured by the rising and setting of some other heavenly body.[16] The basic problematic of ancient astronomy is a situation in which one motion measures another. Units of measure are taken from within the system itself, and one part is a unit of measure for other parts. Let us call this situation a problematic of *internal measurement*. This basic problematic is reflected also in the etymology of the word for "speed." The word τάχος is a paronym derived from the root ταχύς, which means "swift" or "fast" and takes as its proper opposite βραδύς. The term τάχος thus involves comparison in its meaning.[17]

The sort of internal measurement appropriate to astronomy is expressed in both formulations of the Eudoxan continuum principle—the general definition of ratio in *Elements* V, Def. 4, and the Archimedean bisection principle more prevalent in Aristotle's *Physics*. These formulations differ in origin and implication. The bisection principle, ascribed to Eudoxus by Archimedes, is used in the quadrature of the circle. One magnitude is bisected and then its parts are bisected so as to produce a part smaller than

[15] There is a fundamental intuition of time at work in the comparison of speeds, including a presumption of the evenness of time. I use the term *indexical* in the manner of C. S. Peirce and others for whom an index is a sign of something else because it is an effect of that other. The period of revolution of a heavenly body is the effect of its speed.

[16] D. H. Fowler illustrates how astronomical ratios relate periodic phenomena to one another in his article "Ratios and Proportion in Early Greek Mathematics," in Bowen, *Science and Philosophy* (n. 9), p. 108.

[17] François Lasserre suggests that, by the time of Eudoxus, the term τάχος had an unmistakably astronomical connotation (*Die Fragmente* [n. 5], p. 200). If so, the case parallels the appropriation of the word φαινόμενα by astronomy. The two cases together point to the intellectual power of astronomy with respect to other branches of knowledge. It is often noted that, for Aristotle, reputable opinion (ἔνδοξα) functions in dialectic as φαινόμενα do in astronomy. For a review of the extensive scholarship on ἔνδοξα and an interpretation that relates ἔνδοξα to φαινόμενα, see Kurt Pritzl, "Opinion as Appearances: *Endoxa* in Aristotle," *Ancient Philosophy* 14 (1994): 41–50.

any given magnitude. The principle provides a means for refuting the possibility that the areas of two circles are related to one another according to some ratio *other than* the ratio of the squares on their diameters. The principle is used in *Elements* XII, which refers back to *Elements* X.1, where it is stated in terms of taking parts greater than one-half.

The other formulation of the Eudoxan continuum principle comes in Euclid's Book V, said by an unknown scholiast to be the work of Eudoxus. The principle providing for the continuum, Def. V.4, states, "Magnitudes are said to have a ratio to one another which are capable, when multiplied, of exceeding one another." The standard interpretation of this definition is that it defines "being of the same species" for magnitudes and, furthermore, that it excludes the infinitely large and infinitely small, making every quantity of the same species finite and hence comparable to every other. Recent scholarship has given us reason to be cautious about the various meanings and implications ascribed to this definition.[18] Its basic import, however, is to ensure the comparison of any two finite magnitudes, even those having no common measure. Such magnitudes are comparable because their respective multiples produce quantities that can be related to one another as greater and smaller. Each, through multiples of itself, can thus measure the other, by one multiple falling short of and another multiple exceeding their relatum. The formulation of Aristotle that comes closest to Def. V.4 is the one at *Physics* VIII.10, 266b2: "By continually adding to a finite [magnitude], I shall exceed any definite [magnitude] and similarly by continually subtracting from it I shall arrive at something less than it." A magnitude is added to itself, or some submultiple of itself is subtracted from it.[19] It is significant, however, that Aristotle uses the language of addition and subtraction (προστιθεὶς καὶ ἀφαιρῶν), rather than the Euclidean language of multiplication (πολλαπλασιαζόμενα), for it emphasizes that the measuring relation is binary—a relation between the two original members of the ratio. For example, one quantity is added to itself until it exceeds the other, which remains the same.[20]

---

[18] For examples of the standard interpretation of Euclid's Def. V.4, see Heath, *The Thirteen Books of "The Elements"* (n. 9), 2: 120, and idem, *A History of Greek Mathematics* (New York: Dover, 1981) 1: 385; also Morris Kline, *Mathematical Thought from Ancient to Modern Times*, 3 vols. (Oxford: Oxford University Press, 1972), 1: 69. Recent reevaluations of the standard interpretation are found in Wilbur Knorr, "Archimedes and the Pre-Euclidean Proportion Theory," *Archives internationales d'histoire des sciences* 28 (1978): 183–244, esp. sec.VI–VII, and idem, "What Euclid Meant: On the Use of Evidence in Studying Ancient Mathematics," in Bowen, *Science and Philosophy*, (n. 9), pp. 119–63, esp. pp. 124–31. See also Ian Mueller, *Philosophy of Mathematics and Deductive Structure in Euclid's* Elements (Cambridge, Mass.: MIT Press, 1981), pp. 138–45.

[19] See also the similar formulation at *Physics* III.6, 206b7–12.

[20] The obvious difference between Def. V.4 and the bisection principle is that Def. V.4 multiplies some magnitude to produce a new quantity, while the bisection principle relies on the division of an existing magnitude. Also, in Euclid's Def. V.4 each magnitude is multi-

The centrality of binary relation without reference to a third appears even more clearly in the convergence principle, whether anthyphairetic or Archimedean, that Aristotle uses in his arguments about physical things. In an anthyphairetic ratio, the smaller of the two relata measures the larger as many times as possible, the remainder then measuring the smaller of the original two quantities. This process continues until the last remainder measures its predecessor without remainder; or it continues indefinitely, producing progressively smaller parts, the addition of which approaches a limit.[21] In an Archimedean convergence (the bisection principle), progressively smaller parts are produced by bisection of the same part. The addition of these parts also approaches a limit. Convergence of the anthyphairetic or Archimedean variety manifests the principle of binary relation

---

plied and the resultant respective multiples are compared to each other. In bisection, on the other hand, operations are performed only on one magnitude so as to make it converge upon the other.

Definition V.4 and the bisection principle are associated with different possible sources for Eudoxus's mathematics. Def. V.4 is linked with the discovery of incommensurables because incommensurability is the concern of *Elements* V. Indeed, the definition of ratio is often seen as subordinate to Def. 5, which defines proportionality in terms suitable to a general approximation of irrational quantities. The bisection principle, on the other hand, is associated with the geometric problem of quadrature of curvilinear figures. Each of these possible sources highlights a different aspect of the proportionality theory generally identified as Eudoxan. The problem of incommensurability directs attention to the structure of the continuum as the necessary background for handling irrational quantities. Quadrature problems utilize a method of convergence in the context of a *reductio* argument using proportionality. Most scholarship on this topic in the *Physics* treats the topological issue, the structure of the continuum. Nevertheless, in his actual arguments about motion, Aristotle himself uses as his continuum principle almost exclusively the principle of convergence; and he uses it, following the pattern of quadrature proofs, in *reductio* arguments involving proportionality.

There is difference of opinion as to which formulation of the continuum principle, Def. V.4 or the Archimedean convergence principle, was formulated earlier and which belongs to Eudoxus rather than Euclid. Hans-Joachim Waschkies, in *Von Eudoxos zu Aristotles* (Amsterdam: B. R. Grüner, 1977), places emphasis on Def. V.4, and in particular on the homogeneity criterion for magnitudes that it contains. He believes that this principle was formulated by Eudoxus and known to Aristotle (pp. 20–33). Knorr, on the other hand, believes that Def. V.4 is a reworking of Eudoxus's ratio principle, which was itself more like Archimedes' lemma of bisection; in Euclid, this reworking served as a ground of proof for the treatment of incommensurability and for proving a Euclidean version of the bisection principle used in quadrature ("Archimedes and Proportion Theory" [n. 18], pp. 205–11). Knorr believes that Aristotle's uses of the convergence principle already represent a first stage in the rethinking of the Eudoxan method of "exhaustion" in the direction of a more general formulation relevant to the Eleatic paradoxes of motion and divisibility (p. 210). Scholars subscribing to the standard interpretation tend to view the two different formulations as implying each other, and therefore have not clearly distinguished them historically. See, for instance, Heath: "[Def. V.4] is also practically equivalent to the principle which underlies the method of exhaustion now known as the Axiom of Archimedes" (*History* [n. 18], 1: 385).

[21] Anthyphairetic ratios played a role in the development of the medieval theory of proportion. Concerning anthyphairetic formulations in Arabic and medieval texts, see John Murdoch, "The Medieval Language of Proportions: Elements of the Interaction with Greek Foundations and the Development of New Mathematical Techniques," in *Scientific Change*, ed. A. C. Crombie (New York: Basic Books, 1963), pp. 237–71, especially 253–55.

more clearly than Def. V.4, because in each method of convergence, no presumption of indefinite finite extendability is involved. There is a definite finite range for measurement, set by the whole of the larger relatum. Measurement is then internal to the binary relation.

Aristotle uses internal measurement to compare the speeds of bodies in the sublunary sphere.[22] One example of his application of internal measurement is a passage in *Physics* IV.8 famous as a source of his so-called mechanics. Aristotle says that the same body moving through two media of different densities will move more quickly through the finer medium, and the times will be proportionate to the densities of the impeding bodies (216a6–7). The possibility of forming a proportionality of times and resistances is important because of what happens to one side of the proportionality when the null resistance of the void enters into consideration: in this case, there cannot even be a ratio between the void and a plenum (215b20). Aristotle expresses the reason as follows:

> However, there is no ratio to measure the extent to which the void is exceeded in density by body, just as what is nothing has no ratio to any number. Four exceeds three by one, and it exceeds two by more than one, and it exceeds one by still more; but there the sequence of ratios ends. The excess of four over nothing cannot be expressed by any ratio, because it has to be possible to split the greater number into the less number and the remainder, which would mean that four would be the sum of the remainder and nothing.[23]

Aristotle says this would be like a line's exceeding a point by some amount. The null resistance of the medium cannot enter into a ratio because it is like the point, no collection of which can make a line. His reference to splitting the greater number into the lesser and a remainder most readily describes the beginning of an anthyphairetic convergence. His point is that only a finer density, not the void, can be related to the density of air in this situation, because no matter how many times nothing is added to itself, it will

[22] I have appropriated the term "internal measurement" from an article by Robert Sokolowski, "Measurement," *American Philosophical Quarterly* 24/1 (1987): 71–79. Sokolowski gives an account—of internal measurement as descriptive of how *modern* scientific theory works—which differs in some respects from the description I give of the Eudoxan ratio theory as an internal measurement system. Sokolowski says that different *parts* of a situation, when idealized, are brought into relation so that one measures the other. In a Eudoxan ratio, on the other hand, one of the "parts" of the situation of measurement actually functions as a whole to be approximated through its measurement by the other. For the Eudoxan ratio, idealization is less important as a background condition for mathematical science, and the method of measurement is linked to the empirical situation that makes one part of the system the "whole" for the other. In spite of this difference, the contrast of internal and external measurement that Sokolowski develops may prove to be very useful for understanding ancient mathematical science, once we recognize that ancient ratio theory is an internal measurement system. For a version of this contrast applied to the conception of time periods in Eudoxus and Aristotle, see below.

[23] *Physics* IV.8, 215b12–18. The translation is by Robin Waterfield, *Aristotle: Physics* (Oxford: Oxford University Press, 1996).

never produce something. This formulation highlights the "exhaustion" of the larger quantity by additions of the smaller, as in *Physics* VIII.10. Even from this point of view, the convergence of one member of the comparison upon the other is the key element in the substitution of a finite for a null quantity.

As Aristotle expresses it, this convergence is a division of the larger by the smaller to produce a portion measured by the smaller and a remainder. Hence, it is a taking away of the larger by measuring, where the measuring unit is the other member of the ratio. The internal measure is applied, in this passage, to a proportionality between times and densities, but Aristotle makes clear, after his next argument in the chapter, that what he is really aiming at is the comparison of movements (216a8–11). In the final analysis, it is speeds that are correlated to the difference in densities: ἐχέτο δὴ τὸν αὐτὸν λόγον ὅνπερ διέστηκεν ἀὴρ πρὸς ὕδωρ, τὸ τάχος πρὸς τὸ τάχος (215b6–7). The mutual measurement involved in the convergence principle, whether anthyphairetic or Archimedean, is very well suited to the comparison of faster and slower motions where it is their relation to one another that matters, not their measurement by some other standard. It is also most appropriate to the measurement of limited quantities whose dimensions are finite and fixed.

That it is limited quantities of fixed dimensions that are measured draws attention to the absence of a notion of indefinite extendability in Aristotle's uses of ratios. This absence is especially noticeable in his account of the potential infinite in *Physics* III.6, where an Archimedean version of convergence is used. The infinite resides in the possibility of continual divisions of the *same part* of some magnitude. A magnitude is continuous because any part just produced can be divided to produce another smaller quantity like itself. This suggests that Aristotle's conception of infinite divisibility is a kind of successor function and is very "process-dependent." The next operation always returns to the part produced by the immediately preceding cut. This fact about infinite divisibility keeps the meaning of the continuum linked to things already known or defined.

Another thing of interest in *Physics* IV.8 is the idea of one *density* converging upon another. Aristotle does not hesitate to compare the relation between degrees of density to the relation between faster and slower speeds. He does not use the term μέγεθος to refer to the degrees of density or speed. It would appear that he sees no need, in this context, for a presumption of the homogeneity of magnitude for all proportionals.[24] It is simply that each type of thing constitutes a natural continuum. Whatever

[24] Elsewhere, Aristotle discusses the homogeneity of magnitude as an aspect of mathematics (*Posterior Analytics* I.5, 74a17–25). For a discussion of this issue, see Ian Mueller, "Homogeneity in Eudoxus's Theory of Proportion," *Archive for History of Exact Sciences* 7 (1970): 1–6.

division can be made in one medium can be matched in the other. The reasoning is proportional and the methodology need not involve geometrical abstraction. This is a difference between the mathematics of the continuum, which involves abstraction, and the mathematics of continua, which may include features of continuous things in its formulation.

It is worth noting that the reasoning in *Physics* I.7, where Aristotle argues that every natural substance has a substratum, is strikingly similar. Both *Physics* I.7 and IV.8 rely on reasoning to a fourth proportional, the presumption of which, in Eudoxan proportionality theory, is licensed by the continuum principle applied to the relata of each ratio. The homogeneity of all four proportionals is clearly excluded from consideration in the ratios of I.7, where Aristotle claims that different substrata are related proportionally to plants and animals, statues, houses, or things formed by growth or alteration (190b1–10). Nevertheless, he does apply a version of homogeneity to the relation of matter and form, making them extremes within a genus of change (I.5, 188a31–b26). Reasoning to a fourth proportional relies on a presumption of homogeneity within each ratio. Once the extremes of change are admitted as homogeneous, the extension of proportional reasoning to the relation of matter and form is not different in kind from the proportionality of times and distances in *Physics* IV.8.

Besides being an example of internal measurement, *Physics* IV.8 shows Aristotle's emphasis on speeds and their comparison in terms of time, his use of proportional reasoning comparing dissimilar types of natural entities, and his preference for a definition of the continuum that involves no inference to indefinite extendability. All these things point to some venue other than the abstraction of pure mathematics for the mathematical treatment of motion in the *Physics*.

The difficulty with linking these aspects of Aristotle's treatment of motion to the mathematical treatment of speeds in astronomy is that we know so little about Eudoxus's *On Speeds*. This work was an early treatise in the tradition known as the "Little Astronomy," which included later treatises by Autolycus (*On the Moving Sphere*, *On Risings and Settings*) as well as by Euclid, Theodosius, and Aristarchus.[25] Alan Bowen and Bernard Goldstein suggest that later ancient interpretation in the tradition of the Great Astronomy, Ptolemy's *Syntaxis*, placed emphasis on elements of the earlier tradition that do not accurately reflect its problems and approaches. They contrast Simplicius's emphasis on the explanation of planetary motion with what they take to be the greatest contribution of Eudoxus, the system of homocentric spheres.[26] I would submit that a similar forgetfulness occurred

[25] For a complete list and an evaluation of the evidence, see Heath, *Aristarchus of Samos* (n. 3), pp. 317–21.

[26] Alan C. Bowen and Bernard R. Goldstein, "A New View of Early Greek Astronomy," *Isis* 74 (1983): 330–40.

with respect to the significance of speeds as the basic problem of fourth-century astronomy. There is a need to reconstruct the context of questions addressed and expectations to be answered for astronomy in the fourth century B.C. Fortunately, there is some textual evidence from the philosophers that shows how the issue of speeds set the agenda for answers sought and also determined the factors to be quantified for celestial motion and, by extension, for sublunary motion. In evaluating this evidence, astronomy's problematic of internal measurement is crucial, because another feature of internal measurement will be the key to astronomy's influence on Aristotle. I will call this feature the *principle of the indexical whole.*

We have seen that, in general, a ratio based on convergence functions by internal measurement. One finite quantity takes parts of another by a series of operations. Each operation of making cuts is related to the previous operation by a rule that is always the same. Accordingly, one of the two relata starts as a kind of whole in which the other makes cuts. Such a system has no need of reference to an external standard for its unit. This aspect of internal reference is a necessity in astronomy, as we have seen. What we shall find is that, for astronomy, this problematic brings with it a standard of measure determined from within the situation of comparison of speeds. The standard is some movement in the heavens that can function as a whole with respect to other movements. Divisions of this whole serve to determine variations in speed. What we wish to see first, however, is how this principle of the indexical whole arises from the problems given in experience by τὰ φαινόμενα.

Eudemus of Rhodes reports that it was Plato who set the task for astronomy as preserving the appearances of the motion of the planets by the hypothesis of regular and ordered motions and that Eudoxus first met the challenge with his system of homocentric spheres.[27] Discussions of saving the appearances sometimes neglect precisely which phenomena are to be preserved in the motions of the heavenly bodies. Plato's own account in the *Gorgias* makes it clear:

> And if someone asked about astronomy, when I said that this craft too achieves everything by speech, and asked, "And what is the speech of astronomy about, Socrates?", I would say it is about the movement of the stars [ἄστρων], the sun, and the moon, how they are related in speed to one another.[28]

For Plato, the task of preserving appearances is not primarily to make predictions or even to describe the locus of the movement, but rather to account for the *differences* in the speeds of the fixed stars, sun, moon, and

---

[27] Lasserre, *Die Fragmente* (n. 3), F121.

[28] *Gorgias* 451c5–9. The translation is by Terence Irwin, *Gorgias* (Oxford: Clarendon Press, 1979).

planets.[29] The problem of speeds was a matter of accounting for which heavenly bodies ran ahead and which behind in the common direction of their movement, and when, if ever, these relations changed, either through one body's passing another (πάροδος) or by a change of direction (τροπή). In the *Phaedo*, Plato describes the data of astronomy in terms of these relations. Recounting his disappointment with Anaxagoras, Socrates says he expected to learn from him concerning the sun, moon, and other heavenly bodies not only their relative speeds, their turnings, and other affections (τάχους τε πέρι πρὸς ἄλληλα καὶ τροπῶν καὶ τῶν ἄλλων παθημάτων), but also in what way it was better that they should act and undergo as they do (98a3–5). The indexical celestial events are here portrayed as the products of apparent relative speeds. In the *Timaeus*, it becomes clear that resultant speeds are not the only ones to be treated in astronomy. In the case of the sun, moon, and planets, these speeds are explained by the speeds of circles moving in a direction opposite to the movement of the celestial sphere. The movements of these circles constitute the movement of the Different. The circles are seven in number, and their speeds are related, Plato says, by ratio (36b–d).[30] The τάχη of Eudoxus's title undoubtedly also refer both to phenomenal speeds and to the component speeds that produce them.

Aristotle's *De caelo* II.10 most clearly reflects the influence of Eudoxus's *On Speeds*, as Simplicius's commentary on the chapter shows. In chapter 8, Aristotle has laid down the principle that the speeds of the stars are in proportion to the sizes (τοῖς μεγεθοῖς) of the circles on which they travel (289b16). Here, distance is a factor in speed, but it is a ratio of the linear distances of the heavenly bodies from the earth that is productive of relative curvilinear speeds. He reiterates that principle in chapter 10, attributing to the mathematicians the comparison of the speeds of the stars based on their distance from the swiftest circuit. The swiftest is the largest, the circuit of the heavens itself. This principle echoes Plato's principle from *Laws* X that points on a rotating disk move at speeds relative to their distance from the circumference. But Aristotle ascribes the differences in speed to composition of movements: each heavenly body, besides moving at the same speed as the heavens, also moves on its own circle, which goes in a direction contrary to the movement of the outermost sphere and makes the heavenly body it carries thus move with a composite motion. The planets nearest the outermost sphere are most influenced by it and move at the

[29] Lasserre makes this point, claiming that the fundamental problem of ancient astronomy was the problem of speeds (*Die Fragmente* [n. 3], p. 200).

[30] See F. M. Cornford's commentary in *Plato's Cosmology: The "Timaeus" of Plato* (Indianapolis: Bobbs-Merrill, 1975), pp. 73–93. The movement of the seven circles is divided into groups of three and four. The three circles of the outer planets, besides sharing in the movement of the Different, also move in an opposite sense, slowing down the movement of the Different for their own circles; see Cornford, p. 86.

slowest pace on their own circles (291b4).[31] The system of differing composite movements is a function of the relative positions of the heavenly bodies (πῶς ἔχει πρὸς ἄλληλα τοῖς ἀποστήσιν) in comparison with the distance of the sphere of fixed stars. This array or ordering (τάξις) of the heavenly bodies Aristotle characterizes as the proper subject of inquiry of the astronomers (291a29–32).

In his commentary on this passage, Simplicius assumes that Eudoxus's system of homocentric spheres is Aristotle's model, and he presents the model in terms of the issue of relative speeds, referring specifically to Eudoxus's *On Speeds*.[32] He treats successively the motions of the sun, moon, and five planets. In every case, the first movement belonging to the heavenly body takes a time equal to the motion of the fixed stars (ἰσοχρόνιος κόσμου, ἰσοχρονίως τῇ τῶν ἀπλανῶν).[33] The differences in their times of movement arise from the other circular movements they undergo. Concerning the three movements of the sun, Simplicius says:

> And if the two spheres, the middle [in time of speed] and the least, were themselves unmoved, the sun's movement around would occur in a time equal to the turning of the cosmos. But as it is, since they turn back in the opposite direction, the revolution from one dawn to the next lags behind [ὑστερεῖ] the forementioned time.[34]

This passage shows that the problem of speeds in the heavens lies in explaining the retardation of movement (ὑπολείψις) and "following behind" (ὑστερεῖ)—by reference to indexical events, like the rising of certain fixed stars. Having a common direction is intrinsic to both the problem and the solution. All the heavenly bodies move in the direction of the sphere of fixed stars and at its speed, even when the movements of their own circles carry them backward at the same time. These backward movements, besides making some move slower, can produce change of direction, overtaking, and lagging behind. In the case of the sun, the lag is judged in relation to the rising of particular fixed stars. For the planets, the index is the planet's return to the same point on the zodiac. Each planet moves with the motion of four spheres. Concerning the second and third spheres, Simplicius reports:

---

[31] On moving slowest because of being under the influence of the fastest circuit, see Simplicius, *In de caelo*, 471.21–472.4. In Eudoxus's system, each heavenly body inside the celestial sphere has one sphere productive of its own movement that is moving at the same speed as the celestial sphere. He does not presume that the celestial sphere communicates its own movement to each planetary set. Plato and Aristotle both seem to think that the celestial sphere communicates its motion to the sun, moon, and planets. On this topic, see Lasserre, *La naissance* (n. 3), pp. 227–30.

[32] Lasserre, *Die Fragmente* (n. 3), F124, 69.10.

[33] Ibid., F124, 69.17, 71.12–13.

[34] Ibid., 69.16–70.2. Translations from Simplicius are my own.

> and the second, having its poles in the first, makes its revolution in the opposite direction [ἔμπαλιν], from West to East, around the axis and poles of the circle through the middle of the zodiac, in the time in which each of them [the planets] appears to complete the circle of the zodiac. Thus, in the case of the star of Hermes and that of Hespherus, he says that the circuit of the second sphere is completed in a year, in the case of that of Ares, 2 years, in the case of that of Zeus, 12 years, and in the case of that of Chronos, which the ancients called the star of the Sun, 30 years.
>
> The situation for the remaining two spheres is as follows. The third [sphere for each planet] has its poles on the great circle going through the middle of the zodiac and conceived as being in the second sphere of each. The third sphere turns from South to North in the time each [planet] takes from one appearing to the next, going through all the positions in relation to the sun—which [time] the mathematicians [ὁι ἀπὸ τῶν μαθημάτων] call the synodic period [διεξόδου χρόνον]. This is different for each planet. Thus, too, the revolution of the third sphere is not equal in time to that of the whole but, as Eudoxus believed, takes 19 months for the star of Aphrodite, 110 days for that of Hermes, 8 months and 20 days for that of Ares, and for each of the stars of Zeus and Chronos, approximately 13 months.[35]

Simplicius goes on to say that the fourth sphere for each planet, the one that actually bears the planet, moves in the opposite direction to the third but moves in an equal time, its function being to keep the planet near the zodiac and to cause the variations in speed that account for retrogradation. It is the second sphere that provides for the planet's traversing (πάροδον) of the whole zodiac and the third that accounts for the planet's movement in latitude during its course through the zodiac.[36] The indexical whole is the great circle of the zodiac, with some point on it serving as both beginning and end of a complete traversing.

One thing evident throughout the passages quoted from Simplicius's commentary is the centrality of a benchmark time period set by the movement of one of the heavenly bodies over the indexical whole. For the planets, the index is the sun's completion of the same course through the zodiac. Comparisons are made by means of distances covered in equal times. This means of comparison is a result of the necessity that speeds be known relatively in astronomy. From the modern standpoint, however, speeds could be relative and yet the unit for measuring them still be translatable to other contexts. The unit is a yardstick that can be laid alongside something else to produce a sum or a division. Our conception of "a time" is such a translatable unit of measure. In Eudoxan astronomy, however, the fundamental unit of measurement remains one of the things being measured. The

---

[35] Ibid., 71.13–72.9. To avoid periphrasis, I follow the ancients' generalized use of the term τὸ ἄστρον, "star," to mean "heavenly body," some of which are fixed (ἐνδεδεμένα) and some of which are wanderers (πλανητά).

[36] Ibid., 72.20, 73.3–7.

aim was *not* to know speed in terms of a universal unit. As Plato and Aristotle describe it, the aim is to account for differences in speed, and for the particular array of heavenly bodies at any given time. The indexical whole is a unit inseparable from this context of specific relations.[37] This is an aspect of the incomplete abstraction of astronomy that, when extended to sublunary nature, encourages a contextual treatment of natural principles. Let us consider some indications in the *Physics* that Aristotle did extend the problematic of astronomy to rectilinear natural motion.

In turning to Aristotle's account of sublunary motion in the *Physics*, the first thing to notice is that, in spite of the availability of the heavens as a timepiece for sublunary occurrences, he takes an approach to the times of movement in the sublunary sphere that is similar to the handling of time in *On Speeds*. In his treatise on time (*Physics* IV.10–14), he does not discuss the heavens as a standard of time until the last chapter, where he thinks it appropriate to give a reason why time seems to be the motion of the heavenly sphere: it is because its motion is the most regular and other motions are measured by it (223b18–23), not because of the superiority of the heavens on cosmological or metaphysical grounds. More to Aristotle's purposes than the forging of a cosmological standard is his treatment of time as a measure in chapter 12 of the treatise on time. Even here, however, he points out that not only does time measure motion, but motion in its turn measures different times. This is explainable on the model of astronomy that Simplicius provides. What happens in equal times is measured by that time, but also some occurrence, taken as a whole, measures the different times that different bodies take to complete the occurrence.

Some of the fruits of the approach to measuring movement outlined in the treatise on time are evident in *Physics* IV.8, when Aristotle argues against the possibility of motion in the void. The passage from this chapter treated earlier explained why the void cannot be brought into ratio with some density. The passage to be considered now makes explicit the substitution of a finite density for the null density of the void in a proportionality. Aristotle supposes that a body *A* traverses a certain distance in a void in time *G*. He then supposes that the same body covers the same distance, this time in a plenum like air, in a longer time, *E*. But if this is the case, then there is some other, finer plenum through which *A* will move in the shorter time *G*. For as the two times *E* and *G* are related, there will be some other density corresponding to *G* that is related to the density of air. But *A* covers the void in time *G*. So it traverses either the void or some plenum in the same time, which is impossible (215b22–216a4). There is more than one

[37] As Alan Bowen describes it, a similar state of affairs prevailed in ancient harmonics, where the basic phenomenon to be explained was "the relation between two pitches and not pitch itself." See Alan C. Bowen, "The Foundations of Early Pythagorean Harmonic Science: Archytas, Fragment I," *Ancient Philosophy* 2 (1982): 95.

layer of Eudoxan influence in this *reductio* argument. Within the argument, Aristotle uses the principle that a finite quantity may always be divided into a smaller finite quantity of any size. The "outer" layer of Eudoxan influence is the framing of the argument in terms of the same object's covering equal distances in equal times.[38] If the body does not alter in weight, the only reason for its traveling faster is a difference in the medium through which it travels (215a25–29).[39] It is important that traveling faster means taking less time to complete the same occurrence. This is in fact the definition of "faster" (θᾶττον) that Aristotle gives at the end of Book IV, in chapter 14 of the treatise on time: "By the term 'moving faster,' I refer to what changes sooner [πρότερον] into its subject, moving with an even motion the same distance [as another]" (222b34–223a1). This passage states, for sublunary motion, the principle of the indexical whole that is so important in astronomy. Aristotle even treats this proposition as a general principle, saying that it applies to both circular and rectilinear motion, and similarly also to other types of change (223a2–4).

In *Physics* VIII.8, Aristotle quite explicitly carries this principle over from circular motion into his treatment of rectilinear motion. Circular motion and rectilinear motion are, for him, the two simple motions.[40] But rectilinear motion cannot be continuous (συνεχής). Aristotle means by this that it cannot go on indefinitely without a break or gap that realizes a point of termination. Circular motion is the only continuous motion. Circular motion is "from itself to itself"—that is, its starting and ending points are the same. What is more, in circular motion the moving thing is always in different places, the requirement for continuity in motion (264b21–22). For rectilinear motion to go on indefinitely, on the other hand, it would have to be often in the same places. Aristotle says that rectilinear motion would have to be proceeding in opposite directions at the same time, and therefore passing the same point going in contrary directions, in order to make a claim on continuity. This is because it would have to *change direction* yet not realize any point of rest that finishes its motion in the first direction. If the universe were not finite, rectilinear motion could continually be in another place without reversing direction. But the only continuous motion of a finite sort is circular motion. The upshot is that, unlike circular motion, every rectilinear motion must come to an end, and come to an end in another. It proceeds "from itself to another" (264b19).

---

[38] Knorr believes that this idea was a definition or axiom of Eudoxus's *On Speeds*, and that Eudoxus would have established as a theorem that distances traversed are in proportion to time; see Wilbur Knorr, "Infinity and Continuity: The Interaction of Mathematics and Philosophy in Antiquity," in *Infinity and Continuity in Ancient and Medieval Thought*, ed. Norman Kretzmann (Ithaca, N.Y.: Cornell University Press, 1982), p. 120.

[39] Aristotle adds a *ceteris paribus* clause: ἐὰν τἆλλα ταὐτὰ ὑπάρχῃ(215a28).

[40] *De caelo* I.2, 68b26–69a30

This passage is striking in its derivation of the reference points for sublunary motion from considerations governing both circular and rectilinear motion. From these considerations, the passage provides a reason, based on the finiteness of the universe, for why every sublunary motion must end. The criterion of completeness is the central thing in arriving at this conclusion. While, in circular motion, completion is a return to the same point, in rectlinear motion, completion must be a termination in some other point. The reference to a completed whole is itself a requirement for judging speeds in the internal measurement system of a Eudoxan ratio. In astronomy, it is the return to the same point on a circle that yields the criterion for judging speed. When transferred to rectilinear motion, the criterion of completeness means conceiving a movement as a whole, including starting point and ending point. In this way, the principle is established that the indexical wholes in rectilinear motion will be movements that have genuine completions, in some way comparable to a return to the same point on a circle. Once this criterion of completeness is transferred to rectilinear motion, a holistic approach to the analysis of sublunary motions is ensured.

One consequence of applying internal measure to sublunary motion is the conception of sublunary motion in terms of τὸ πρότερον and τὸ ὕστερον. These terms are crucial to the comparison of speeds in astronomy, where moving bodies share a common direction of motion. Differences in speeds are judged by a heavenly body's running ahead or lagging behind, passing or turning back. These terms are also the most appropriate for describing the relation between points within a continuum ordered according to a Eudoxan ratio. Taking the bisection principle as an example, a smaller quantity which is some part of a whole is made to approach the size of the whole by means of the addition of successive bisections of the smaller. Each bisection adds one-half of the part produced by the immediately preceding bisection. The additions establish a series of cuts in the larger quantity.[41] These cuts are ordered like the series of natural numbers. When the quantity being divided is an indexical whole, the series is ordered in the direction of the termination point of movement over the whole. This ordering of the range making up the indexical whole of a Eudoxan ratio applies also to the stages of change in rectilinear motion. In sublunary motion, priority and posteriority will be judged by reference to points on the course covered. Τὸ πρότερον is farther from the termination point, because in the natural order of how the end must be accomplished, it is a point that a moving body en route to the termination must pass *before* it reaches another point, τὸ ὕστερον. Speeds are judged in terms of times taken to complete a whole—or, derivatively, in terms of parts of a whole covered in the same

[41] In Physics III.6, Aristotle calls this series of additions the infinite by addition (206b3-9), meaning a potential infinite whose additions will never exceed the whole being divided..

time. Accordingly, stages of change are ordered by priority and posteriority in relation to the end of the change.

In *Physics* VI.6, Aristotle makes this point, developing to the fullest extent the implications of the continuum principle for the nature of sublunary motion. There, he says that every thing that "is changing" previously "has changed." His pointed use of present and perfect tenses is significant, because the perfect tense signifies, for Aristotle, the presence of the τέλος of a change.[42] He means that anything in the process of change has also achieved some part of the end. Using generation as the species of change, he says that everything in process of coming-to-be has come to be before, though it has not always come to be the final product of the generation, "but sometimes something else, for example one of the stages of the change [οἷον τῶν ἐκείνου τι—literally, one of those belonging to that], for example of a house, the foundation" (237b9–13). He makes the relation of stages of change to their end part of the import of the continuum idea. In the passage just quoted, he says this relation of coming-to-be and having-come-to-be belongs to "as many things as are divisible and continuous" (237b11), and he emphasizes that because of this relation of parts of a change to the end of change, "straightway something infinite inheres in the thing coming-to-be or passing-away" (237b13–14). When a change and the underlying course it covers are divided at any point, the part exists in an ordered relation of priority with respect to the later termination point of the course.

This passage makes explicit the principle of holism—that parts of a change must be understood in relation to the completion of the change. The passage has important implications for the meaning of measurement for Aristotle. George Molland has said that an object's being continuous presents difficulties for measurement, because "if there are no ultimate constituents of a continuum, there are, as it were, no inbuilt markers."[43] But ordering in relation to an end, which characterizes Aristotle's holism in this passage, yields any number of different series of markers, understood as ordinals—e.g., first, second, third. There is no ultimate fineness of discrimination, but there is a structure of marking.

---

[42] On the implications of tense markers in Aristotle, see J. D. Ackrill, "Aristotle's Distinction between *Kinesis* and *Energeia*," in *New Essays on Plato and Aristotle*, ed. Renford Bambrough (London: Routledge & Kegan Paul, 1965), pp. 121–41; and Terry Penner, "Verbs and the Identity of Actions," in *Ryle*, ed. O. P. Wood and G. Pitcher (Garden City, N.Y.: Anchor Books, 1970), pp. 393–460. There has been a great deal of research in both philosophy of action and linguistics on the family of distinctions that correspond roughly to Aristotle's understanding of tense markers. See, for instance, Alexander Mourelatos, "Events, Processes, and States," *Linguistics and Philosophy* 2 (1978): 415–34, and Robert Binnick's encyclopedic *Time and the Verb: A Guide to Tense and Aspect* (Oxford: Oxford University Press, 1991).

[43] Molland, "Aristotelian Holism" (n. 8), p. 229.

This point also bears upon Aristotle's definition of time, in *Physics* IV.11. Aristotle says that time is the number of motion with respect to before and after (ἀριθμὸς κινήσεως κατὰ τὸ πρότερον καὶ ὕστερον) (219b2). This definition is usually understood as laying the foundation for a view of time as a metric. By "time," Aristotle means periods of time that serve as units to measure motion. Sarah Waterlow expresses this view:

> *chronos*, so far as he is concerned with it, is not time as such (if by "time" is meant something like succession or successiveness or the possibility of these): rather, it is time *qua* quantifiable. For that is how he explains *chronos*, i.e. as "the number" and also "the quantity of movement in respect of before and after."[44]

Waterlow thinks that Aristotle's account of time gives insufficient weight to the issue of temporal succession. She interprets τὸ πρότερον and τὸ ὕστερον in terms of distance from the present, whether with respect to the past or future. I submit that this view and others similar to it presume that Aristotle is working with the kind of measuring system in which time periods are units that can be laid along another quantity to measure it completely in terms of the unit. The units themselves, as units, have no particular internal structure. Viewed from the standpoint of internal measurement, however, the reference to prior and posterior in the definition of time suggests that Aristotle means by ἀριθμός *ordinal* number, and that the ordinal must be understood in relation to a whole ordered by a ratio. In this case, it is *movement* that measures another movement, as in astronomy. Times of completion are indexical of comparison of movements, and even a period of time has an internal structure of priority and posteriority established by the relation of its parts to the indexical whole.

Finally, it is important to notice that Aristotle's incorporation of Eudoxan thought produces a genuine scientific theory of motion. If rectilinear motion can be measured, it is because motion is (1) defined by starting and ending points that are different, and (2) characterized by parts existing in an ordered relation to one another and to their end. In this theory of motion, ascribing direction or progressivity to movement does not depend on the teleological view of nature elaborated in *Physics* II. Indeed, it would be interesting to investigate to what extent Aristotle's view of final cause is built upon or takes as its point of departure a theoretical foundation of internal measurement.

The way in which Aristotle involves ends with infinite divisibility, in *Physics* VI.6, refocuses attention on the indexical whole and how it is cho-

---

[44] Sarah Waterlow [Broadie], "Aristotle's Now," *Philosophical Quarterly* 34 (1984): 109–10. For instances of a similar interpretation, see Julia Annas, "Aristotle, Number, and Time," *Philosophical Quarterly* 25 (1975): 97–113; Michael J. White, "Aristotle on 'Time' and 'A Time,'" *Apeiron* 22 (1989): 207–24; and idem, *Continuous and Discrete* (n. 2), pp. 74–86.

sen. In astronomy, the starting and ending point of a body's circular movement is determined in accordance with a ranking of heavenly bodies determined by experience. The sphere of fixed stars is most regular in its movements. The sun has the greatest effect on other events, both celestial and meteorological. The way to judge movements in relation to the sun and fixed stars must be with reference to the horizon, which is determined by the place of the observer. All these facts about astronomy show that the requirement for indexical wholes links astronomy to experience and makes it more difficult to separate spherics from celestial phenomena. It is likely that Aristotle thinks indexical wholes for sublunary motion similarly come out of the rankings of experience—for instance, by noticing self-terminating movements that culminate in functional units that can be defined. The centrality of indexical wholes to Eudoxan method in astronomy suggests, however, that Aristotle would not seek to abstract his use of ratios in sublunary motion from the wholes provided by experience. The abstraction remains incomplete, and the movement is defined by its end.

THE LATIN SOURCES OF *QUADRANS VETUS*, AND WHAT THEY IMPLY FOR ITS AUTHORSHIP AND DATE

†WILBUR R. KNORR

The medieval tract on the quadrant called *Quadrans vetus* or *Quadrans secundum modernos*, or simply *Quadrans*, has given rise to a vigorous interpretive debate. Among the issues disputed are the author's name and the date of composition: Was he magister *Robertus* (or *Johannes*) Anglicus, *in* (or *de*) Monte Pessulano (that is, Montpellier)? Was it written after the mid-thirteenth century, or perhaps as much as a century earlier? The classical pronouncements by eminent scholars like Paul Tannery, Pierre Duhem, José M. Millás y Vallicrosa, and Lynn Thorndike have been scrutinized in recent studies by Emmanuel Poulle, Steven K. Victor, and Nan L. Britt Hahn.[1]

In reconsidering these issues my approach will be through a source analysis of *Quadrans vetus* based on literary parallels. Partial efforts of this sort have already been undertaken by Victor and by Hahn. A textual methodology has also been exploited by Millás, who revealed how accounts of a form of the "quadrant with cursor" can be found already in the tradition of practical geometry and astronomy in the eleventh and twelfth centuries. But none of these accounts has attempted to establish the *specific* affinities of *Quadrans vetus* to the immediate precursors within its own genre—that is, to thirteenth-century Latin tracts such as those by Sacrobosco and Campanus. When this is done, I will show, the answers to the other issues become clearer; in particular, the question of the date is resolved. Furthermore, one comes to realize how very extensively *Quadrans vetus* depends on these literary sources.

[1] These editions and studies will be cited continually throughout the present discussion: P. Tannery, "Le traité du quadrant de maître Robert Anglès (Montpellier, xiiie siècle)," 1897; reprinted in idem, *Mémoires scientifiques*, vol. 5 (Toulouse: Privat; Paris: Gauthier Villars, 1922), pp. 118–97 (edition of medieval Latin and Greek texts, with commentary); idem, "Le traité du quadrant" (1899), in ibid., pp. 198–203. P. Duhem, *Le système du monde*, vol. 3 (Paris: Hermann, 1915; repr. 1958), esp. chap. viii: "L'école de Montpellier: Robert l'Anglais," pp. 291–98; J. M. Millás y Vallicrosa, "La introducción del cuadrante con cursor en Europa," *Isis* 17 (1932): 218–58; idem, *Estudios sobre Azarquiel* (Madrid and Granada: Consejo Superior de Investigaciones Científicas, 1943–50); L. Thorndike, "Robertus Anglicus," *Isis* 34 (1942–43): 467–69; idem, "Who Wrote *Quadrans Vetus*?" *Isis* 37 (1947): 150–53; E. Poulle, "Les instruments astronomiques de l'Occident latin aux xie et xiie siecles," *Cahiers de civilisation médiévale x–xii siècles* 15 (1972): 27–40; S. K. Victor, *Practical Geometry in the High Middle Ages*, Memoirs of the American Philosophical Society 134 (Philadelphia: American Philosophical Society, 1979); N. L. Hahn, *Medieval Mensuration: Quadrans Vetus and Geometrie Due Sunt Partes Principales*, Transactions of the American Philosophical Society 72, no. 8 (Philadelphia: American Philosophical Society, 1984).

In the first section, I will review the debate and briefly critique the arguments proposed for the various positions. In the next two sections I will take up the comparison of *Quadrans vetus* with two of the principal related tracts, namely, the Sacrobosco commentary by Robertus Anglicus, and the quadrant tract by Campanus. In the conclusion, I consider what these findings imply in regard to our view of the *Quadrans vetus* and its place in the thirteenth-century academic context.

1. *QUADRANS VETUS*

In his edition of the *Quadrans vetus* in 1897, Tannery departed from the general consensus among prior bibliographers in maintaining the author's name to be Robertus Anglicus rather than Johannes.[2] From what he takes to be the earliest datable copy (MS Cambridge, Univ. Lib., Ii 3.3) he infers its composition "avant 1276."[3] Accordingly, he takes it as given that the author of *Quadrans vetus* is the same as the Robertus Anglicus to whom is due a commentary on Sacrobosco's *De spera*, made at Montpellier in 1271. However, he distinguishes the date of composition of the tract from the date of the instrument's introduction into the West, for he finds an explicit allusion to this very form of quadrant in the tract on Arzachel's *Saphea* by Guillelmus Anglicus, "civis Massiliensis," in 1231.[4] Indeed, on the basis of astronomical tables associated with *Quadrans vetus*, Tannery affirms with "une certaine vraisemblance" that this specific form of quadrant came to the West in the second half of the twelfth century—although he embraces the possibility, if lesser probability, of an even earlier date, in the first half of the twelfth century.[5] He further raises the possibility that Robertus's epithet "Anglicus" might be a family name (e.g., Anglès or Langlois), rather than a designation of English nationality.[6]

Duhem adopts Tannery's views on the name Robertus, the identification with the Sacrobosco commentator, and the resulting date.[7] But he rejects Tannery's attempt to convert the epithet "Anglicus" into a family name, for he finds a specific interest in and experience of English affairs in several passages of the commentary.

---

[2] Tannery, "Le traité du quadrant" (1897), pp. 126–36.

[3] Ibid., pp. 131, 135. But Tannery later recognized that the Cambridge manuscript ought instead to be dated ca. 1299, and so he discarded this dating argument: "Le traité du quadrant" (1899), pp. 201–2. In its place he proposed a revised date of before 1274, on the basis of tables in another manuscript, in the collection of Boncompagni. Since the location of this manuscript is now unknown, Tannery's inference cannot be checked.

[4] Tannery, "Le traité du quadrant" (1897), pp. 141–42.

[5] Ibid., pp. 147–48.

[6] Ibid., pp. 137–38.

[7] Duhem, *Système du monde* (n. 1), pp. 292–93, 298.

Similarly, Millás accepts Tannery's positions on the author's name as Robertus and the composition's date as the latter part of the thirteenth century.[8] But he emphasizes that the literary tradition of writing on the quadrant and closely related devices (most notably, astrolabes) extends back as far as the eleventh century. He calls attention to works attributed to Gerbert, Hermannus Contractus (perhaps by misattribution), the pseudo-Gerbertian *Geometria incerti auctoris* edited by N. M. Bubnov, and others.[9] These would provide the generic background for the accounts of the construction and uses of the quadrant, as set out in the thirteenth-century tracts.

In his 1943 paper on Robertus Anglicus Thorndike endorses most of Tannery's positions, although he agrees with Duhem in rejecting Tannery's "gratuitous attempt to metamorphose <Robertus> into a Frenchman."[10] A few years later, however, in his paper of 1947, he pointedly criticizes Tannery's arguments for the name Robertus as "weak and far-fetched" and "dubious," and even "cavalier," and recommends instead the greater likelihood of the name Johannes.[11] On the date, however, he accepts Tannery's placing of it prior to 1276, even though he never seems entirely sure of the grounds for this dating.[12] In this Thorndike shows himself unaware of Tannery's subsequent revision of the dating argument in the 1899 note. More strikingly, he seems not to realize that by switching the name of the author to Johannes, thereby prohibiting his identification with the Sacrobosco commentator Robertus, there would be no reason for construing "before 1276" as "ca. 1276."

Poulle acknowledges that Thorndike's critique rendered Tannery's arguments "insoutenables," but he proposes certain modifications of his own: that the writing might be dated as early as the latter part of the twelfth century, that is, to around the time when the quadrant in this particular form was first introduced. Nevertheless, he recommends: "L'édition de Tannery étant cependant tout à fait classique, il est commode de laisser à cet auteur inconnu le nom de Robertus Anglicus."[13]

In her edition of *Quadrans vetus* Hahn reviews the various arguments on authorship and dating. From a much more extensive survey of the extant

[8] Millás, *Estudios sobre Azarquiel* (n. 1), pp. 377–78.

[9] Millás, "Introducción del cuadrante" (n. 1).

[10] Thorndike, "Robertus Anglicus" (n. 1).

[11] Thorndike, "Who Wrote *Quadrans Vetus*?" (n. 1).

[12] See Thorndike, "Robertus Anglicus" (n. 1), p. 468n: "the reasons ... remain a mystery to me." Realizing later that Tannery had relied on the Cambridge manuscript, he nevertheless admits: "I do not know why the catalogue dated the MS. in 1276" (Thorndike, "Pierre of Limoges," *Isis* 36 [1945–46]: 4n). Yet he accepts this dating of the Cambridge manuscript in L. Thorndike, *The Sphere of Sacrobosco and its Commentators* (Chicago: University of Chicago Press, 1949), p. 63n; and in about a dozen entries in L. Thorndike and P. Kibre, *Catalogue of Incipits of Medieval Scientific Writings in Latin*, rev. ed. (London: Mediaeval Academy of America, 1963).

[13] Poulle, "Les instruments astronomiques" (n. 1), pp. 37–38.

manuscripts, she sustains Thorndike in favor of the name Johannes, although she recognizes that the decision "will have to be an educated guess, as there is no evidence that settles the question."[14] On the date she notes, as had Millás before her, a close structural affinity to the treatise on the quadrant in the *Libros del saber* commissioned by Alfonso X of Castille in 1276–77. By taking the Spanish work as the source for *Quadrans vetus* she infers 1277 as a *terminus post quem*, while she derives a *terminus ante quem* as 1284, from what she takes to be the earliest datable copy (MS Milan, Bibl. Ambros., H 75 sup.).[15]

In a stern review of Hahn's edition, Poulle iterates in the strongest terms his allegiance to Tannery.[16] The name Robertus, he affirms, "meme si c'est a tort," has become attached to the work through Tannery's edition, and so, in the absence of definitive support for the name Johannes (which Hahn herself admits), he prefers to treat the work as anonymous. On the date, he discounts Hahn's use of manuscript dates for establishing limits; but against dating it after 1277 (which Hahn argues from literary comparisons with the *Libros del saber*) he recalls Tannery's use of the 1276 dating of the Cambridge manuscript (which, as we have noted, Tannery himself came to recognize to be an error). Poulle subsequently has reaffirmed his support of Tannery's alternative hypothesis, that the tract may have been written much earlier—that is, toward the end of the twelfth century or the beginning of the thirteeenth.[17]

On the questions of both the date and the author's name, positions have been defended through surveys of the extant manuscripts. Poulle has wisely cautioned against using manuscript datings, in view of the notorious inconsistencies and uncertainties in the judgments of cataloguers and paleographers; yet even he has been misled, following Tannery's inference from the misdated Cambridge manuscript. Accordingly, I have compiled an inventory of what can with reasonable confidence be taken to be most of the earliest copies—that is, those from the late thirteenth to the early fourteenth century (see Appendix 1)—the determination being made on the basis of a range of criteria, including contents (e.g., where *Quadrans vetus* occurs in the vicinity of dated compositions) and dated annotations, as well as the style (sometimes, the identity) of graphical forms and scribal hands that might define related clusters of manuscripts.

---

[14] Hahn, *Medieval mensuration* (n. 1), p. xviii.

[15] Ibid., pp. xx, xxiii–vi, xl. The Milan manuscript, however, must be dated *after* 1284, for the same reason that compels the adjustment of the date of the Cambridge manuscript (sc., not 1276, but after 1276): each holds a dated colophon that must be the date of composition of the tract to which it is attached, not the date of the scribal copy (see n. 18 below).

[16] Emmanuel Poulle, "Two Medieval Texts," *Journal for the History of Astronomy* 15 (1984): 135–38.

[17] Emmanuel Poulle, "L'astronomia," in *Federico II e le scienze*, ed Pierre Toubert and Agostino Paravicini Bagliani (Palermo: Selerio, 1994), pp. 122-37, esp. pp. 127-28.

None of the early copies is explicitly dated.[18] Whatever dates can be assigned inferentially lend little assistance in dating *Quadrans vetus*, however, since only one copy seems even to admit the possibility of a date before 1280, namely MS Vatican, Pal. lat. 1414, and that with no high degree of confidence.[19] This evidence thus merely confirms what we would have accepted in any event, that *Quadrans vetus* was written some years before 1288, when Profatius Judeus (Jacob ben Machir) issued his *Quadrans novus* (extant in the Latin translation made by Armengaud Blasius at Montpellier in 1290)—for the latter work can be assumed to extend (if not supersede) the former, which. like it, was composed at Montpellier.[20]

As for the name, one ought to be aware that Tannery presented no case at all in support of his preference for Robertus. For having listed the attributions given in thirty-five copies (of which eleven were Paris copies he examined personally, but the rest were known to him only through catalogs or correspondence with curators or colleagues), he neither summarizes the data nor formulates his inference therefrom, but merely *assumes* the issue as settled: immediately after the final entry he writes, "En résumé, on peut trouver au moins une trentaine de manuscrits du Traité de *Robertus Anglicus*"—the issue being not to determine the correct name, but rather to reveal the wide circulation of the treatise.[21] It seems that Tannery *presupposes* the name Robertus, through the connection with the Sacrobosco commentator of that name, although he does not offer a word of evidence that might link the commentary to *Quadrans vetus* (see the next section).

---

[18] The two ostensible exceptions: MS Cambridge Ii 3.3 includes a colophon dated to 1276 (fol. 139v), and MS Milan H 75 sup. ends with a colophon dated to 1284 (fol. 73r)—but both dates must refer to the composition of the associated tracts, as one finds by comparison with other copies of the same tracts. The Cambridge copy must be dated after 1292, and most likely to 1299, as both Tannery and Hahn have recognized. The Milan copy is probably assignable to the late 1280s, or early 1290s, for I have identified several other manuscripts, datable to about then, where the same scribal hand appears; see nn. 33 and 36 below, and also Wilbur Knorr, "Solar Tables in the Tradition of 13th Century Quadrant Tracts" (forthcoming).

[19] Complicating the matter is the fact that the Vatican copy is composite, and the early date would apply only to the first part, whereas *Quadrans vetus* falls in the later second part (see n. 36 below). In sect. 3 below I consult this manuscript for my text of Campanus's *Practica quadrantis*.

[20] This must be confirmed, however, by textual comparison of the works, which has not yet been undertaken. But F. S. Pedersen has observed that parts of the subsequent adaptation of *Quadrans novus* by Petrus de Sancto Audomaro (St. Omer) incorporate whole sections from the *Quadrans vetus*: see Pedersen, *Petri Philomenae de Dacia et Petri de S. Audomaro Opera Quadrivialia*, Corpus Philosophorum Danicorum Medii Aevi, 10 (Copenhagen: Gad, 1983–84), pp. 580–81. This establishes a *terminus ante quem* of 1293 for *Quadrans vetus*.

[21] Tannery, "Traité du quadrant" (1897) (n. 1), p. 132.

To summarize, then, as Tannery does not: of his 35 manuscripts, the name Robertus is given in 4,[22] Johannes in 9, and Arnaldus de Villanova in 1;[23] 19 are anonymous, 1 ambiguous,[24] and 1 misattributed.[25] Tannery could accept these data as confirming his preference for Robertus only through his conviction that all appearances of the name Johannes were "simplement en rapport avec la chance de confusion des abréviations *Ro.* et *Io.*"[26]

Thorndike, adding to Tannery's inventory three copies listed by Boffito and Melzi d'Eril and eight more from his own research, summarizes thus: 14 or 15 with Johannes, 4 or 5 Robertus, 2 Grosseteste, 1 Arnaldus de Villanova, 24 or 25 anonymous.[27]

Hahn tabulates the names in the full repertory of known manuscripts thus: 20 with Johannes, 19 Robertus (of which 6 intend Grosseteste), 2 Robertus *and* Johannes, 50 anonymous, 6 uncertain.[28] This survey, however, includes many copies to which her access was indirect (e.g., through catalogs). Of the 65 she consulted directly, the figures are thus: 12 with Johannes (11 in titles, 1 in colophon), 6 Robertus (4 in titles only, one of these being to Grosseteste; 1 in colophon only; 1 in both title and colophon), 1 Ricardus (in a colophon), the rest anonymous.

One sees how strongly these figures depend on selection. When the full ensemble of almost one hundred manuscripts is taken into account, the two names seem in a statistical dead heat (20 for Johannes vs. 19 for Robertus). But even this result is deceptive, in that it includes—at least, in Hahn's summary—data from catalogs that may be dubious. A more important shortcoming, I believe, is that all these lists give equal weight to all copies, whether early or late. For such a work with a substantial tradition of extant copies over several centuries, however, none of the later copies is likely to preserve variants that are unrepresented in the earlier ones. Thus, the figures based on the later copies, reflecting merely the rate of propagation of variants copied from the early ones, should lack authority in determining original readings. Yet in Tannery's list only one of the copies with the

[22] Of these, the name in manuscript no. 5 (MS Paris, Bibl. nat., lat. 14070) is written only in a second scribal hand; and in no. 15 the name is expanded as Robertus Greathead Lincolniensis.

[23] No. 26 (MS Munich, Bayer. Staatsbibl., lat. 353), as listed in the Munich catalogue; but inspecting the manuscript itself, I find no attribution for *Quadrans vetus*.

[24] Tannery judges the marginal title in no. 3 (MS Paris, Bibl. nat., lat. 7437) difficult to decipher, being either "Io." or "Ro." (Hahn, however, reads the same unambiguously as Johannes); but he finds a later marginal note with the phrase "quadrantem Roberti" that leads him to construe it as "Ro."

[25] No. 31 is actually the *Practica geometrie* with almost the same incipit as *Quadrans vetus*, as Tannery later realized: see "Traité du quadrant" (1899) (n. 1), pp. 202–3.

[26] Tannery, "Traité du quadrant" (1897) (n. 1), p. 131.

[27] Thorndike, "Who Wrote *Quadrans Vetus*?" (n. 1), pp. 151–52.

[28] Hahn, *Medieval Mensuration* (n. 1), p. xviii.

name Robertus (MS Paris, BN lat. 14070) is early (late thirteenth century), the rest being from the mid-fourteenth century or later; Thorndike's list adds only one such early copy with Robertus (MS Vatican, Pal. lat. 1414). By contrast, several of their copies with Johannes are assignable to before ca. 1300.

More precisely, if we consult the early copies—that is, those assignable to the late thirteenth or early fourteenth century, as I have listed them in Appendix 1—a clearer pattern emerges. I have classed these under four headings. (i) Of four that lack the supplementary solar tables, three being datable probably to the 1280s and one (MS Yale, Cushing 11) to ca. 1310: 2 have Robertus, 2 are anonymous. (ii) Of fifteen with supplementary tables of the primary type and datable to the late thirteenth century (specifically, 1280s or early 1290s): 2 have Johannes, 0 Robertus, 11 are anonymous, 2 unknown. (iii) Of fifteen with supplementary tables based on the *Almanac* of Guillelmus de S. Clodoaldo (St. Cloud), and hence datable after 1292 (specific assignments varying from 1292–95 to 1304–7): 6 have Johannes, 0 Robertus, 9 are anonymous. Thus, of the 34 copies in these three groups: 8 have Johannes, 2 Robertus, 22 are anonymous, 2 unknown.

In group (iv) I assemble those other copies that Tannery and Hahn state to be from the thirteenth or fourteenth century (presumably, on the basis of catalogs or other indirect information), but that I have not myself examined. I thus cannot confirm these estimates or make them more precise (e.g., whether middle or late fourteenth century). Thus, I am not inclined to accord great weight to this evidence. But for the record, of these 19 copies: 4 have Johannes, 2 Robertus, 1 Lincolniensis (that is, a mistaken form of Robertus), and 12 are anonymous.

At first glance, then, the testimony of the earliest manuscripts would favor the name Johannes over Robertus almost overwhelmingly. But the matter is more complex. First, it is clear that the manuscripts form interrelated subgroups, as my numbering scheme is intended to indicate. Thus, in (i) the two copies with Robertus (MSS Paris, BN lat. 14070, and Vatican Pal. lat. 1414) display other agreements—in particular, their variant numbers in the short table of longitudes (typically inserted into the text of *Quadrans vetus* as sect. [29] in the numbering scheme of Tannery and Hahn), which they share with only one other manuscript (namely, MS London, BL Harley 3735) among the 65 copies collated by Hahn.[29] Further, they often display common variants, such as the omission of "zodiaci" in [14] (ed. Hahn, p. 26, line 1). Moreover, in MS Paris, BN lat. 14070, the words in its title of *Quadrans vetus*, "magistri Roberti Anglici" (fol. 76), do

[29] Ibid., pp. 38–39.

not come from its exemplar, for they are written in a fifteenth-century hand, the same as several other such headings and marginalia in this manuscript.[30]

Indeed, of Hahn's 6 manuscripts ascribed to Robertus, 4 (namely, the Paris and Vatican copies just mentioned, plus MS Naples, Oratoriana di Girolamini, XI, fourteenth cent.; and MS Oxford, Bodl. Lib., Canon. ital. 157, fourteenth cent.) frequently share variants.[31] Similarly, the manuscripts in (iii, a–c) form a coherent group, suggestive, in my view, of their production by one and the same atelier (see sect. 4 below). Thus, one might infer that the ascription to Johannes in six of them reflects a single shared prototype. Inspection of Hahn's apparatus confirms this in some cases—e.g., conspicuously for MSS London, BL Royal.12.C xvii (*M*) and Oxford, Ashmole 1522 (*Q*) in group (iii, c). Further, these two manuscripts, together with three others in (iii, a–c)—MS Florence, Bibl. laur., plut.18, s.c. 6 (*Z*), MS London, BL Egerton 844 (*j*), and MS Florence, Bibl. naz., Magl. II iii 24 (*u*)—and two later manuscripts, are noted by Hahn to constitute a textual group.[32] But this group displays too much fluctuation in the patterns of variants to allow its descent from a single prototype. Moreover, although the manuscripts in group (ii, a) can be seen as a precursor from the same atelier as those in (iii, a–c), the Milan copy (Hahn's *F*), which is ascribed to Johannes, and the London copy BL Harley 3735 (*k*, mentioned above) often share variants with the Robertus group *JNSt*. Indeed, the same scribal hand appears in the Vatican and Harley copies, as well as in ten other manuscripts that I have identified, including several in groups (ii, a) and (iii, a–c).[33] On the other hand, in (ii, b) the Cambridge manuscript Univ. lib. Ii 1.15, assigned to Johannes, cannot be affiliated with the main groups (ii, a) and (iii, a–c).

---

[30] Note, in particular, in this same hand the comment "anno 1443 ... in 2° ciclo," in the top margin of fol. 24, its copy of the *Kalendarium* of Grosseteste.

[31] One can inspect Hahn's apparatus (*Medieval Mensuration*, [n. 1]) for the sigla *J, N, S, t* corresponding to the Paris, Naples, Oxford, and Vatican copies, respectively. These are often joined by the Harley (*k*) copy, as also MS Venice, Bibl. marc., VIII 79, 14th c. (*U*), and MS Paris, Bibl. nat., lat. 7416B, 14th c. (*V*)—all three being anonymous, and the last being one of the four copies collated by Tannery in his edition ("Traité du quadrant" [n. 1], p. 133). For instance, for "ponantur" in [19] and [20] *JSUVkt* (et al.) write "scribantur"; for "in fine versus S" in [20] they write "in ultimis duobus spatiis"; for "Marcii" in [26] they write "in quo scribatur Marcius"; in [40] *JNSUVt* add "eternus." Other agreements of this sort occur in [9], [19], [27], [28], and [40]. Note that the variants stated here are not restricted to these manuscripts.

[32] Hahn, *Medieval Mensuration* (n. 1), p. lxxxi.

[33] Compare the hand in MS Vatican, Pal. lat. 1414, fols. 183v–208 with MS London, Royal 12.C.ix, fols. 62–179, in group (i); the following in (ii, a): MS Milan, H 75 sup., fols. 70–73 (including the colophon dated to 1284, on fol. 73); MS London, Harley 3735, fols. 6–16v, 64–72v (the copy of *Quadrans vetus*), 74–84 (datable to 1293); MS Basel, Oeffent. Bibl. der Univ., O II 7, fols. 41–57v, 66–165v (ca. 1292); and in group (iii, b): MS Copenhagen, Add. 447, 2°, fols. 38–83v (including its copy of *Quadrans vetus*, fols. 77–83v, and the supplementary tables, fols. 46v–48v; datable to 1296–99).

In summary, the manuscripts with Robertus display a certain coherence as a subgroup, suggestive of descent from a common prototype.[34] By contrast, the manuscripts with Johannes display a more diverse lineage, indicating textual mixing, and even some coincidences of scribal hand with the Vatican manuscript (Pal. lat. 1414) in the Robertus group. One could thus propose that the occurrences of the name Robertus derive from a single mutation (of an original Johannes), subsequently propagated in later copies. By contrast, it would be more difficult to account for the occurrences of Johannes in such a manner, were Robertus the original reading, in view of the greater diversity of the associated portion of the manuscript tradition.

Ultimately, the entire tradition of the name Robertus seems to descend from a prototype now most closely represented by the Vatican (Pal. lat. 1414) copy. What prevents me from dismissing the attribution as a simple scribal error, however, is that this manuscript has in one respect a unique claim to authenticity: its special variant of the short table of longitudes is the only one, among all those collated by Hahn, that is correct; from it the prevalent form of the table arose through scribal error.[35] Further, the Vatican copy is one of the oldest extant copies, perhaps even the oldest.[36]

So, the picture remains confused. Both names, Johannes and Robertus, are attested from the earliest phase of the tradition of *Quadrans vetus* discernible through the extant manuscripts. Conceivably, both are correct: it might be possible that a tract compiled by one Robertus Anglicus was subsequently edited by one Johannes Anglicus, both in Montpellier—the latter,

[34] Note that neither the Vatican (*t*) nor Paris (*J*) manuscript could be such a prototype, nor could either have been copied from the other; for each displays unique variants that isolate it from all other copies. For instance, in [5] *t* omits "a quo puncto tu vis protrahere lineas rectas iuxta" and later in [5] omits "limbus" (see other such unique readings in [6], [26], and [31]); similarly, in [5] *J* omits "ducantur due linee, et includent infra se superius"; in [14] *J* changes "Tunc sciendum est quod" to "Et," and again in [14] *J* changes "respondent" to "tunc erunt"; and in [40] *J* omits "declinationem solis."

[35] I set out the analysis of the table in Knorr, "Solar Longitude Tables" (n. 18). Note that the Vatican copy (MS Pal. lat. 1414, fol. 169) attaches the common, but incorrect, values as a marginal variant. The Paris (MS BN lat. 14070) and London (MS BL Harley 3735) versions of the table combine the Vatican values with the common form into a single three-place table.

[36] On its last page, fol. 224r, it has an annotation stating 2 (or 12?) November 1266 as the date of copying, and 62 Florentine florins as the price. Since the line is written in a fifteenthth-century hand, however, the source of the information is unclear, and the bibliographer, Ludwig Schuba, who hypothesizes that it possibly refers to a bookseller's document, seems disinclined to accept it; see Schuba, *Die Quadriviums-Handschriften der Codices Palatini Latini in der Vatikanischen Bibliothek* (Wiesbaden: L. Reichert, 1992), pp. 195–96, 199. The Vatican manuscript, however, is composite. Beginning with *Quadrans vetus* at the quire starting on fol. 168, the scribal hands in the second part (fols. 168–225v) differ from those in the first part (fols. 1–167v), with the exception only of a single short quire (fols. 86–89v) that is seen to be a later insertion (cf. fols. 183v–208). Thus, the year 1266 could perhaps apply to the first part of the manuscript, other than the inserted short quire; but the second part must be later, since the dominant hand appears in other manuscripts from the 1280s and 1290s (e.g., MS Milan, Bibl. Ambros., H 75 sup.; MS London, BL Harley 3735; and at least ten others); see n. 18 above.

for instance, supplying the supplementary tables (which are lacking from the Vatican and Paris copies), and making a fair number of purely linguistic changes. This hypothesis would entail, one must admit, a remarkable coincidence in the names. If only one of the names is correct, however, then the other need not have arisen from mere scribal error (as Tannery supposed), but rather as an editorial correction for what was taken to be an error or omission in the prototype. For this same reason we can accept "Anglicus" as correct (even though the epithet appears in only a very few manuscripts), and prefer "in" over "de" as the preposition with "Montepessulano": for Anglicus would not casually be supplied, if absent from the copyist's exemplar; and "in" is a very unusual variant, compared to the altogether common "de."

It remains to be seen, then, under what circumstances either name, Johannes or Robertus, could be thought to be a correction for the other. This will emerge in the next section.

2. Robertus Anglicus and the Commentary on *De spera*

The commentary on Sacrobosco's *De spera*, completed as "an introduction for the scholars studying in Montpellier" by one "Robertus Anglicus" in 1271, consists of fifteen *lectiones* that hold closely to the order and content of Sacrobosco's treatise.[37] At certain points the commentator elaborates longish digressions occasioned by the text, some of which introduce technical matters.

One of these, in *lectio xi*, takes its cue from Sacrobosco's definition of the "hour" as the rising time of half of a zodiacal sign, and his observation that every day and every night thus consists of twelve such hours.[38] After a few remarks on the equal and unequal systems of hours, he proceeds to an account of time measurement by means of instruments. He first discusses the form of a mechanical clock—then not yet invented, but under consideration by certain *artifices horologiorum*—that could, through appropriate daily adjustments, keep track of the equinoctial motion precisely, whence it "would be a very accurate timekeeper and be better than an astrolabe or any other instrument of astronomy for taking the hours."[39]

He then describes a simple device, a form of sundial, by which one can mark off the time of noon on any day via the shadow of a pointer. Next he

[37] For a critical edition of the Latin text with English translation and notes, see Thorndike, *Sphere of Sacrobosco* (n. 12), pp. 199–246 (hereafter, ed. Thorndike). The author's name and the place and date appear in the colophon in three manuscripts, although one gives an alternative "Parisius" and another gives an alternative "1272" (ibid., p. 198n).

[38] Ibid., pp. 103–4.

[39] Ibid., p. 180 (cf. L. Thorndike, "Invention of the Mechanical Clock about 1271 A.D.," *Speculum* 16 [1941]: 242–43).

explains how this instrument can be used toward finding one's latitude, "which is the distance of the zenith from the equinoctial," that is, "the altitude of the pole": this is done by marking the elevation of the sun at noon "when the sun is in the first degree of Aries or ... Libra" by means of "a quadrant or astrolabe" and then subtracting this figure from 90 degrees, the remainder being the latitude.[40]

An account to the same effect occurs in *Quadrans vetus*. Indeed, it displays strong verbal coincidences with the passage in Robertus's commentary. I give both texts, indicating by italics the phrases common to both:[41]

> Commentary: Utilitas autem istius instrumenti non est modica, nam per istum circulum invenitur certissime *latitudo regionis, que est distantia zenith ab equinoctiali,* que quidem distantia, cum sit equalis *altitudini poli,* ex consequenti etiam patet *altitudo poli* in omni regione. Nam *sole existente in* primo gradu *Arietis vel* in primo gradu *Libre,* cum umbra solis ceciderit super lineam diametralem circuli, notetur *altitudo solis* cum quadrante vel astrolabio, deinde *altitudo* solis *subtrahatur ab 90 gradibus et residuum erit latitudo regionis.* Hoc est distantia zenith ab equinoctiali vel elevatio *poli* sine aliquo errore. (ed. Thorndike, p. 181)

> Quadrans vetus: [41] Si autem vis scire *latitudinem regionis id est distantiam cenith ab equinoctiali* sive *altitudinem poli,* quod idem est, *altitudinem solis* accipies in meridie *sole existente in* principio *Arietis vel Libre.* Et illam *altitudinem subtrahe de 90 gradibus, et residuum erit latitudo regionis* sive altitudo *poli.* (ed. Hahn, pp. 54–55)

One can see that Robertus's passage is hardly other than a transcript of the passage in *Quadrans vetus*, with a few obvious additions and accommodations to his own context. For instance, his line "cum umbra solis ceciderit" describes the use of the shadow dial (cf. "utilitas istius instrumenti" in the first line) that he has just described, whereas *Quadrans vetus* merely assumes as known the operation of finding noon (although no explicit account of this operation appears in the work).

In its next section [42] *Quadrans vetus* describes an alternative method, accommodated to any day of the year, rather than just the equinox. It then goes on to describe yet another method, employing observations of a suitable circumpolar star. Here one takes two sightings of its meridianal transit, the "major" and the "minor," and computes the angle halfway between them, which equals the altitude of the pole, or, equivalently, the observer's latitude. The same method is given in Robertus's commentary, immediately

[40] Ed. Thorndike, p. 181.

[41] In this I overlook differences of merely syntactic form (e.g., *subtrahatur* vs. *subtrahe*), or minor lexical variants (e.g., *ab* vs. *de*). In my quotations from Hahn (*Medieval Mensuration* [n. 1]; hereafter, ed. Hahn), I make a very few minor changes, such as writing *sol* in lower case, to conform to the conventions in Thorndike and other editions, as cited below.

following the passage just quoted. Again, I give the lines from Robertus and *Quadrans vetus*, indicating their agreement by italics:

> Commentary: Aliter tamen, si quis vellet, posset *accipere altitudinem* poli in quacumque regione ita quod respiceret *de nocte aliquam stellam notabilem*, que sit prope *polum* de maiori Ursa vel de minori, et suspendat astrolabium super virgam unam vel quadrantem respiciendo stellam illam, donec fuerit *in maiori sua altitudine*, et tunc notetur numerus graduum altitudinis stelle in limbo. Deinde expectet idem donec *eadem stella* fuerit in maiori suo descensu et similiter accipiatur numerus graduum altitudinis stelle eadem hora. Tunc *minor* numerus *subtrahatur de maiori* et residuum dividatur in duas medietates equales, et una illarum *medietatum addatur* numero *minori, et illud quod inde provenit erit* vera *altitudo poli*. (ed. Thorndike, p. 181)

> Quadrans vetus: [43] Vel sic: *Accipe altitudinem alicuius stelle* fixe *notabilis* iuxta *polum* quando est *in maiori altitudine de nocte*, et altitudinem *eiusdem stelle* quando est in minori altitudine in eadem nocte vel alia, et *minor* altitudo *subtrahatur de maiori et medietas* differentie *addatur* altitudini *minori, et quod inde provenit est altitudo poli* sive latitudo regionis. (ed. Hahn, pp. 56–57)

Here again the commentator transmits the *Quadrans* passage almost verbatim, with occasional elaborations (e.g., "notetur numerus graduum ... in limbo," which would already be obvious to the reader of *Quadrans*), and arbitrary variants (e.g., "numerus" for "altitudo;" "in maiori suo descensu" for "in minori altitudine"). Note that the arithmetic operation of finding the mean elevation is described in identical terms in both (namely, subtracting the lesser from the greater, taking half the difference and adding this to the lesser), where it would be trivial to supply an equivalent variant (e.g., half the sum of the two values, as Sacrobosco expresses the same in his quadrant tract).

One can argue a priori that the passage in *Quadrans vetus* must have chronological priority, as the source for Robertus's commentary, rather than vice versa. For the discussion of how to find the latitude is well integrated into the structure of *Quadrans vetus*, as one of a series of basic applications of the quadrant. By contrast, this is only a digression in the commentary. It would thus be entirely reasonable that a commentator consult a special tract on the quadrant as the occasion recommended. Indeed, his phrase "cum quadrante vel astrolabio" signals dependence on such a tract, since a parallel passage in Sacrobosco's *De spera* itself mentions only the astrolabe, not the quadrant.[42] Conversely, the author of a technical tract

[42] In the commentary, in an earlier passage on measuring the altitude of the pole as part of the operation of measuring the circumference of the globe, note the phrase "per ambo foramina pennularum que sunt in quadrante vel in astrolabio" (ed. Thorndike, p. 158), which has its parallel in Sacrobosco's "sumpto astrolabio ... per utrumque mediclinii foramen" (ibid., p. 85). It is odd, however, that the commentator of Montpellier here volunteers the latitude for Paris, "et inveniet Parisius altitudo poli 48 graduum," where Sacrobosco says nothing to

like *Quadrans vetus* would hardly consult a general work like the Sacrobosco commentary as the source for a discussion like this.

But this order can be established on textual grounds. For the present passages have a prior source in the practical geometry, *Artis cuiuslibet consummatio*.[43] The operations of finding the equinoctial meridianal altitude of the sun, the latitude from it, and the same alternatively, from the altitude of the celestial pole, appear in its Book II, chaps. 5, 7, and 8:

> Artis: [II 5] Equinoctialem *solis altitudinem* investigare ... quando *sol est in* primo gradu *Arietis vel Libre* ... , sumatur altitudo meridiana per precedentem et illa erit equinoctialis.... [II 7] Per altitudinem equinoctialem *latitudinem regionis* invenire. *Altitudo* equinoctialis inventa ut docuimus *subtrahatur a 90 gradibus, residuum est latitudo regionis et poli altitudo.* Secundum hoc *48 graduum est latitudo regionis parisiensis,* scil. *distantia ab equinoctiali* linea.... [II 8] Poli altitudinem inquirere. Sumatur *alicuius stelle* vicine *polo* maxima *altitudo* ... per meridionalem lineam, *quando erit in* directo eius. Similiter sumatur minima *altitudo eius* que *subtrahatur a maiori.* Excessus *medietas addatur minori. Et* illa *erit altitudo poli.* (ed. Victor, pp. 230, 234, 236)

The italics indicate verbal agreements between the *Artis* and *Quadrans vetus*. Several of these are also agreements between Robertus's commentary and *Quadrans vetus*—but the latter holds "latitudo regionis sive altitudo poli" (end of [41]), parallel to *Artis* (end of II 7), where the commentary reads "latitudo regionis ... vel elevatio poli"; and *Quadrans* [43] has "quando est in," parallel to the "quando erit in" of *Artis*, where the commentary describes the whole operation ("suspendat astrolabium ... donec fuerit in").[44] Moreover, the convergences linking the tracts constitute almost the entirety of these passages as they appear in *Quadrans vetus*, whereas in the cases of *Artis* and the commentary each contains substantial elaborations of its own. For instance, in the second passage, the commentator's "accipere altitudinem ... ita quod respiceret de nocte aliquam stellam notabilem ... prope polum ... donec fuerit in maiori sua altitudine," corresponds (by way of expansion, but retaining all the key terms) to a single line in *Quadrans vetus*: "accipe altitudinem alicuius stelle fixe notabilis iuxta polum quando est in maiori altitudine de nocte"; the latter in its turn corresponds to a single line of *Artis*: "sumatur alicuius stelle vicine polo

---

occasion the remark. But a comparable line appears in *Quadrans vetus* [44], just after the passages cited: "altitudinem poli ... que est in Montepessulano 44 graduum fere et Parisius 48 graduum" (ed. Hahn, pp. 58–59).

[43] I will cite from the edition by S. Victor, *Practical Geometry* (n. 1).

[44] There is only one agreement of the commentary with *Artis*, not shared by *Quadrans*: their "primo gradu," where *Quadrans* has "principio." Both terms, however, are standard synonyms for the initial point of a zodiacal sign (for instance, "principio" occurs in Sacrobosco's quadrant tract, "primo gradu" in Campanus's, in the analogous discussion). Indeed, Robertus writes "in principio Arietis vel Libre" only a few pages later (ed. Thorndike, p. 189); clearly, then, he knows both expressions.

maxima altitudo ... quando erit in directo eius," but with notable differences in vocabulary.[45] If the *Quadrans* passage were based on the commentary, one would have to impose an implausible assumption: that its compiler somehow managed to retain only those lines also found in the *Artis*.

The passages in *Quadrans vetus* and the commentary, however, resemble each other far more closely than either of them does the *Artis*, both in the structure of the account and in the special phrases employed. For instance, they both write "latitudo regionis ... distantiam zenith ab equinoctiali" at the start of the first passage, where the *Artis* situates the comparable phrase (but without "zenith") only later in II 7. Further, at the end of the second passage, where the *Artis* writes only "illa erit altitudo poli," both the commentary and *Quadrans* write "quod inde provenit est altitudo poli." These comparisons indicate that the passage in *Artis* must be the source consulted by the author of *Quadrans vetus*, which in its turn has been taken over and elaborated in the commentary.

Tannery supposed that Robertus Anglicus wrote both the commentary and *Quadrans vetus*, although he offered no support for his hypothesis. The appearance of this extract in the commentary might seem to provide such support. That is not necessarily the case, however, since authors sometimes quote from works other than their own. In fact, I have been unable to discern any significant verbal usage shared by the two. Among the conspicuous stylistic features of *Quadrans vetus*, for instance, is its strong preference for second-person active forms; e.g., in [41] cited above: "si vis scire latitudinem"; "altitudinem solis accipies"; and "subtrahe"; and in [43]: "accipe." By contrast, the commentary favors third-person expressions, with the consequent emphasis on passive forms; thus, it substitutes for the words just cited "invenitur ... latitudo"; "patet altitudo" and "notetur altitudo"; "subtrahatur" in the first passage; and "si quis vellet, posset accipere" in the second. Of course, both works employ both types of idioms in one place or another. In his computation of the dimensions of the earth in *lectio iv*, for instance, the commentator adopts a profusion of first- and second-person forms: "nota quod," "debeo removere," "accipe," "divide," "non potes accipere," "accipias," "habeo," "aufer," and so on.[46] Presumably, these dictions are a carryover from the technical source consulted by the commentator,[47] but it remains significant that in the commentary's ex-

---

[45] Note the similar progression in the second passage, from "excessus medietas addatur minori" in *Artis*; to "medietas differentie addatur altitudini minori" in *Quadrans*; to "una illarum medietatum addatur numero minori" in the Commentary.

[46] See Thorndike, *Sphere of Sacrobosco* (n. 12), p. 159.

[47] I have not identified this source. But a parallel to Robertus's account appears in the Sacrobosco commentary of Michael Scot at the corresponding place, and it likewise adopts second-person forms (e.g., "accipe," "subtrahe," "divide," etc.); the computation agrees line for line with that in Robertus, leading to the same figure for the diameter of the earth

tract from *Quadrans vetus* (as just quoted), second-person forms are converted to third-person. Furthermore, in the parallel passages in the *Artis*, third-person or impersonal forms happen also to be favored: e.g., the free-standing infinitives "investigare," "invenire," and "inquirere," and the passives "sumatur altitudo," "altitudo inventa," and so on.[48] In *Quadrans vetus* these are converted to second-person forms, until the end of [43], where the passives "subtrahatur" and "addatur" are retained from the *Artis*. It is thus all the more striking that the entire passage in the commentary is recast into passive forms, beyond the two retained from *Quadrans* in the closing line.

Where the overall style of *Quadrans vetus* is concise and direct, the commentary tends to be prolix and formalistic. Marking the latter, for instance, is the ubiquity of constructions with gerundives, such as "sciendum," "dicendum" (with rare alternative, "dico"), and "notandum" (with common alternative, "nota"), one or another of which stands near the beginning of practically every paragraph. Such terms are rare in *Quadrans vetus*: "sciendum" once (in [14]), "incipiendum" once (in [15]), and "videndum" once (in [24]); "dicendum" occurs four times (once each in [11], [32], [48], and [64]), but even here, the last three occurrences may be influenced by one of the sources consulted, the *Practica geometrie* (see below, sect. 4).

Admittedly, the works fall into different expository types, scholastic commentary vs. technical treatise, such that each will likely reflect the linguistic habits typical of its respective genre. But that need not exclude common idioms, such as sequential adverbial phrases to set off new topics.[49] Thus, one finds that "consequenter" is notable in *Quadrans vetus*, as initiator of chapters [8], [10], [11], [13], [15], [18], [25], [30], [48], and [58], where all these occurrences can be argued as due to the author, rather than any of his sources.[50] Otherwise, "postea" is about as common, both as an initial marker (e.g., at [16], [19], [20], [23], [26], [27], etc.) and internally; while the familiar "deinde," "unde," and "tunc" figure only rarely as initiators, but often as internal markers of sequence. On the other hand, the term "item," which is commonplace in general prose, is almost absent from *Quadrans vetus*; it occurs only four times ([6], [53], [62], [63])—of which

---

(namely, 80181 stades plus a half and a third of a stade); but the texts are not in notable verbal agreement; see ibid., pp. 300–1.

[48] The *Artis* generally subscribes to a concise, impersonal style; but second-person forms are hardly rare in it: e.g., "habebis" in II 6, "probabis" in II 7, "scias" and "scies" in II 19, etc., and imperatives like "mensura," "pone," "aspice" in II 30, etc.

[49] I have analyzed such phrases as indexes of style in certain medieval geometric authors, such as Johannes de Tinemue; see the summary in Wilbur R. Knorr, "John of Tynemouth alias John of London," *British Journal for the History of Science* 23 (1990): 293–330.

[50] For these purposes, it seems best to focus on the first part of the tract, the account of the composition of the quadrant in [3]–[31], because the second part, on applications of the quadrant, is heavily influenced by sources (see sect. 4 below).

the last two are copied from a source (*Practica geometrie* [7] and [8]); and the second may be similarly influenced by another source, the *Artis*.[51]

By contrast, "item" is frequent in the commentary, occurring at least eight times in *lectio i* alone, and more frequently thereafter. Of course, "postea," "deinde," "unde," and "tunc" are present, if not with conspicuously high frequency, but I find "consequenter" only once (sc., ed. Thorndike, p. 173). Further, a striking phrase is "ad evidentiam" plus genitive (e.g., "ad istius diffinitionis evidentiam," "for making this definition evident," in *lectio i*, ibid., p. 145); such forms occur at least twenty times in the 56 pages of the commentary, with a few additional occurrences of comparable phrases like "ad intelligentiam" (twice: ibid., pp. 159, 179), "ad auctoritatem" (p. 192), and "ad maiorem introductionem" (p. 198). Indeed, idioms with "ad" proliferate more generally in a variety of forms on every page of the commentary, as if to signal this as the author's hallmark. To be sure, none of these phrases is unique; what is distinctive is the frequency.[52] By contrast, phrases of this form are absent from *Quadrans vetus*.

Despite his commitment to the arid scholastic format, the commentator manages occasionally to inject notes of a more personal kind, such as the unique report about research on the mechanical clock (ed. Thorndike, p. 180), his defense of the climate of Britain (ibid., p. 187), and his relaying of stories either heard by himself, like that of the enchanter in England (ibid., p. 191), or taken from the classical literature (ibid., pp. 191–92). The only personal note I find in *Quadrans vetus*, on the other hand, is the author's advice that the sighting holes must be quite narrow to get an accurate measurement ([57]; cf. sect. 3 below).

These items reflect different writing styles and personalities, and so appear to indicate that the commentary and *Quadrans vetus* are due to different authors. At the very least, they offer no encouragement at all for the hypothesis of common authorship as proposed by Tannery and accepted by Duhem, by Millás, and, at one time, even by Thorndike.

If the works are by different authors, it would seem implausible that they happened to have the same name (sc., Robertus Anglicus). But, of course, we already encounter a set of ostensibly unlikely coincidences: both authors are surnamed "Anglicus," are active "in Monte Pessulano," and have credentials as "magister" in the discipline of astronomy, although in

[51] In *Artis* II 30, the line "Item per astrolabium… "occurs just after a section paraphrased in *Quadrans* [53].

[52] I find "ad evidentiam," but only once, in the 95-page commentary on Sacrobosco by Michael Scot (ed. Thorndike, p. 338), in the presence of variants like "ad maiorem huius scientie declarationem" (p. 251) and "ad auctoritates" (p. 275). Thus, the particular idioms favored by Robertus Anglicus are effectively absent. Other forms, such as the gerundives, and common expressions with "ad" (e.g., "ad primum," "ad tertiam rationem," "ad obiectum," etc.) appear frequently in Michael's commentary, so they may be viewed as generic in scholastic discourse.

different ways: for (to paraphrase the opening of *Quadrans vetus*) if astronomy "has two parts: *theorica* and *practica*," the commentary falls under the first category, *Quadrans vetus* under the second.

If the author of *Quadrans* was named Robertus, however, there would be no evident reason for the appearance of the variant Johannes, for these coincidences with the Sacrobosco commentary would only reinforce the name, rather than confuse it.[53] On the other hand, if the correct name is Johannes, we can recognize the potential for just such a confusion with the alternative Robertus. For the *Quadrans vetus*, first issued at Montpellier, must still have been circulating there around 1271, when Robertus quoted from it in his commentary. In that academic environment, a commentator on Sacrobosco, like Robertus, would be expected to be well known, as a contributor of scholastic lectures and disputations within the philosophical curriculum. By comparison, practical studies in geometry and astronomy would be marginal, and adepts, such as the author of *Quadrans vetus*, would not have the standing or repute of the scholastic masters. In this context, a copyist or editor of a manuscript of *Quadrans vetus* would more likely be aware of the name and work of the commentator than of the practitioner, and so might supply the name Robertus to the heading of an otherwise anonymous copy, or even "correct" the name Johannes in his exemplar to Robertus. This could happen all the more readily, if the editor knew that *Quadrans vetus* was cited in the commentary.

Such an explanation assumes the emendation by a knowledgeable copyist, rather than a mere scribal error. The Vatican copy, MS Pal. lat. 1414, corroborates this assumption in an odd way. As I have said, it is an early copy (probably from the 1280s), the primary extant witness to the name Robertus. In its title for *Quadrans vetus* (fol. 168) the name Montpellier is written "Monte Pessimo," doubtless a misreading of the abbreviated form "Monte Pessu<la>no"; but the slip seems too good to be inadvertent—as if the scribe knew the place and did not like it much.

It would take only one such substitution of Robertus for Johannes to account, via propagation in a chain of subsequent copies, for the small number of appearances of that name.[54] This "emendation" would be that much easier, if it happened that Robertus and Johannes were near-contemporaries in Montpellier. I will establish this in the next section.

---

[53] I suppose one could conceive of a confusion with Johannes de Sacrobosco, to whom also is due a tract on the quadrant. But this is not done in any of the early manuscripts of *Quadrans vetus*.

[54] I emphasize, however, that the Vatican manuscript was not itself the primary locus of this mutation; for, as I have noted, it contains a few idiosyncratic variants and deletions that isolate it from all other copies (see n. 34 above).

3. THE *PRACTICA QUADRANTIS* OF CAMPANUS

The *Practica quadrantis* of Campanus, although generically related to the *Quadrans vetus* of Johannes Anglicus (as I shall henceforth name its author), nevertheless displays several notable differences. Whereas the latter begins at once with an account of the construction of the quadrant, Campanus prefaces that with a long account of purely astronomical phenomena, namely, how the sun's motion in the ecliptic correlates with the varying length of daylight over the year. Then, in his account of the construction of the quadrant, Campanus, unlike the other, omits any practical comments, such as the choice of materials or the use of compass and ruler. Further, in a departure surprising for this author noted for compilations of special astronomical tables, Campanus omits any discussion of the use of the solar tables in conjunction with the quadrant, whereas this is prominently discussed in *Quadrans vetus*. Campanus's account of the geometric applications covers only the measurement of altitudes and so omits the sections on lengths and on solids that appear in *Quadrans vetus*.

In their sections on astronomical and altimetric applications, however, where the tracts overlap, the parallels are substantial. Indeed, as I intend to show now, the correspondences between the tracts by Campanus and Johannes Anglicus are in major part verbatim. In the passages cited below, I present an eclectic text of Campanus, based on three early copies: my principal witness is MS Vatican, Pal. lat. 1414, fols. 208–11, a manuscript that I believe dates from the 1280s;[55] I have collated this in part against MS Oxford, Bodl. Lib., Laud misc. 644, fols. 211–13, from ca. 1276;[56] and MS Vatican, Urb. lat. 1428, fols. 93–95v, perhaps ca. 1270.[57] A partial listing of other manuscripts is set out in Appendix 1.

First, in the section on the construction of the quadrant, two of the features are described in similar terms by Campanus and Johannes Anglicus. The sights are described as "two small tables" set "over the side of the quadrant," and "perforated" with "openings" "at equal distances" from the base:

---

[55] See n. 36 above.

[56] In MS Oxford, Laud misc. 644, fols. 125–27, there is a copy of the tract on eclipses (inc. *Quicumque voluerit facere almanac*; second chap. inc. *Cum in quolibet mense cuiuslibet anni coniunctionem vel preventionem volueris invenire*), that in other copies (notably, MS Cambridge, Ii 3.3; see n. 18 above) is dated in its colophon to 1276; further, an annotation after the colophon of its copy of the Humeniz Canones (fol. 176v) explicitly cites the cycles for 1277 (emended by erasure of an indeterminable date earlier in the same decade).

[57] I estimate the date on the basis of its close affinity in contents, annotations, and graphical features to MS Tournai, Bibl. de la ville, 87, which was copied in 1268; although the Tournai manuscript was destroyed by fire in the aftermath of an air raid in 1940, notice of it survives in the ample catalog description by Paul Faider and Pierre van Sint Jan, *Catalogue des manuscrits conservés a Tournai* (Gembloux: Duculot, 1950), pp. 86–90.

Campanus: Posuerunt quoque *duas tabulas parvas perforatas super latus quadrantis* in quo sunt centra horarum, unam earum equidistare alteri, et ambas perpendiculariter insistere lateri quadrantis cui infiguntur, et *foramina* earum fecerunt equalia et *equidistantia* a radice ipsarum. (MS Vatican, Pal. lat. 1414, fol. 209v)

Quadrans vetus: [5] Et *super latus quadrantis* dextrum *due tabule parve* erunt equidistantes ab extremitatibus lateris AC; et sint *perforate* parvis *foraminibus equidistantibus* linee AC descendenti. (ed. Hahn, p. 16)

Quadrans vetus speaks of the "tables" as being "equidistantes ab extremitatibus," that is, "equally distant from the ends," although such a precisely symmetrical disposition is not requisite. By contrast, Campanus speaks of the same tables as "unam earum equidistare alteri," that is, "each of them parallel to the other," which indeed is a factor in the proper arrangement. It may be, then, that *Quadrans vetus* has misunderstood Campanus's term "equidistare."

For comparison, I quote the parallel from the *Quadrans* of Sacrobosco:

Sacrobosco: Quadrante itaque formato *infigantur* due pinne perforate de cornu vel ebore vel metallo quocumque in latere quadrantis ex parte linee in qua assignantur *centra* arcuum *horarum* ... itaque *foramina* pinnarum *equaliter distent* a linea marginali. (MS New York, Pub. Libr. 69, fol. 72)

One sees that Sacrobosco's term "pinne" differs from the "tabule" in the other two tracts, which together omit any such specific allusion to the materials as he gives (that is, "of horn or ivory or metal"). By italics I indicate phrases that Campanus shares with Sacrobosco: that the tables are "attached" (*infiguntur*) to the side of the quadrant on which the "centers of the hour <arcs>" lie, and the "openings are equidistant" from the base, or marginal line. *Quadrans vetus* happens to agree on this last (sc., "foraminibus equidistantibus"), but not on either of the other phrases.

Similarly, Campanus and Johannes Anglicus describe in identical terminology the plumb line, as being a "silken thread" or "perpendicular" with a "lead" weight "at its end" and a bead marker, or "pearl," movable up and down its length:

Campanus: Et *posuerunt clavum* unum parvum in angulo quadrantis cui appenderunt *filum* unum parvum *de serico* habens *in extremitate* altera *plumbum* an<n>exum a<d> modum *perpendiculare*[58] et in filo illo *posuerunt ma<r>garitam* unam percussam que possit *per* ipsum *filum moveri* a summo eius usque deorsum. (MS Vatican, Pal. lat. 1414, fols. 209v–210)

Quadrans vetus: [31] Deinde in puncto A *ponatur clavus* subtilis cui alligemus *filum de serico* subtili, *in* cuius fili *extremitate ponatur perpendiculum de*

[58] One would expect *perpendicularem* ("in the perpendicular manner") or *perpendiculi* ("in the manner of a perpendiculum").

> *plumbo* vel alterius rei ponderose. Et *margarita* perforata *moveatur super filum* de loco ad locum. (ed. Hahn, p. 42)

Not only does Johannes share Campanus's nomenclature, as indicated in the italicized phrases, but he also displays a close synonymy in his variants: e.g. his "in puncto A ... cui alligemus" corresponds to Campanus's "in angulo ... cui appenderunt," his "clavus subtilis" to Campanus's "clavum parvum," his "margarita perforata" to "margarita percussa," his "de loco ad locum" to "a summo eius usque deorsum." The correlative passage in Sacrobosco diverges from *Quadrans vetus*, with, as before, a few coincidences with Campanus (indicated by italics):

> Sacrobosco: Deinde fiat *perpendiculum, s. filum, dependens a clavo* fixo *in angulo* concursus linearum marginalium in cono quadrantis *habens plumbum in extremitate* et curriculum in medio, s. nodulum mobilem sursum et deorsum pro voluntate artificis. (MS New York, Pub. Libr. 69, fol. 72r–v)

Campanus's phrase "from its top downward" (*a summo eius usque deorsum*) may be instigated by Sacrobosco's "up and down" (*sursum et deorsum*). But specific parallels of this sort are infrequent in Campanus and Sacrobosco, so I think that they are only coincidental here, reflecting a general nomenclature for the quadrant, and that Campanus's tract is independent of Sacrobosco's.

In both of these passages, as elsewhere, Campanus describes the features of the quadrant in third-person plural active verbs (e.g., "posuerunt," "fecerunt"), as if describing the habitual practice of instrument-makers. Similarly, Sacrobosco alludes explicitly, here and elsewhere, to the *artifices*. This suggests that both writers base their descriptions on actual instruments, from which they infer the procedures of the artisans.

I turn now to the set of astronomical exercises described by Campanus. The first of these shows how to find the solar altitude: one positions the quadrant toward the sun so that its rays pass through the sights; then, where the plumb line falls on the edge of the quadrant, one can read off the angle of altitude:

> Campanus: *Si vis scire altitudinem solis in omni hora* diei *per quadrantem, appone quadrantem soli et* fac *radium* eius intrare *per ambo foramina duarum tabularum* ita quod perpendiculum sponte descendat et adhibeat limbo quadrantis, et nota locum perpendiculi *in gradibus limbi*. Quot gradus *abscindit perpendiculum* ex limbo, tot gradibus ascendit <sol> in illa hora. (MS Vatican, Pal. lat. 1414, fol. 210)

> Quadrans vetus: [33] *Si* igitur *velis scire altitudinem solis in omni hora per quadrantem, pone* punctum A *quadrantis* versus *solem* et punctum C versus te, *et* dimitte *radium* solis transire *per ambo foramina duarum tabularum* perforatarum. Et vide quem gradum *ex gradibus in limbo abscindat perpendiculum*, et respice numerum graduum sumptum superius, et numerus graduum ostendet tibi altitudinem solis. (ed. Hahn, p. 46)

As I indicate by italics, the two writers agree practically verbatim on phrases distributed throughout the passage. Johannes tends to vary the verbal forms (e.g., "velis" for "vis," and "pone" for "appone"), and introduce synonyms (e.g., "dimitte radium solis transire" for "fac radium eius intrare; vide quem gradum" and "respice numerum graduum" for "nota locum"). Here, as we have already seen in [31] cited above, and as we will see in other passages below, the variant in *Quadrans vetus* inserts specific letters (e.g., "punctum A") correlated with its diagrams, whereas Campanus's tract has no diagrams. Further, *Quadrans vetus* here omits any parallel to Campanus's comment on the situation of the plumb line ("ita quod perpendiculum ... quadrantis"). Presumably, the remark could be considered superfluous. Nevertheless, even where the verbal agreement is quite loose, as in the last line in the passage, the technical agreement is precise.

In their next sections, *Quadrans vetus* [34] and Campanus (MS cit., fol. 210) describe how to locate the ecliptic position of the sun by means of the cursor. This entails simply lining up the plumb line with the current date and reading off the corresponding mark of the ecliptic. The passages fluctuate between complete verbatim agreement and a more paraphrastic correspondence

Next, in its paragraphs [35]–[38], *Quadrans vetus* takes up these same operations as executed alternatively via the solar tables, where it displays a loose textual parallelism with Sacrobosco. By contrast, Campanus is silent. Immediately after this, *Quadrans vetus* [39] describes how to determine the solar declination via the cursor of the quadrant, and here it returns to close verbal agreement with what is the very next section of Campanus (MS cit., fol. 210). In both, one uses the plumb line first to correlate the equinoctial point with an angular mark on the limb of the quadrant, then to match the current date with a second angular position; the difference of the two angles is the solar declination on that day. The passages conform in a close paraphrastic agreement, sometimes verbatim. One is struck, however, by the absence from Johannes's text of "declinacio eius 7trionalis vel meridionalis" and a corresponding line at the end, distinguishing the northerly and southerly declinations; for this distinction is critical in later problems, such as finding the latitude (sect. [42]) and the hour (sect. [47])—where, however, Johannes seems to be working with a source like Sacrobosco.

Quadrans vetus next (sect. [40]) describes how to find the declination alternatively, by means of the solar tables, where it comes again into parallelism with Sacrobosco. Similarly, in sections [41]–[43], its account of how to find one's latitude via the solar altitude or via circumpolar stars, *Quadrans vetus* follows partly Sacrobosco, and partly the *Artis* (cf. passages cited in sect. 2 above).

In the next sections, describing how to find the hour of day, Johannes returns to agreement with Campanus, at precisely the point where he had momentarily departed from Campanus after [39]. The two accounts first explain how to set the cursor in the position corresponding to one's latitude, which both call the "situs perpetuus" of the cursor:

> Campanus: *Si vis scire horam diei in omni regione* in qua fueris per quadrantem, scito longitudinem [! latitudinem] *regionis* illius quot graduum fuerit, *et computa tot gradus in limbo quadrantis incipiendo a latere quadrantis super quod infixe sunt tabule perforate. Et ubi ipse numerus terminabitur,* pone perpendiculum et gira *cursorem* quo usque *principium Arietis sub ipso perpendiculo cadat,* et hic *situs cursoris erit perpetuus in quadrante in illa regione.* (MS cit., fol. 210)

> Quadrans vetus: [44] *Si* autem *vis scire horas diei in omni regione,* vide *latitudinem regionis* sive altitudinem poli, ut dictum est, que est in Montepessulano 44 graduum fere et Parisius 48 graduum. *Et computa tot gradus in limbo quadrantis, incipiendo a latere quadrantis super quod infixe sunt tabule perforate. Et ubi numerus ille terminabitur,* move *cursorem* donec *principium Arietis cadat sub ipso perpendiculo* directe, et talis *situs cursoris quadrantis* [11 MSS add: *perpetuus*] *erit in regione illa.* (ed. Hahn, pp. 58–59)

The correspondence in this passage is more consistently literal than in the ones we have considered before. Moreover, even the discrepancies are often close synonyms. Of Johannes's additions, most are of the usual minor sort already seen. But the addition of "que est in Montepessulano ... et Parisius ... " is noteworthy, since this particularizes the use of the instrument to Montpellier and Paris, precisely where Campanus notes its adaptability to any latitude (sc., "quot graduum fuerit"). Johannes's precedent for the line appears to come from a Paris source, such as the *Artis* II 7 (cf. quotation in sect. 2 above).

Quadrans vetus [45] and Campanus (MS cit., fol. 210) then describe an alternative method for setting the cursor: observing the meridian altitude of the sun and then moving the cursor until the present date lines up with the plumb line. In contrast to the preceding passage, the correspondence here is somewhat loose, although the linkage is still clear.

Next, to find the hour one first sets the plumb line against the present day and moves the bead to where it crosses the line for the sixth hour; then, taking the present altitude of the sun, one takes the hour from the point where the bead of the plumb line cuts the hour lines. Johannes's account in [46] agrees almost verbatim with Campanus (MS cit., fol. 210), except for his characteristic synonyms, amplifications, and technical equivalents. While Johannes's last line ("si enim cadat ... et sic deinceps") lacks a parallel in Campanus, it may be influenced by the corresponding account in Sacrobosco (cf. MS New York, Pub. Libr. 69, fol. 76r–v).

The next section, *Quadrans vetus* [47], describes how to find the hour, alternatively, if the quadrant lacks the cursor. The procedure is the same as before, save that the noon declination is now to be determined from the solar tables. This passage seems to be modeled in part on [46], and in part on [42]; it displays no direct verbal affinity to either Campanus or Sacrobosco.

The next sections of Campanus set out a few more exercises on time measurement. No parallel appears in *Quadrans vetus*, which has completed its astronomical applications with the discussion in [47] and now begins its section on altimetry. Campanus also presents a series of altimetric problems. Since these utilize the shadow square of the quadrant, however, he first makes a comment on how the values of *umbra versa* and *umbra recta* can be converted one to the other.[59] This section corresponds to a later paragraph of *Quadrans vetus*, where the correspondence is firmly literal:

> Campanus: *Si autem ex umbra versa volueris* elicere *rectam, per numerum punctorum umbre verse divide 144, et quod exierit erit numerus punctorum umbre recte.* E converso quoque *invenies umbram versam per rectam,* videlicet si *diviseris 144 per numerum punctorum umbre recte, exibit tibi in divisione numerus punctorum umbre verse.* (MS Vatican, Pal. lat. 1414, fol. 210v)

> Quadrans vetus: [54] Sed ad hoc quod scias omni hora accipere umbras, oportebit te mutare umbram rectam in umbram versam et e contrario. [55] *Si autem vis ex umbra versa* habere numerum punctorum umbre *recte, per numerum punctorum umbre verse divide 144, et* illud *quod exierit* post divisionem *erit numerus punctorum umbre recte.* [56] Si vis *invenire umbras versas per rectam, divide 144 per numerum punctorum umbre recte,* et *exibit tibi in divisione numerus punctorum umbre verse.* (ed. Hahn, pp. 75–77)

But for the occasional synonym (e.g., Johannes's "habere" for Campanus's "elicere"), the correspondence here is almost completely verbatim. Note also how Campanus's "e converso" is echoed in Johannes's "e contrario" in the prefatory sentence [54] that is lacking in Campanus.

In *Quadrans vetus* the first two altimetric exercises, in [49] and [50], parallel Sacrobosco: to measure the height of a tower by contriving that the quadrant assume a 45° angle toward its summit; and to effect the same when the tower and the observer are not situated in the same plane (cf. MS New York. Pub. Libr. 69, fols. 77v–78, 79).

Campanus has no such special cases, but considers as his first example how to find the height of a tower by means of the shadow square of the quadrant (MS Vatican, Pal. lat. 1414, fol. 210v). Although *Quadrans vetus* does not address the tower measurement in this way, in a later chapter [78] it describes how to measure the depth of a well (*puteus*) in quite similar terms. But this chapter owes primarily to another source, the *Artis* II 34, which itself lacks those parallelisms with Campanus.

---

[59] The rule, that *umbra recta* Θ = 144 / *umbra versa* Θ, follows from the trigonometric equivalents: for Θ, the angle between the horizontal and the observer's sightline toward the summit, *umbra versa* $\Theta = 12 \tan \Theta$, and *umbra recta* $\Theta = 12 \cot \Theta$.

My sense, then, is that in [78] Johannes, as he paraphrases the *puteus* passage from *Artis*, imports idioms he elsewhere has already taken from Campanus. For we have no grounds to suppose that any form of Campanus's tract included a section on stereometry, to set a precedent for this last part of *Quadrans vetus*. Indeed, Campanus ends his tract with the remark that one can employ the same methods toward the measurement of depths and planes, but "let these things briefly said suffice for the sake of an introduction on the *practica quadrantis*."[60] This makes explicit the *absence* of these procedures from his tract.

Although, as noted above, Johannes appears to have turned to a model like Sacrobosco for his first exercises on measuring towers, his next example in [51] now corresponds very closely to a variant method next presented by Campanus (MS cit., fol. 210v). In both, the procedure requires one to sight the top of the tower and then note where the plumb line cuts the shadow square. The verbal correspondence is close throughout, and in the last sentence precisely so. At certain points Johannes inserts explicatory phrases "sicut prius"; "respiciendo ... perpendiculum"; "post divisionem"), in effect glossing Campanus. One such added phrase, "scilicet in dextro latere," corresponds to the particular orientation Johannes has chosen for its construction, and is consistent with his diagram (cf. the facsimiles in plates 2–4, Appendix 2). Clearly, the quadrant could be made just as well with the *umbra recta* on the left, although Johannes's configuration is more facile for right-handed observers.[61] Campanus's account leaves unstated which side is the *umbra recta* and whether it lies to the right or the left of the axis of the quadrant.

Johannes agrees with Campanus in reducing the second case (where the plumb line cuts the *umbra versa*) to the first, via the reciprocal relation of the *umbra versa* and *umbra recta* (as set out in *Quadrans vetus* [54–56] and the parallel in Campanus, quoted above). The first case corresponds to the situation where the sightline is inclined to higher than 45°, whence the altitude being measured is greater than the observer's distance from it. But one could just as easily have reversed the order: starting with the second case, operating with the number of the *umbra versa* directly, then referring the other case to this. Alternatively, one could treat the cases independently, as Sacrobosco does (cf. MS New York, Pub. Libr. 69, fol. 78r–v). But the option followed by Campanus and Johannes is critical for the subsequent problem of measuring the inaccessible tower (see below). Both now com-

---

[60] Cf. MS Vatican, Pal. lat. 1414, fol. 211, with "causa introductionis breviter dicta" added in MS Oxford, Laud misc. 644, fol. 213.

[61] To sight with the right eye, the observer will place the instrument at the right side of his face, and will thus read the markings more conveniently if they are engraved on the side of the quadrant that faces toward his left. This is consistent with the orientation assumed by Campanus and Johannes.

plete the procedure: the observer's distance from the tower is to be multiplied by 12, the product then divided by the number previously taken from the shadow square, whence the height of the tower is found. Here Johannes in [52] does little other than transmit Campanus's account (MS cit., fol. 210v) verbatim.

The next problem in *Quadrans vetus*, sect. [53], presents an alternative solution via shadows, when the sun happens to have an elevation of 45°. Campanus has no parallel; but the model for this passage is found in the *Artis* II 30. There follow the rules for converting *umbra recta* to *umbra versa,* and conversely, in [54–56], which as we have seen above are a transcript of an earlier paragraph of Campanus. The next exercise in *Quadrans vetus* [57–58] treats of measuring an inaccessible tower via the shadow square, and corresponds closely to the next example in Campanus (MS cit., fols. 210v–211). Both accounts begin by directing the quadrant toward the summit of the tower and noting the number on the shadow square. Johannes adheres to Campanus's phrasing, with typical substitutions. Campanus's "in loco consideracionis" is expanded by Johannes into "in loco in quo tu stas in hora considerationis." Johannes also introduces an explicit lettering (sc., "signum" becomes "signum D") corresponding to the appearance of diagrams in *Quadrans vetus*—a feature absent from Campanus, as we have already noted. A nontrivial discrepancy appears in Johannes's remark that the sighting holes should be kept "very narrow" (*valde stricta*) in order to avoid error. As Campanus lacks a parallel, Johannes appears to be volunteering this observation on the basis of his own experience using the quadrant.

Both accounts now complete the procedure by prescribing that the observer take another measurement at a second position. The distance between the two sites is then to be multiplied by 12 and divided by the difference in the corresponding shadow markings. The result, when added to the observer's own height, is the height of the tower:

> Campanus: Deinde *elonga te a turri aut appropinqua ei secundum* situm *linee recte*, et considera *iterum* summitatem *turris* per ambo foramina, *et quere* etiam *numerum punctorum umbre recte ad hunc situm secundum, et pone signum in hoc loco* consideracionis, *et mensura* quid sit *inter duo signa*, et *subtrahe minorem* duorum *numerorum* punctorum *umbre recte de maiori, et serva differentiam* inter eos, *et distanciam inter duo* signa *multiplica per 12*, et *productum divide per differentiam* punctorum umbrarum *prius* servatam, *et* ei *quod exierit superadde quantitatem tue altitudinis, et quod* aggregabitur *erit altitudo turris*. (MS cit., fols. 210v–211)

> Quadrans vetus: [58] Consequenter *elonga te a turri vel appropinqua ei secundum lineam rectam*, et *iterum* respice altitudinem *turris, et quere numerum punctorum umbre recte ad hunc situm* in quo *secundo* stabis, *et pone signum* C *in illo loco, et mensura* quot sunt pedes *inter duo signa*, et retine illud;

> postea *subtrahe numerum minorem umbre recte de maiori, et servetur differentia; et distantiam inter* dicta *duo* loca *multiplica per 12 et productum divide per differentiam prius* acceptam, *et illi quod exierit adde quantitatem tue altitudinis et quod* remanebit *erit altitudo turris.* (ed. Hahn, pp. 79–80)

Note that this measurement *requires* that the readings be taken off the *umbra recta* (with conversion, if needed), rather than the *umbra versa.*

The entire passage in *Quadrans vetus* is a virtual transcript of Campanus. The replacement of Campanus's "Deinde" by "Consequenter" is interesting, in view of Johannes's general preference for the latter term (see section 2 above). Otherwise, the paraphrasing is minor: Johannes's "respice altitudinem" for Campanus's "considera summitatem"; his "situm in quo secundo stabis" for Campanus's "situm secundum"; etc. Johannes also adds "retine illud." Besides this, he displays a very slight tendency to abridge. But over all, this passage displays as clearly as any we have considered the close verbal agreement between the two works.

With this example on measuring the inaccessible tower Campanus ends his tract, but for a single concluding sentence naming additional kinds of applications.[62] *Quadrans vetus*, however, adds another case, where the tower is viewed from a valley (sect. [59]), and then goes through a series of alternative solutions of the same altimetric problems "sine quadrante,"—that is, by use of rods and mirrors—in sects. [60–63]; these sections correspond closely, and often verbatim, to the *Practica*, sects. [14] and [6–8], respectively (in the numbering adopted by Hahn). *Quadrans vetus* then continues into the examination of problems in planimetry and stereometry, where it follows now the *Practica* (sects. [9–10]), now the *Artis*, and perhaps also Sacrobosco (see the summary below). In all these sections, its fidelity to the indicated sources matches what we have encountered in its handling of the materials from Campanus.

In comparing the preceding passages, I have tended to explain the version in *Quadrans vetus* as an adaptation of the parallel in Campanus. While the verbal correspondences clearly indicate a direct textual relation between them, it remains to argue explicitly the chronological priority of Campanus over *Quadrans vetus*. This is evident from the following considerations:

First, *Quadrans vetus* includes effectively verbatim parallels of the entirety of two large blocks of text in Campanus: a series of astronomical problems, appearing in order as sections [33–34], [39], and [44–46] in *Quadrans vetus*; and a series of altimetric problems, appearing in order as its sections [55–56], [51–52], and [57–58]. Other passages of *Quadrans vetus* appearing before or after these, or interspersed among them, correspond to other sources, such as Sacrobosco, or the two tracts in practical geometry: the *Practica geometrie* (or *Geometrie due sunt partes princi-*

---

[62] See n. 60 above.

pales), and the *Artis cuiuslibet consummatio*. When compared one to another, however, these other works—Campanus, Sacrobosco, and the two cited practical geometries—display no discernible verbal agreement with each other. If we were to take *Quadrans vetus* as a source for Campanus, we would expect to find passages where Campanus corresponded verbally to one or another of these works. But this is never seen to be the case. On the other hand, if we take Campanus to be prior to *Quadrans vetus*, the pattern of correspondence we have seen between them would follow simply from the compiler's introducing, as he saw fit, materials from other sources to amplify what he took from Campanus.

Second, in a few subtle discrepancies, *Quadrans vetus* presents incorrect variants (e.g., "ipsa" in [46], "quadrantis" in [51], "sumpte" in [52]) where the reading in Campanus is correct ("ipse," "quadrati" and "sumptum," respectively). These variants, if they are not merely scribal errors in *Quadrans vetus*, afford priority to Campanus; for the author of the original text would not have written what we find in *Quadrans vetus*.

Third, among the items absent from Campanus that appear in *Quadrans vetus* (on the precedent of Sacrobosco) is its account of applications making use of the solar tables. But Campanus, whose earliest work includes compilations of astronomical tables, would hardly have omitted such items from his quadrant tract, had they been so well established in his source.

Fourth, one of the marked improvements of *Quadrans vetus* over other accounts of the quadrant (for instance, Sacrobosco) is its account of the cursor. For where Sacrobosco explicitly, but mistakenly, assigns equal intervals to the month spaces and equal intervals for the signs, Johannes states clearly the principle of their variable spacings and gives values for their widths: 11½° for Aries, 8½° for Taurus, etc. (cf. [15]: "quia tanta est declinatio solis in Ariete," and similarly in [17]). Further, he gives values for the initial points of each month, as specified in an accompanying abbreviated table of longitudes (cf. [22]: "videatur in tabula in quo gradu Capricornii incipiat Januarius)."

In this regard, Campanus is only marginally better than Sacrobosco; for although he is not culpable of Sacrobosco's error of asserting equal spacings, nevertheless, he does not say what the spacings are, or even what the principle of spacing is. He says only that "makers of quadrants ... set 6 signs in the ascending half circle for the sun" (*artifices quadrantis ... posuerunt 6 signa que sunt in medietate circuli solis ascendente*: MS Vat. Pal. lat. 1414, fol. 209), and next to it they divide the band for the corresponding half year into five whole and two half months (*et hec medietas anni continet sex menses, 5 videlicet integros et duas medietates duorum*), such that the "last half" (*ultimam medietatem*) of December corresponds to the motion of the sun from the initial point of Capricorn to its middle (*ad medium eius*). Continuing in this loose way, describing each month as running from the "middle" of the corresponding sign to the middle of the next, Campanus differs little from the imprecise account in Sacrobosco. Thus, he could hardly have known the *Quadrans vetus*, with its detailed account of the division of the cursor.

The comparisons with Campanus provide a *terminus post quem* for the composition of *Quadrans vetus*, for the quadrant tract can be classed with Campanus's earliest work. One copy, in MS Tournai 87 (now lost), was written out in 1268.[63] According to the summary compiled by F. S. Benjamin, Jr., Campanus's edition of Euclid's *Elements* can be placed between 1255 and 1259, while his *Theorica planetarum* is datable between 1261 and 1264.[64] Further, Campanus's astronomical tables, now extant in a variety of forms and alternative canons, had already been issued early in the 1260s, for they are cited in Petrus Peregrinus's tract on the astrolabe (ca. 1263).[65]

None of the evidence cited by Benjamin supports assigning an earlier date to any of these writings.[66] Since, further, Campanus died in 1296, it would be difficult to assign to *any* of his writings a date much earlier than 1255, or to his birth a date before the 1220s. I am inclined to place his quadrant tract somewhere in the earliest indicated interval, from 1255 to 1260, perhaps before his own efforts on astronomical tables—for he omits any reference to the use of solar tables with the quadrant, despite the existence of a standard set of tables for this purpose, which already circulate in the 1260s in conjunction with his quadrant tract (as in MS Tournai 87 of 1268). At the least, Campanus is not yet familiar with the basic literature on the quadrant, notably the tract by Sacrobosco, where a full account of the use of the solar tables is given.

Accordingly, *Quadrans vetus* is bracketed between Campanus's *Practica quadrantis* (ca. 1255–60) and the Sacrobosco commentary by Robertus

---

[63] See n. 57 above.

[64] F. S. Benjamin, Jr., and G. J. Toomer, *Campanus of Novara and Medieval Planetary Theory* (Madison, Milwaukee and London: University of Wisconsin Press, 1971), pp. 4–5.

[65] Ibid., pp. 5, 15–16. Benjamin adopts the date 1261 for Petrus's astrolabe tract. But Poulle notes that Petrus only states his use of Campanus's solar table for 1261; Petrus is also aware of a special form of Arzachel's *saphea* ("*cum novello*") that could be known in the Latin West only after the translation of 1263: Emmanuel Poulle, "L'astrolabe médiévale d'après les manuscrits de la Bibliothèque nationale," *Bibliothèque de l'École des chartes* 112 (1954): 92–93.

[66] Benjamin considers possible reasons for assigning the composition of Campanus's tables to 1232 or to 1254, based on the headings of the initial tables in some of the manuscripts; *Campanus of Novara* (n. 64), pp. 3, 16–17. But all that one could securely infer from these tables, I believe, is that Campanus compiled them during the first *cycle* so listed, which would be 1232–60 in the former instance, and 1254–73 in the latter. The reasons for commencing at these particular points have to do with the natural breaks in the cycles. That is, 1232 and 1261 correspond to the Arabic years 630 and 660, respectively, as tabulated in Campanus's principal reference, the Toledo Tables. Similarly, 1254 is the beginning of a 19-year lunar cycle, as indeed is recognized in the heading in the fifteenth-century copy (MS Vatican, lat. 3118, fol. 5v) that suggests this as near the date of Campanus's composition (ibid., p. 16n).

Anglicus (1271). Presumably, some modest interval must have elapsed between the first appearance of Campanus's tract and Johannes's wholesale appropriation of it, whence the *Quadrans vetus* would most plausibly be assigned a date around the middle 1260s. I will venture a more specific date below.

4. THE WRITING OF *QUADRANS VETUS*

In the preceding survey of the quadrant tracts of Johannes Anglicus and Campanus, I have also referred to parallels in other thirteenth-century works. These comparisons are summarized in the following table, where I use the following code letters:

Qv = *Quadrans vetus* (section numbers as in the editions by Tannery, "Traité du quadrant," cit. n. 1; and Hahn, *Medieval Mensuration*, cit.n. 1);
Sb = Sacrobosco, *Quadrans*;
Ts = *Tabule solis* (inc. *Per istas quatuor tabulas*; cf. Knorr, "Solar Tables," cit. n. 18);
Cm = Campanus, *Practica quadrantis*;
Ar = *Artis cuiuslibet consummatio* (book and chapter numbers as in the edition by Victor, in *Practical Geometry*, cit. n. 1);
Pr = *Practica geometrie* (inc. *Geometrie due sunt partes principales*; section numbers as in the edition by Hahn, in *Medieval Mensuration*, cit. n. 1);
? = unknown source (if any), or uncertain assignment.

TABLE OF SOURCE CORRESPONDENCES

Qv	Source	Subject
1	*Pr* 1–2	preface: parts of geometry
2.	?	introductory statement on part I
3	?	on the quadrant and its uses
4–7	?	basic form of the quadrant, limbus, parts
8–10	Sb?	hour lines
11–28	?	cursor
29	?	short table of longitudes
30	Sb?	shadow square (start)
31	Cm?	shadow square (end), plumb line

Table of Source Correspondences, continued.

Qv	Source	Subject
32	?	introduction on uses of the quadrant
33	Cm	altitude of the sun
34	Cm	position of the sun in the ecliptic
35–38	Sb/*Ts*	same, via tables
39	Cm	declination of the sun
40	Sb/*Ts*	same, via tables
41	Sb/*Ar* II.5–7	latitude, via the sun at equinoxes
42	Sb	latitude, via the sun generally
43	*Ar* II.8	latitude, via the stars
44–46	Cm	finding the hour via hour lines
47	?	same, via tables (cf. Sb)
48	?	introductory statement on altimetry
49	Sb	height of a tower, seen at 45°
50	Sb	same, with observer in a different plane
51–52	Cm	height of a tower via the shadow square
53	*Ar* II.30	same, via shadows at 45°
54–56	Cm	converting *umbra recta* and *umbra versa*
57–58	Cm	height of an inaccessible tower
59	*Pr* 14	same, measured from a valley
60–61	*Pr* 6	height, via a rod, *sine quadrante*
62	*Pr* 7	same, via shadows
63	*Pr* 8	same, via a mirror
64	?	introductory statement on planimetry
65	Sb?	length of a field, via the quadrant
66–67	*Pr* 9	same, via a rod
68	*Pr* 10	same, via a mirror
69–70	*Ar* I.22, 21	the area and circumference of a circle
71–73	?	area of triangles (cf. *Ar* I 2–4)
74–75	*Ar* I.10, 9	area of a square and a rectangle
76	*Ar* I.15	area of a pentagon
77	?	introduction on "depths and volumes"
78–79	*Ar* II.34–35	depth and volume of a well (*puteus*)
80	*Ar* III.1	volume of a square-based column
81–83	*Ar* III.7	volume of a barrel (*modius*)
84–86	*Ar* III.8	volume of a tun (*dolium*)
87	*Ar* III.12	volume of a quadrangular *vas*

In most of the places where *Quadrans vetus* is listed as following Campanus's *Practica quadrantis*, our examination of the parallels has revealed a strongly literal correspondence. The same applies even more to its parallels with the *Practica geometrie*, where in addition to exploiting the preface, the *Quadrans vetus* takes over a whole block, consisting of six of the sections of the *Practica*, to constitute a like block of itself, with but a single interruption and only one change of order.[67] In the case of the *Artis* the dependence is evident in frequent coincidences of phrases and whole lines, although this occurs in the context of extensive editing.[68]

With Sacrobosco, however, although the correspondences in content and technique are close, the verbal parallels are usually so loose that there is a real question whether in fact *Quadrans vetus* is exploiting it directly, rather than through an intermediate version not yet identified. Sacrobosco must underlie those items involving the use of tables (that is, *Qv* [35–38] and [40]), since we have no clear evidence that any other form of solar tables was available prior to the tradition of the ones presupposed to be by Sacrobosco (which actually accompany some copies of Sacrobosco's tract). These tables, which I call the *Tabule solis*, are an edited excerpt, with canon, from the Humeniz Tables (in the Paris edition of 1239).[69] Presumably, *Quadrans vetus* could consult both Sacrobosco and the *Tabule solis* for its own accounts in [35–38] and [40] explaining how to use the tables. Similarly, sections [41–43] appear to interweave Sacrobosco's accounts on finding latitude with the corresponding account in *Artis*. Of the several methods of determining heights, the first, in [49–50], utilizing the quadrant, seems to follow Sacrobosco; the second, in [51–52], using the shadow square, follows Campanus; the third, in [53], alternatively via shadows, follows the *Artis*; and the fourth, in [60–63], describing three procedures by rod, then by shadows, then by mirror, follows the *Practica*.

Of the sections marked only by "?" some would seem prima facie to be original to the author of *Quadrans vetus*—in particular, the bridge statements introducing each new section, at [2], [32], [48], [64], and [77], for these are instigated by the peculiarities of ordering he adopts. Even so, there may be an influence from the *Practica*, for the idioms with

[67] Hahn has already noted most of these correspondences, except for that of *Qv* 59 to *Pr* 14 (see *Medieval Mensuration* [n. 1], pp. lvi, lix).

[68] Victor describes as "very close," "close," or "similar" the correspondences of *Ar* III 7, 8, and 12 to the listed sections of *Qv* (*Practical Geometry* [n. 1], pp. 349n, 353n, 361n). Although Hahn accepts *Ar* only as a minor source for *Qv*, she admits the connection relative to *Ar* I 22, and "possibly" also that to II 34–35 (*Medieval Mensuration*, pp. lix–x). All the other parallels listed between *Ar* and *Qv* are additional observations of my own.

[69] For details on the relation of the *Tabule solis* to the Humeniz Tables and to Sacrobosco see, respectively, Knorr,"Solar Tables" (n. 18), and Wilbur R. Knorr, "Sacrobosco's *Quadrans*: Date and Sources" (in progress).

"dicendum" in *Qv* 32, 48, and 64 (otherwise absent from *Qv*) seem an echo of *Pr* 3.

For the other uncertain sections we can hold out the possibility of yet another source exploited by our author, although I think it at least as plausible that his own hand is responsible. In *Qv* [47] the phrase "quod post subtractionem vel additionem remanserit" occurs in precisely the same context as it did in *Qv* [42], where it was a paraphrase of Sacrobosco's "quod post augmentationem vel diminutionem provenerit"; presumably, then, *Qv* [47] is the author's adaptation of his own earlier section. Similarly, in the sections on triangles, *Qv* [71–73], one can conceive that the text, though verbally different from the parallels in *Artis*, nevertheless proceeds from the author's free editing of that work. The most significant of the unidentified sections occur in the first part, on the construction of the quadrant. Where parallels in technique or wording can be noted with Sacrobosco and Campanus, the specific verbal correspondence is so loose that one might suspect an alternative source. Only in the instance of the long account of the composition of the cursor in *Qv* [11–28] does it differ substantially from the procedure described in Sacrobosco and Campanus, and in this *Qv* is distinctly superior to both the others.

It is appropriate that *Quadrans vetus* draws so liberally from Campanus's quadrant tract, which around the mid-thirteenth century was one of only a few specimens in a genre that would subsequently proliferate. But the particularly close relation of *Quadrans vetus* to the *Practica* is harder to comprehend, since the latter was not particularly notable and several widely circulating alternatives were available, such as the *Artis* (early thirteenth century), the *Practica* of Hugo de Sancto Victore (twelfth century), the pseudo-Gerbertian *Geometria* (eleventh or twelfth century), and numerous adaptations.[70]

Most strikingly, by making his own preface an extract from the preface to the *Practica*, the author of *Quadrans vetus* not only has ceded his privilege to define his own work—a remarkable abdication, even for a medieval writer—but has done so against its genre. For a tract on the quadrant would call for comments on the discipline of practical astronomy, not geometry. In the work itself, moreover, the block taken over from the *Practica* is initiated by the phrases "sine quadrante" and "si non habeas quadrantem" ([60], ed. Hahn, p. 83), stipulations that seem odd in the context of a work *about* the quadrant. The author seems to have striven to create relevance for the materials from the *Practica*, and to have done so without precedent in the quadrant tracts by Sacrobosco and Campanus.

---

[70] See the summary of this genre of writing compiled by Victor in *Practical Geometry* (n. 1).

This suggests to me the hypothesis that both works might be by the same author. The *Practica*, as his own prior effort, could provide geometric exercises taken over, in many cases virtually verbatim, as the nucleus of a larger work, the *Quadrans vetus*—conceived as an expanded edition superseding the *Practica*. The later work retains the core of problems in *Pr* 6–16. But the next portion of the *Practica (Pr* 17–20), describing the construction and use of the *gnomonicum instrumentum* (that is, a form of shadow square), is replaced in *Quadrans vetus* by the elaborate account of the quadrant and its applications, based on the principal treatments then available—namely, the tracts by Sacrobosco and Campanus. This expansion has the effect of reducing the *Practica*-based segment to merely a set of alternative solutions. But the connection is still evident in the preface, taken from the *Practica*, which announces, now incongruously, an exposition of practical geometry. To complete its three-part scheme of altimetry, planimetry, and stereometry, *Quadrans vetus* augments its survey of altimetry, as based on Sacrobosco, Campanus, the *Artis*, and the *Practica*, with two series of planimetric and stereometric rules. For this, the sketchy treatment in the *Practica* is replaced by an ampler account, based on the *Artis*.

This hypothesis of common authorship seems to fit well the structure of *Quadrans vetus* and its relation to its sources. But it encounters certain discrepancies.[71] For instance, some of the exercises are situated differently within the three-part subdivision of geometry; e.g., the measurement of the depth and volume of a well coming under stereometry in *Qv* [78–79], but under planimetry in *Ar* II. 34–35. Further, while it may be difficult to discern a "style" in a work like *Quadrans vetus*, which scavenges so freely from its sources, the verdict of a stylistic analysis is ambiguous: both works subscribe to a spare technical manner, where the precedent tracts tend to be more diffuse—but in some of its habitual phrases the *Practica* differs from the *Quadrans vetus*; e.g., to initiate chapters "consequenter" is notable in *Qv*, but absent from *Pr*; while *Pr* favors the phrases "et si" or "quod si," which are not prominent in *Qv*.

Thus, I can present common authorship only as a possible hypothesis. It is striking, nevertheless, that *Quadrans vetus* effectively supplants the *Practica*, in that among the 105 copies of the former and 20 copies of the

[71] Hahn (*Medieval Mensuration* [n. 1], p. xxiii) takes the *Practica* to be an early work, perhaps from the twelfth century, and this, if correct, would of course preclude its being by the author of *Quadrans vetus*. Hahn derives her date from M. Clagett, *Archimedes in the Middle Ages*, vol. 3 (Philadelphia: American Philosophical Society, 1984), pp. 215–17); but Clagett's argument, based on points of terminological usage, secures the early thirteenth century only as a *terminus post quem* for the *Practica*. On the other hand, the manuscripts of the *Practica* do not date back that far, the oldest datable copy (which I take to be MS Oxford, Bodl. Lib., Ashmole 341) coming from the early 1270s (see the listing in Hahn, pp. 199–200). On codicological grounds, then, the *Practica* and *Quadrans vetus* seem not to be far apart.

latter listed by Hahn only 4 manuscripts contain both, and these are not among the earliest copies.[72] The nearly identical prefaces of the two works may have led the copyists to take *Quadrans vetus* as a revised version of the *Practica*. But it seems to me that the author of *Quadrans vetus* may actually have had this intent.

Of the two issues debated about *Quadrans vetus* with which I began, those of its date and the name of its author, the first has now been definitively settled. For *Quadrans vetus* must come after the works it exploits: the Humeniz Tables (1239), the *Tabule solis* (prob. ca. 1245), Sacrobosco's *Quadrans* (prob. 1245–50), and Campanus's *Practica quadrantis* (prob. 1255–60). Further, it must precede the work that cites it: the commentary on *De spera* of Robertus Anglicus (1271). This places *Quadrans vetus* sometime in the 1260s.

I can venture a specific proposal. The short table of solar longitudes in *Qv* 29, I have found, is drawn from the first bissextile table of the *Tabule solis*, with an augment of precisely 10' in each entry.[73] A difference of this kind would be occasioned by the aim to adjust the reference table to the time of the extract. The parameter of 2' per bissextile cycle was familiar among thirteenth-century astronomers, such as William of St. Cloud.[74] Thus, the 10' of the *Quadrans vetus* table would accrue in five cycles, or twenty years. Further, a certain date does appear prominently in the context of the *Tabule solis*, namely in the epigram that closes the *Compotus* of Sacrobosco: expressed ambiguously, it could mean either 1244 or 1256, and each can be supported from medieval annotations.[75] If our compiler took 1244 as the date of the *Compotus* and the adjacent Sacrobosco works, such as his *Quadrans* and its related tables, then the increment of twenty years brings us to 1264. This date, if tentative, seems to me more likely than either of the other two bissextile years, 1260 or 1268, as the date of the

[72] Hahn sets out inventories of known manuscripts of both tracts in *Medieval Mensuration*, pp. 196–200. The copies containing both tracts are MS Vatican, lat. 3114, 13th–14th c.; MS Oxford, Bodl. Lib., canon. ital. 157, 14th c.; MS New Haven, Yale Medical School Lib., Cushing 11, early 14th c.; and MS Vienna, Öst. Nat. Bibl., lat. 5184, 15th c. The Vatican copy, where the two tracts are in different sections and different hands, may be composite. The New Haven manuscript includes copies of six quadrant tracts (Sacrobosco, Campanus, *Quadrans vetus*, and three versions of the *Quadrans novus*), besides the *Practica*, among other tracts in astronomy.

[73] The analysis is set out in Knorr, "Solar Tables" (n. 18).

[74] It is adopted by William in his *Almanac* of 1292, for instance, and results by approximation from the parameter of 1° per 131 years that he and other Latin scholars like Roger Bacon assign to Asophus (sc., al-Sûfî); see Bacon's *Compotus*, cap. 3, in *Opera hactenus inedita Rogeri Baconi*, ed. R. Steele, fasc. 6 (Oxford: Clarendon Press, 1926), p. 14.

[75] Cited, with analysis, by O. Pedersen, "In Quest of Sacrobosco," *Journal for the History of Astronomy* 16 (1985): 175–221 (specifically, pp. 186–88). Neither 1244 nor 1256 corresponds to the writing of the *Compotus*, which by internal evidence is datable to 1235. I argue elsewhere that 1256 is the meaning of the debated line, referring to Sacrobosco's death; see Knorr, "Sacrobosco's *Quadrans*" (n. 69).

Quadrans vetus table. Further, the table is explicitly cited for the purposes of calibrating the cursor: "videatur in tabula ... " ([25], ed. Hahn, p. 33)—whence we can accept it as an intrinsic part of the text, rather than an addition by a later editor. This would set the composition of *Quadrans vetus* around 1264, probably within the bissextile interval 1264–67.

As for the name of the author, we have seen that Tannery's preference for Robertus rests on nothing other than fiat, since the titles in his listed manuscripts display a greater number with the name Johannes. The same pattern favoring Johannes is evident in the surveys by Thorndike and Hahn, and I have pointed out that among the two dozen earliest manuscripts, it is even more prominent (Johannes appearing eight times, Robertus only twice, the rest being anonymous).

But if this falls short of a definitive proof, we need not be compelled to treat both names as equally likely.[76] The ambiguity arose very early in the tradition of *Quadrans vetus*—that is, certainly by the 1280s, whence copies with each name are extant (e.g. MS Vatican, Pal. lat. 1414 for Robertus; MS Milan, H 75 sup. for Johannes). As Thorndike stresses, what one must explain is why there is an ambiguity at all. Tannery's remark about scribal confusion (of "Ro." for "Jo.") is in principle symmetric, and so does not sustain Robertus over Johannes; further, Thorndike and Hahn have doubted whether such a confusion is even plausible.

Recognizing that *Quadrans vetus* was written at Montpellier only a few years before the Sacrobosco commentary, in which it is quoted, we obtain the cause of the confusion: that the author of the commentary, Robertus Anglicus, might by error be taken as the author of *Quadrans vetus*. That the two works are by different authors is arguable from their lack of stylistic affinity. But the commentator, as a master in the scholastic curriculum of philosophy and the arts, would be better known than the author of a quadrant tract, a master specializing in practical technical studies; and thus the name Robertus might be written in place of Johannes—the more familiar name for the less familiar—by an editor or scribe situated within the Montpellier academic context. Conversely, if Robertus were the correct name from the start, one could not easily account for the appearance of the variant Johannes, and even less for its predominating over Robertus in the subsequent tradition of the work.

In Tannery's assessment, the *Quadrans vetus* attests to an author writing "dans une langue claire et relativement correcte," but of little originality, and, worse, culpable of some serious lapses in technical understanding.[77] The harshness of the last judgment, however, follows from Tannery's own

---

[76] Or even less to adopt Robertus in deference to the judgment of Tannery, as Poulle feels inclined to recommend.

[77] Tannery, "Traité du quadrant" (1897) (n. 1), p. 149.

misunderstanding of the provenance of the associated supplementary tables, which he misdates to the twelfth century—while we have seen that these tables relate to 1239, and the short table of longitudes to ca. 1264. Thus, far from being anachronistic items attached by the author, through ignorance of the astronomical phenomena involved, they reflect his aim to bring the reference tables up to date. In other respects, nevertheless, Johannes Anglicus does sometimes propagate errors in his sources.[78]

The lack of originality is even more acute than Tannery imagined. For we have seen from the source analysis how *Quadrans vetus*, at least in its sections on applications, is hardly other than a collage of items from Sacrobosco, Campanus, and the two practical geometries. With a date as late as the 1260s, the composition of *Quadrans vetus* cannot be coordinated with the first introduction of this form of the quadrant, since clear references occur as early as 1231 in the astrolabe tract of Guillelmus Anglicus,[79] and the tracts by Sacrobosco and Campanus precede *Quadrans vetus* in the detailed description of the instrument. Yet in its account of the construction of the quadrant, in particular of the cursor, *Quadrans vetus* is sound, where Sacrobosco is incorrect and Campanus incomplete. Two anonymous tracts likewise correctly describe the cursor: *Ad quadrantem faciendum* (MS Paris, BN lat. 7475, fols. 60v–63v) and *Cum quadrantem componere volueris* (MS Paris, BN lat. 7416B, fols. 62–63v). But neither of these suggests any textual connection to *Quadrans vetus*, and neither is clearly earlier than it.[80]

Another feature that differentiates *Quadrans vetus* from all these other quadrant tracts is its ploy of integrating diagrams into its expositions, both for the construction of the quadrant and for its applications. It is remarkable that the tracts of Sacrobosco and Campanus lack diagrams. Indeed, I know of only one early copy of Campanus's tract that presents a diagram of the quadrant, namely that in MS Vatican, Urb. lat. 1428, fol. 95v (see the reproduction in Appendix 2). This figure, of course, lacks the letters keying its elements to the text, but otherwise it closely resembles the form typically held in *Quadrans vetus*. The Vatican manuscript appears to be relatively

[78] I find three such instances: the wrong formula for the area of the pentagon in [76], taken from *Artis*, where the *Practica* offers a correct, but nonarithmetical, alternative; the misleading account of the mirror procedure in [68], taken from the *Practica*; and the incorrect rule in [81–82] for the *modius* (in the form of the truncated cone), also taken from *Artis*.

[79] The passage is cited by Tannery, "Traité du quadrant" (1897) (n. 1), pp. 141–42. See also Millás, "Introducción del cuadrante" (n. 1), p. 251; idem, *Estudios sobre Azarquiel* (n. 1), pp. 449–54; and Hahn, *Medieval Mensuration* (n. 1), pp. xxx–xxxii.

[80] Both anonymous tracts are mentioned in passing by Tannery, "Traité du quadrant" (1897), p. 139. MS Paris, BN lat. 7475 dates from ca. 1272 (not, as Tannery states, 1371, which is a much later annotation); MS Paris, BN lat. 7416B dates from the early fourteenth century. I know of no other copies of either tract.

early in date, perhaps ca. 1270;[81] thus, we cannot affirm for certain that *Quadrans vetus* innovated the quadrant diagram. But at the least, it has incorporated a feature only recently associated with tracts of its kind.

Of course, the appearance of geometric figures, with lettering keyed to the text, is found in practical geometries, such as the *Artis* and the pseudo-Gerbertian *Geometria*.[82] It is odd, however, that even as *Quadrans vetus* employs lettered diagrams in [66] against its model from *Practica* [9], in its borrowing from the *Artis* (e.g., II 34, as paraphrased in [78]) it omits such key letters despite their presence in the source.

Ultimately, what sets *Quadrans vetus* apart from all other tracts on the quadrant is the extent of its circulation. Whereas the tract of Sacrobosco survives in eighteen copies so far identified, the tract of Campanus in forty or more, and other variants in only single copies each, the *Quadrans vetus* of Johannes Anglicus exists in at least nine dozen.[83] The popularity of this work of such modest literary and technical merits owes, I believe, to a stroke of luck: that it was chosen as a study text in the elementary astronomical curriculum for the University of Paris late in the thirteenth century—for the two dozen manuscripts that I list in Appendix 1 under groups (ii, a) and (iii, a–c),[84] seem to have that provenance. Specimens of an evolving anthology of introductory works, originally constructed around the tracts by Sacrobosco early in the thirteenth century, they represent an impressive expansion of astronomical study at Paris in the three decades of the 1280s, 1290s, and 1300s.[85] The cited copies, in fact, form such a coherent ensemble, in terms of content and graphical format (including the manner of executing and decorating figures and of ornamenting initials), that I would assign them to a single book-producing atelier.[86]

The virtues that recommended *Quadrans vetus* to the designers of this corpus can only be a matter for speculation. Presumably, its author intended to consolidate and supplant his principal sources—the quadrant tracts by Sacrobosco and by Campanus, and the two practical geometries, the *Artis*

---

[81] In contents it seems very closely related to MS Tournai 87, copied in 1268; the latter, however, had no figure for Campanus's tract (see the description by Faider in *Catalogue des manuscrits* [n. 57], p. 89).

[82] The *Practica* presents figures, but these lack reference letters keyed to the text.

[83] To the 105 listed by Hahn, I add 4 additional copies: MS Vienna, Öst. Nat. Bibl., lat. 2367; MS Copenhagen, Kong. Bibl., Add. 447, 2°; MS Stockholm, Kung. Bibl., X 767; and MS Basel, Öffent. Bibl. der Univ., O II 7. In the Basel manuscript, the copy of *Quadrans vetus* originally present (as indicated in the table of contents) has since been lost.

[84] Among which I include MS London, BL Royal 12.C.ix; and possibly also MS Vatican, Pal. lat. 1414, both from group (i).

[85] O. Pedersen has produced an account of this general class of astronomical manuscripts in "The Corpus Astronomicum and the Traditions of Medieval Latin Astronomy," *Colloquia Copernicana* III, Studia Copernicana, vol. 13 (Wroclaw and Warsaw: Polish Academy of Sciences, 1975), pp. 57–96. But this must be considered a very preliminary study, with much work remaining to be done.

[86] Some aspects of this analysis appear in Knorr, "Solar Tables" (n. 18).

and the *Practica*—and the conciseness and clarity of his exposition seem to have suited the purposes of the arts curriculum, within which the work's technical subject matter was of only marginal interest for the key disciplines of philosophy and theology. But the fact that it was so chosen accounts for the production of almost a quarter of the extant copies at Paris within a span of only about twenty years. From such a start the subsequent proliferation of copies of *Quadrans vetus* can be viewed as part of the transmission of the Paris curriculum of astronomy to other centers at later times

APPENDIX 1: MANUSCRIPTS

CAMPANUS, *PRACTICA QUADRANTIS*, AND *QUADRANS VETUS*

I present here partial lists of manuscripts containing copies of the *Practica quadrantis* of Campanus, and the *Quadrans vetus* of Johannes Anglicus. Most of these manuscripts have been available to me either for direct inspection in the libraries or through microfilm copies. For the others I depend on accounts in secondary sources, which I mark thus: TK = L. Thorndike and P. Kibre, *A Catalogue of Incipits* (n. 12) (column number); Tannery = P. Tannery, "Traité du quadrant" (n. 1), pp. 127–32 (manuscript number); Hahn = N. L. Hahn, *Medieval Mensuration* (n. 1), pp. 3–5 (manuscript letter). For entries with such a key letter, the dating estimates are taken from the indicated source; otherwise, they are my own.

Campanus, "Practica quadrantis"

F. S. Benjamin, Jr., claimed to know of as many as forty manuscripts containing Campanus's tract on the quadrant: see Benjamin and Toomer, *Campanus of Novara* (n. 64), p. 15. I have consulted about one-third that number, all from the thirteenth century or early fourteenth:

- Cambridge, Univ. Lib., Ii 1.17, fols. 72–75v; prob. late 1280s
- Florence, Bibl. Naz., Pal. 639, fols. 46–47v
- Lincoln, Cath. Lib., 148, fols. 102v–109v (defective at end); 13c
- London, BL Egerton 843, fols. 40–43; 13c (TK 1405)
- London, BL Harley 4350, fols. 38–43; 13c
- London, BL Royal 12.C.ix, fols. 35–37v; prob. 1280s
- New Haven, Yale Med. Sch. Lib. Cushing 11, fols. 127–30; early 14c
- New York, private coll. Henry C. Taylor, fols. 18–21v; 13c (described by Thorndike, *Sphere of Sacrobosco*, cit. n. 12, pp. 73–74; present location unknown)
- Oxford, Bodl. Lib. Laud misc. 644, fols. 211–13; ca. 1276
- Paris, BN lat. 7196, fols. 27v–30v; 13c
- Tournai, Bibl. de la ville 87, fols. 116v–121; 1268 (destroyed in 1940; cf. Faider, cit. n. 57)
- Vatican, Pal. lat. 1414, fols. 208–11; prob. 1280s
- Vatican, Urb. lat. 1428, fols. 93–95v; ca. 1270
- Verdun, Bibl. publ. 25; 13c (cf. *Catalogue général des manuscrits des bibliothèques publiques*, V [Paris, 1879], pp. 441–42)

Seven more copies are listed in TK 396 (Inc.: *Debes scire quod circulus solis* ...) and 1405 (Inc.: *Scire debes* ...).

Quadrans vetus

Hahn lists 105 copies of this tract (*Medieval Mensuration*, cit. n. 1, pp. 196–99), to which I can add 4 more (see n. 83). I here list only those copies from the first half-century of its tradition—that is, from the late thirteenth and early fourteenth centuries. For each entry I indicate whether the supplementary tables are present (and in which form), and also the manner of the attribution. Under ii a, I list copies that seem to form a coherent group, probably stemming from the same atelier. Likewise, under iii a, b, and c, I list copies that appear to have come from the same atelier at three different periods. The folio numbers pertain to the text of *Quadrans vetus*, excluding the tables where present.

(i) Copies lacking the supplementary tables:

London, BL Royal 12.C.ix, fols. 23–28; prob. 1280s (tables only in short forms: beginning of month for a single year and every five days in a single year; anon.)

New Haven, Yale Med. Sch. Libr. Cushing 11, fols. 132–37; early 14c (anon.)

Paris, BN lat. 14070, fols. 76–80, late 13c (*Robertus Anglicus*, written in fifteenth-century annotator's hand)

Vatican, Pal. lat. 1414, fols. 168–73; prob. 1280s (title in first hand: *Robertus Anglicus de Monte Pessimo* [!]; colophon in first hand: *Robertus Anglicus de Monte Pessalano*)

(ii) Copies with supplementary tables, as in the *Tabule solis*:

a Basel, Öffent. Bibl. der Univ. O II 7; ca. 1292 (tract now missing, but listed in table of contents, just before "*tabula quedam ad inveniendum locum solis*," namely, the five supplementary tables, extant at fols. 58–60)

Bruges, Bibl. de la ville 522, fols. 51–57; prob. 1280s (anon.)

Escorial, Real Bibl. O II 10, fols. 64–68v; prob. 1280s (anon.)

London, BL Harley 3735, fols. 64–70; ca. 1292 (anon.)

Milan, Bibl. Ambros. H 75 sup., fols. 52–56v; after 1284, prob. late 1280s (*Johannes de Monte Pesulano*)

Paris, BN lat. 7195, fols. 36–41v; prob. ca. 1290 (anon.)

Paris, Bibl. Ste. Geneviève 1043, fols. 43–48v; 1296–97 (anon.)

Princeton, Univ. Libr. Garrett 99, fols. 162v–168v; ca. 1293 (anon.)

Stockholm, Kung. Bibl., X 767, fols. 66–70; ca. 1292 (defective at beginning; attribution unknown)

b Cambridge, Fitzwilliam Mus. McClean 166, fols. 3–8v; 13c (anon.)
Cambridge, Univ. Lib. Ii 1.15, fols. 11–15; prob. 1290s (*Johannes Anglicus in Monte*)
Darmstadt, Hess. Land. Bibl. 2661, fols. 135–43; 13c (anon.)
Munich, Bayer. Staatsbibl. Clm 353, fols. 33v–38; prob. 1290s (anon.)
Paris, BN lat. 7194, fols. 39–44; early 14c (anon.)
Paris, BN lat. 7416B, fols. 51v–57v; early 14c (anon.; tables in same form appear three times: in same hand, at fols. 17–18v, 96–98; in different hand, partial copy at fols. 64r–v)

(iii) Copies with tables based on the *Almanac* of William of St. Cloud (1292):

a Florence, Bibl. Laur. plut. 18 sin. cod. 6, fols. 59v–68 (defective set of tables: 1292–94 only; anon.)
London, BL Egerton 844, fols. 49–55v; 1292–95 (*Johannes in Montepessulano*)
London, BL Harley 3647, fols. 55–60; 1292–95 (anon.)
Paris, BN nouv. acq. lat. 1893, fols. 41v–46; 1292–95 (anon.)
Salamanca, Bibl. Univ. 2662, fols. 41–46v; 1292–95 (anon.)
Vienna, Öster. Nat. Bibl. lat. 2367, fols. 66–74v; 1292–95 (anon.)

b Cambridge, Univ. Lib. Ii 3.3, fols. 56–60; 1299 (lacks supplementary tables, but short table has entries explicitly labeled for 1299; *Johannes in Montepessulano*)
Copenhagen, Kong. Bibl. Add. 447, 2°, fols. 77–83v; 1296–99 (tables appear separately at fols. 46v–48v, canon at fol. 45, being quotation of *Quadrans vetus* [35–38] and [40], in later annotator's hand; anon.)
Oxford, Bodl. Lib. Tanner 192, fols. 67–74v; 1296–99 (*Johannes Anglicus in Monte Pessulano*)
Paris, BN lat. 7298, fols. 55v–59; 1296–99, but misdated as if 1292–95 (anon.)

c Florence, Bibl. naz. Magl. II iii 24, fols. 176v–179v; 1303–07 (*Johannes de Montepessulano*)
London, BL Royal 12.C.xvii, fols. 61–71; 1303–07 (*Johannes in Montepess.*)
Oxford, Bodl. Lib. Ashmole 1522, fols. 70–78; 1303–07 (*Johannes in Montepess.*)
Paris, Bibl. de la Sorbonne 595, fols. 55–61v; 1303–7 (tables appear separately at fols. 65v–67v; anon.)

d Copenhagen, Kong. Bibl. NKS 275ª, 4°, fols. 64–72v; 1292–95 (anon.)

(iv) I list the following manuscripts, mostly from later in the fourteenth century, from information in Tannery and Hahn:

Bruges, Bibl. de la ville 528; 13c–14c (Hahn *B*; anon.)
Erfurt, Wissen. Bibl. Amplon. F 376; 14c, second half (Tannery 21, Hahn *e*; *Johannes Anglicus in Monte.*)
Erfurt, Wissen. Bibl. Amplon. Q 348; mid-14c (Tannery 17, Hahn *f*; colophon: *Robertus Anglicus*)
Erfurt, Wissen. Bibl. Amplon. Q 369; 14c (Hahn *g*; anon.)
Erfurt, Wissen. Bibl. Amplon. Q 386; 14c (Tannery 30, Hahn *h*; anon.)
Florence, Bibl. Laur. plut. 18 sin. cod. 3; 14c (Tannery 24, Hahn *Y*; anon.)
Madrid, Bibl. Nac. 8883; 14c (Hahn *i*; *Johannes Anglicus in Monte.*)
Naples, Bibl. Naz. VIII C 49; ?13–14c (Hahn *r*; anon.)
Naples, Oratoriana di Girolamini XI (C.F.I.26); 14c (Hahn *N*; *secundum Lincolniensem episcopum*)
Oxford, Bodl. Lib. Can. ital. 157; 14c (Hahn *S; Robertus Anglicus de Monte Pessulano*)
Oxford, Corpus Christi Coll. 41; early 14c (Tannery 32, Hahn *P*; anon.)
Oxford, Univ. Coll. 41; 14c (Tannery 23, Hahn *P*; anon.)
Paris, BN lat. 7267; 14c (Tannery 6, Hahn *T*; *Johannes Anglicus*)
Paris, BN lat. 15121; 13 or 14c (with incomplete supplementary tables) (Tannery 8, Hahn *W*; anon.)
Paris, BN lat. 15125; 14c (Hahn *X*; anon.)
Rome, private coll. of Prince Boncompagni, 51 (1892 catalog); 14c (Tannery 13; *Johannes Anglicus in Monte.*)
Rome, private coll. of Prince Boncompagni, 323 (1892 catalog); 13c (Tannery 14; anon.)
Vatican, Urb. lat. 1408; 14c (Hahn *a*; anon.)
Venice, Bibl. marc. VIII 77 (XI 79); 14c (Hahn *U*; anon.)

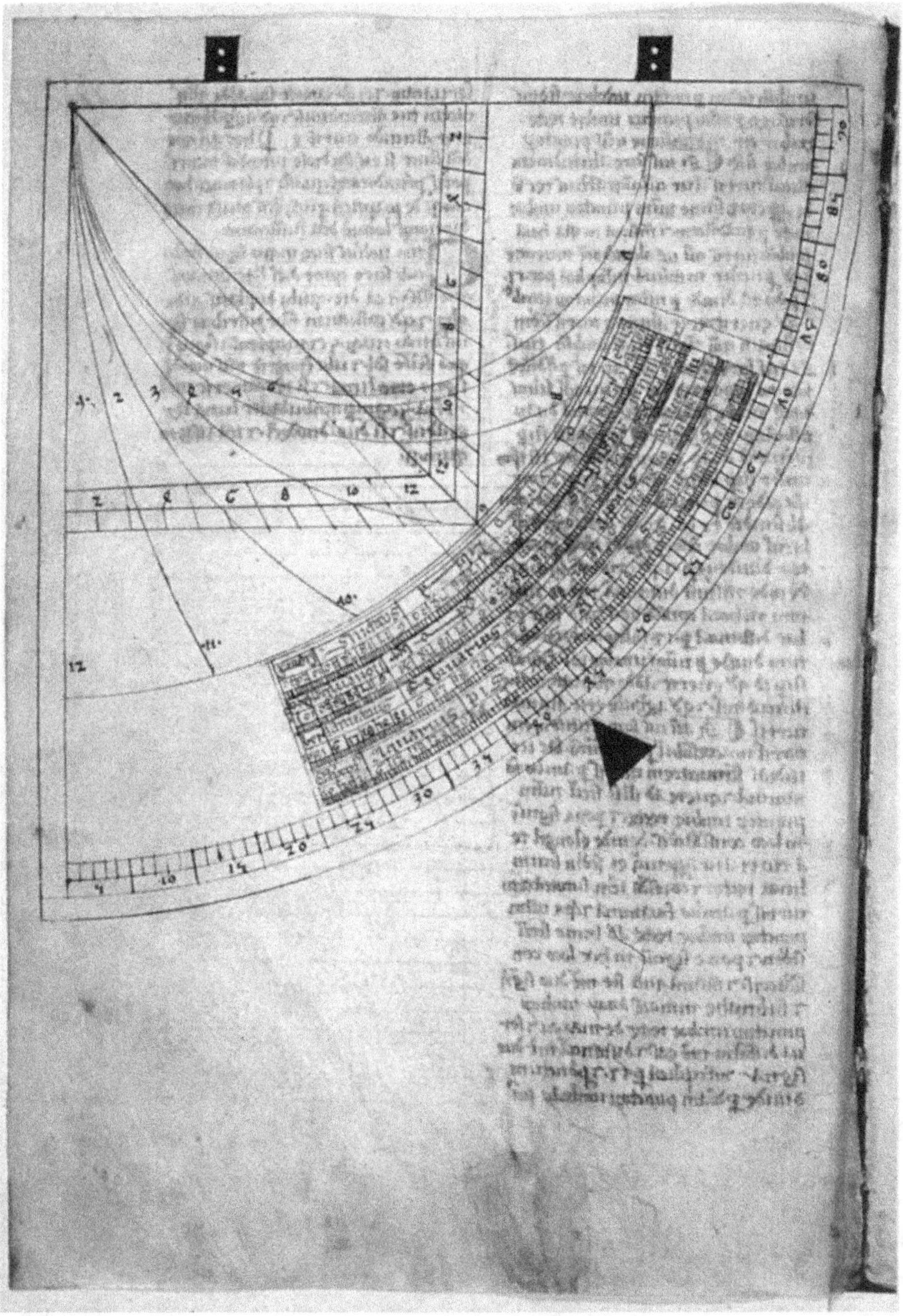

Plate I. Vatican, Biblioteca Apostolica, Urb. lat. 1428, f. 95v. (Photograph by courtesy of the Biblioteca Apostolica Vaticana.)

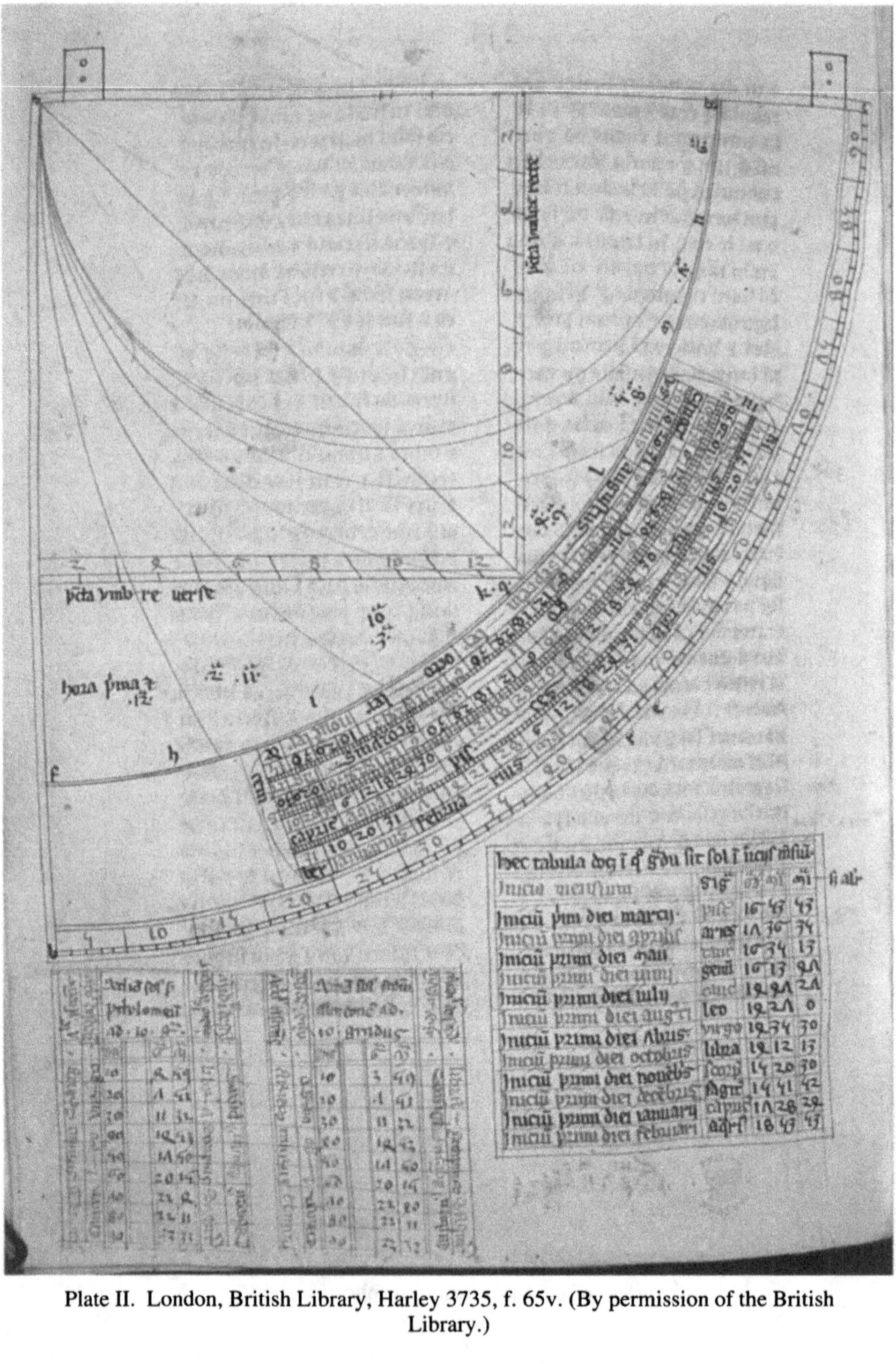

Plate II. London, British Library, Harley 3735, f. 65v. (By permission of the British Library.)

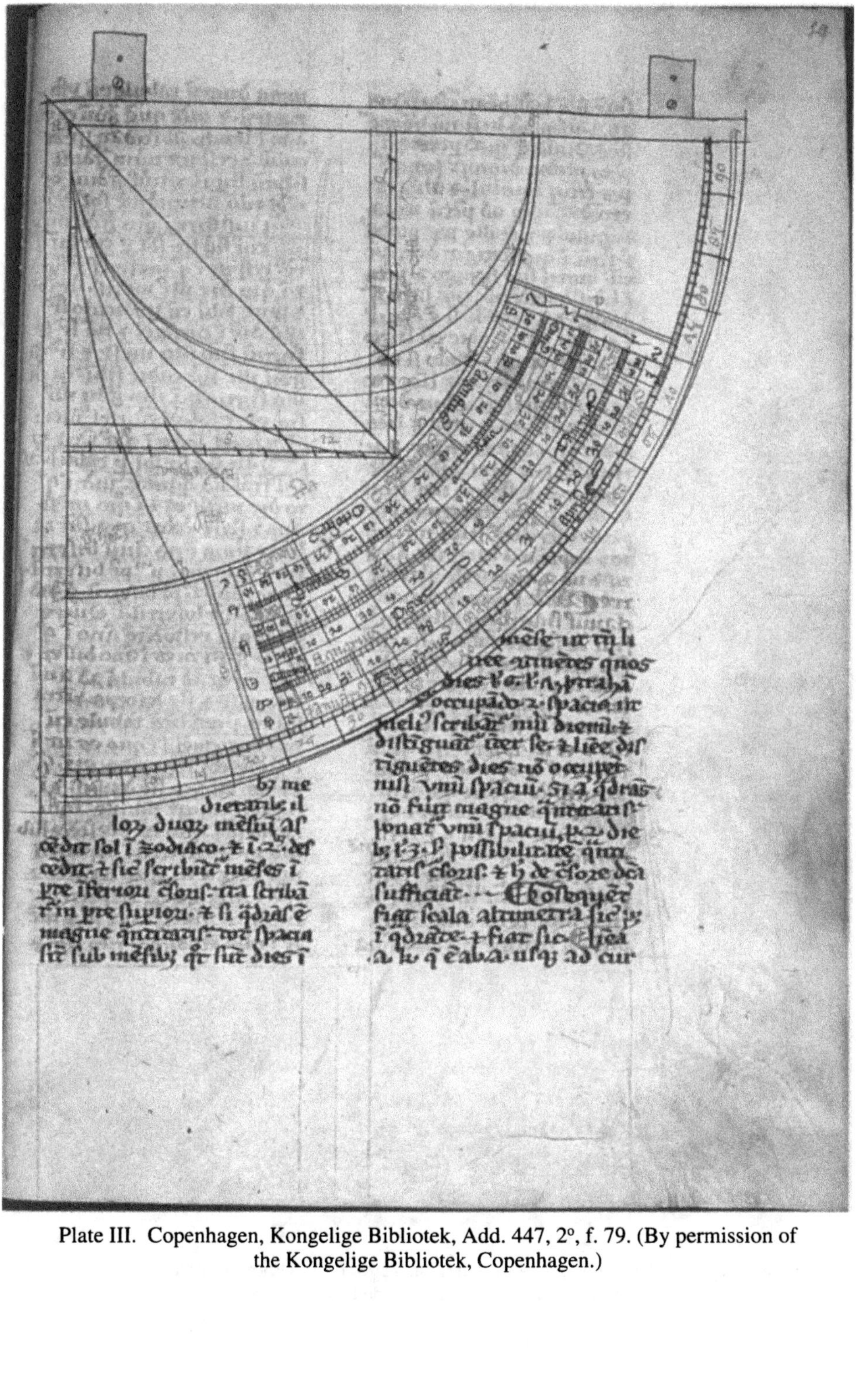

Plate III. Copenhagen, Kongelige Bibliotek, Add. 447, 2°, f. 79. (By permission of the Kongelige Bibliotek, Copenhagen.)

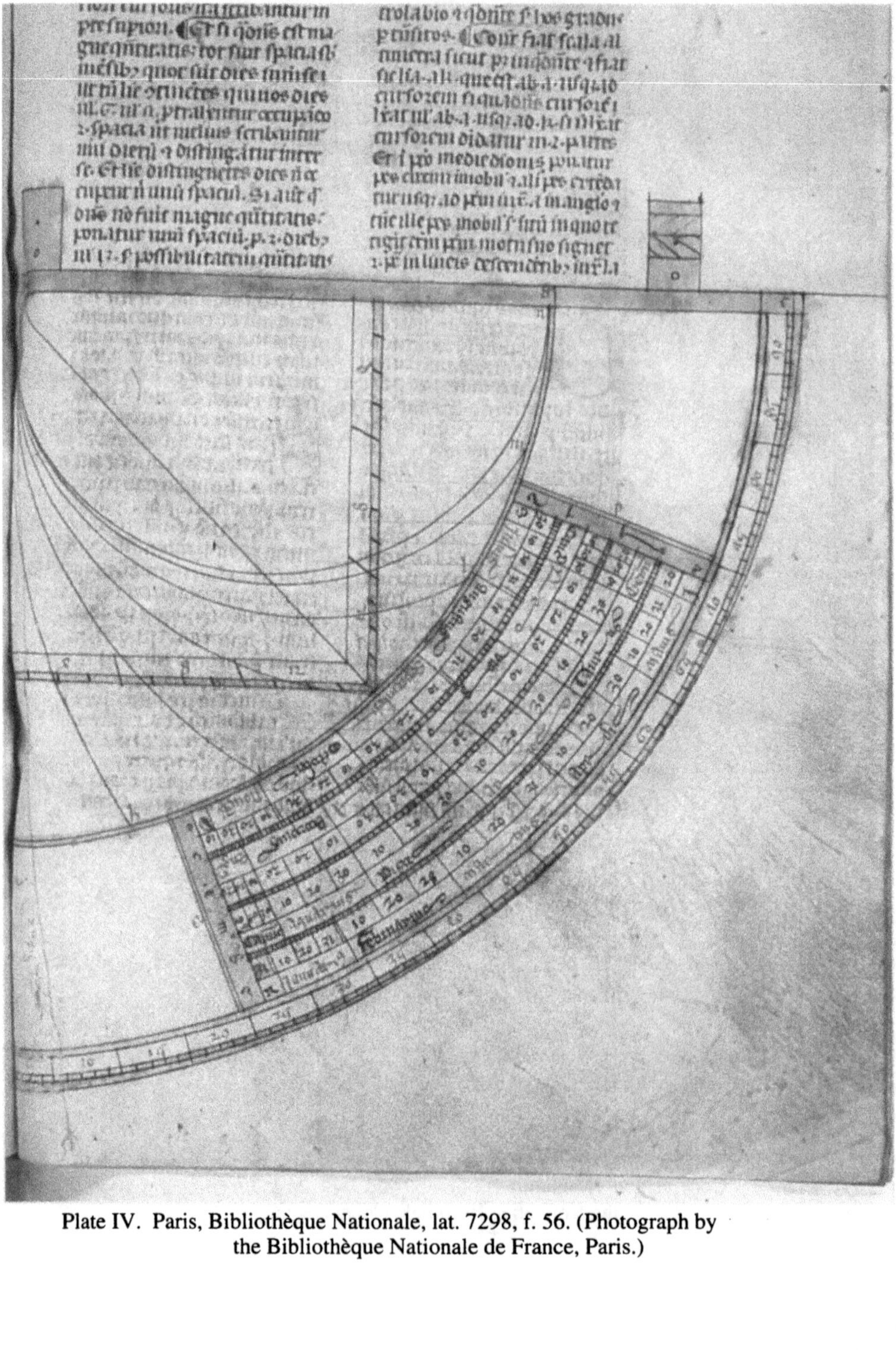

Plate IV. Paris, Bibliothèque Nationale, lat. 7298, f. 56. (Photograph by the Bibliothèque Nationale de France, Paris.)

APPENDIX 2: PLATES

MANUSCRIPT DIAGRAMS OF QUADRANTS

I. Vatican, Urb. lat. 1428, fol. 95v

Plate I is taken from MS Vatican, Urb. lat. 1428, fol. 95v. This is the only manuscript I know of that includes a diagram in the vicinity of the *Practica quadrantis* of Campanus. (The copies of Sacrobosco's *Quadrans* likewise omit diagrams.) In form it very closely resembles that characteristically found with copies of *Quadrans vetus* from the 1280s onward, as in groups (ii) and (iii) in Appendix 1 (from which the specimens in the other plates are taken).

The figure is carefully drawn by means of compass and ruler. The labeling is in the text hand of the manuscript. Engraving on the cursor alternates in rubric and bright blue (the latter being an unusually striking color, seen also in the diagrams and tables of a few other manuscripts—e.g., the Paris, London, and New York manuscripts listed under *Tabule solis* in my "Solar Tables," n. 18). As is typical, the central axis of the cursor (corresponding to 0 Aries and 0 Libra) lines up with degree 45 in the outer scale. The cursor extends 24 degrees to either side, and the spacings for the signs—e.g., of Aries, Taurus, and Gemini (11°, 9°, and 4°, respectively)—conform approximately with those stated, not in Campanus, but in *Quadrans vetus* (sc., 11½°, 8½°, and 4). An uncommon feature of this diagram is the inclusion of the *perpendiculum*, drawn as a red line, with the *plumbum* as a triangle in blue ink; the *margarita* appears as a point about halfway up the line. The hourlines are numbered from 1 to 6 at their upper parts, and from 7 back down to 12 at their lower parts. Each side of the shadow square is divided into 12 parts, every second part being numbered.

II. London, BL Harley 3735, fol. 65v

In Plate II the diagram from MS London, BL Harley 3735, fol. 65v, displays the standard form in *Quadrans vetus*, with many features as in the Vatican copy. The cursor is labeled in alternating rubric and black, the hourlines numbered from 1 to 6, then 7 back to 12, and the shadow square numbered by twos. But the *perpendiculum* is absent. The cursor extends 24 degrees to each side of its central axis, and the first three signs have spacings of 11½°, 8½°, and 4, as specified in the text. But the equinoctial line is set opposite degree 42 (the colatitude of Paris), rather than 45. The labeling here is in the same hand as that of the text.

Visible at the lower right is the short table of longitudes, in the same hand, with the heading: *Hec tabula docet in quo gradu sit sol in iniciis mensium*, which usually appears as [29] of the text. This table, with three places instead of two, is quite uncommon, being found only in MS Paris, BN lat. 14070, and related to that in MS Vatican, Pal. lat. 1414. At the lower left are two copies of a short table of the solar declination. This manuscript, I believe, is the only one having such a table (the usual form being the complete table, set among the supplementary tables, as in the Harley copy); it is in a different hand, but one found in several other manuscripts in groups (ii) and (iii) from the 1290s.

III. Copenhagen, Kong. Bibl. Add. 447, 2°, fol. 79

The figure in Plate III comes from MS Copenhagen, Kong. Bibl., Add. 447, 2°, fol. 79. The form is much as the preceding, but on a somewhat smaller scale. The cursor labeling alternates between rubric and black; the shadow square is now labeled by fours. The cursor is centered on degree 45, and the signs receive spacings of 11½°, 8½°, and 4 to fill out the 24 degrees of each half. The labeling hand differs from that of the text (which happens to be the same as the text and labeling hand in the London manuscript in the preceding plate). Most distinctive of this new hand is the flourishing of the "s" and "r," the use of a vertical stroke to mark capitals (as with "T" and "P"), and the high, closed arch of the "a." Note also the inscription of "*(J)uniu(s)*" at the upper margin of the cursor.

Along the outer margin of the *limbus*, the numeral "75" is in a different hand from the rest—but in the same hand as the figure in the Harley manuscript. The same hand, as said above, is also responsible for the text in the Copenhagen manuscript. Thus, while the figure has been delegated to the second scribe in this case, the first scribe appears to have kept notice of the work.

Light yellow wash overlies the *pinnule*, with the sighting holes at the top, and the borders of the *limbus* and cursor.

IV. Paris, BN lat. 7298, fol. 56

In Plate IV is shown the diagram from MS Paris, BN lat. 7298, fol. 56. This copy is approximately contemporary with the Copenhagen copy (1296–99), and the forms of the diagrams in the two are closely similar. The text of the Paris copy is in a rounded, almost Italianate hand, different from both of the Copenhagen hands. But the diagram labeling in the Paris copy matches that of the Copenhagen diagram hand. In addition to the letters mentioned previously, one should note also the terms *Marcius* (for the "r"), *Aprilis* (for the "l"), *Junius* (for the marginal letters), *Taurus*, *Gem*, *Leo*, *Virgo*, *Libra*, and so on. The Paris hand introduces a second form of flourishing for terminal "s" (with a backward sweep of the upper stroke), as in *Aquarius*, but this is seen also in the Harley copy (e.g., in *Aries*).

Light yellow-greenish wash highlights the *pinnule* and the borders of the *limbus*, cursor, shadow square, and upper side of the quadrant.

The coincidences of style in the details of execution of these three last figures, particularly in the characteristics of the labeling hands, indicate that all three are the product of the same two scribes. By comparisons of the same sort, one can establish the unity of the manuscripts classed in (ii, a) and (iii, a–c) as the products of a small group of scribes associated, it would appear, with a single atelier. The Vatican copy must be associated with an unrelated earlier scribal group.

ROGER BACON'S *DE LAUDIBUS MATHEMATICAE*: A PRELIMINARY STUDY

GEORGE MOLLAND*

As Item 16 in his still-authoritative bibliography of Roger Bacon's works, Andrew George Little included a text *De laudibus mathematicae*, represented by two manuscripts: London, BL Royal 7.F.vii, fols. 72r–77r, and Oxford, Bodleian, Digby 218, fols. 98r–103r.[1] Little began his description of the text by calling it "a different recension of matter which occurs for the most part word for word, in *Opus majus*, Part IV,"[2] and this probably explains why it has hitherto received little attention. Nevertheless, Little went on to make clear that the short work contained material not in the *Opus maius*, and this provides at least a partial justification for the present study.

At present I know of no manuscript copies of this work other than the two noted by Little. Of these, the first is from one of a pair of codices, found in the Royal Collection in the British Library (the other being 7 F.viii), that were in the possession of the Franciscan William Herbert around 1300 and that were lodged in the Franciscan Convent at Hereford; both have annotations by Herbert.[3] They contain several other writings by Bacon and are usually dated to the late thirteenth century. Later they were in the Lumley library, and in the seventeenth century they passed to John Prideaux, Bishop of Worcester, and to the Theyer collection. The Oxford manuscript was probably written a little later; at one time time it belonged to Thomas Allen and was, it seems, bound into a single codex, together with other works (including a fragment of Bacon's *Opus minus*), by

---

*I must acknowledge profusely a wealth of comments, criticisms, and suggestions that I have received from Edith Sylla. Unfortunately I have not been able to take them all on board, and, as is proper, there are points on which for the time being we must agree to differ—but without her counsel this article would have been by far a poorer performance.

[1] A. G. Little, "Roger Bacon's Works, with References to the MSS. and Printed Editions," in *Roger Bacon: Essays*, ed. A. G. Little (Oxford: Clarendon Press, 1914; reprint, New York: Russell and Russell, 1972), pp. 375–426, at pp. 393–94.

[2] Part IV is to be found edited in J. H. Bridges, ed., *The "Opus Majus" of Roger Bacon*, 3 vols. (Oxford, 1897–1900; reprint, Frankfurt: Minerva, 1964), 1: 97–403, although portions of this almost certainly did not form part of the original *Opus maius*. (Subsequent parenthetical volume and page numbers will refer to this edition.)

[3] M. R. James, "The Library of the Grey Friars of Hereford," in *Collectanea Franciscana*, vol. 1 (Aberdeen: University Press, 1914), pp. 114–23, 154–55; G. F. Warner and J. P. Gilson, *British Museum: Catalogue of Western Manuscripts in the Old Royal and King's Collections*, 4 vols. (London: for the Trustees, 1921), 1: 201–3; N. R. Ker, *Medieval Libraries of Great Britain: A List of Surviving Books*, 2nd ed., (London: Royal Historical Society, 1964), p. 100; A. G. Little, "The Lamport Fragment of Eccleston and Its Connexions," *English Historical Review*, 49 (1934): 299–302.

Kenelm Digby.[4] The two versions of the *De laudibus mathematicae* are very close: both break off at the same place, and it is tempting to consider whether the Oxford version might not be copied from the London one; at any rate it does little to correct the latter's faulty readings.

An obvious question that comes to mind concerning the *De laudibus*, but one that cannot be answered definitively, is that of its temporal relation to the *Opus maius* and the other works for the pope. We can assert with some confidence that it was not very much later than they, for in it, as in the *Opus maius* and the *Opus tertium*, Bacon provides an etymological argument for distinguishing true mathematics from false mathematics (comprising magic and deterministic astrology), an argument that he later strongly rejected. This involved the assertion that the Greek root for true mathematics was *mathesi* with a shortened middle vowel, while false mathematics came from *matesi* with a lengthened middle. But in later works, such as the *Communia mathematica* and his introduction to the pseudo-Aristotelian *Secretum secretorum* (probably from around 1270), he derided this as a most stupid opinion and maintained almost the precise opposite.[5]

Bacon refers to the *De laudibus* as a letter,[6] but gives no indication of the addressee (a question to which we shall revert at a later stage). There are signs that it was conceived as a very long letter, for its opening phrase seems clearly to allude to a preceding part on experimental science,[7] and this could tempt one to think that it was intended as part of a draft of the *Opus maius*. It does not include the fulsome salutations to the pope that are characteristic of the *Opus tertium*, but for the most part these are also absent from the *Opus maius*. However, in the *Opus maius*, Part IV is devoted to mathematics, while experimental science is the concern of Part VI, and it was the placing of mathematics after experimental science that led Little to abandon his earlier suggestion that the *De laudibus* was intended as a pref-

---

[4] W. D. Macray, *Catalogi codicum manuscriptorum Bibliothecae Bodleianae pars nona: Codices a viro clarissimo Kenelm Digby, Eq. Aur., anno 1634 donatos, complectens* (Oxford: Clarendon Press, 1883), pp. 231–33; A. G. Watson, "Thomas Allen of Oxford and His Manuscripts," *Medieval Scribes, Manuscripts, and Libraries: Essays Presented to N. R. Ker*, ed. M. B. Parkes and A. G. Watson (London: Scolar Press, 1978), pp. 279–314.

[5] R. Steele, ed., *Secretum secretorum, cum Glossis et Notulis: Tractatus brevis et utilis ad declarandum quedam obscure Dicta Fratris Rogeri, Opera hactenus inedita Rogeri Baconi* (Oxford: Clarendon Press, 1920), p. viii; A. G. Molland, "Roger Bacon's Appropriation of Past Mathematics," in *Tradition, Transmission, Transformation: Cultural Exchange in the Premodern World*, ed. F. Jamil Ragep and Sally P. Ragep with Steven Livesey (Leiden: Brill, 1996), pp. 347–65, at pp. 348–49.

[6] "Quia superfluum esset in presenti epistola illas utilitates iuxta singulas partes mathematice pertractare": *J* [= MS London, BL Royal 7.F.vii] fol. 72ra, *Q* [= MS Oxford, Bodleian, Digby 218], fol. 98ra.

[7] "Post hanc scientiam experimentalem, mathematica est dignior et utilior respectu theologie": ibid.

ace to the second volume of the *Compendium philosophiae*.[8] The plan of the *Opus minus* did include an enumeration of the sciences in the reverse order from that in the *Opus maius*, but if the *De laudibus* was to form part of that work one would not expect so much material identical to that presented in the *Opus maius*. Also this hypothesis would not square easily with the mathematical passages of the *Opus minus* preserved in Vatican MS Reg. Lat. 1317, fols. 103r–107v.

One argument for a relatively late date for the *De laudibus* is that it includes the ninefold division of mathematics, whereby the subject was first divided between those things common to the whole subject, and those proper to its particular branches. The latter were then divided into eight parts: the four traditional subjects of the quadrivium—arithmetic, geometry, astronomy, and music—each of which had a theoretical and a practical subdivision.[9] This separating out of the common things was strongly insisted upon in Bacon's *Communia mathematica*, but is far less prominent in the works for the pope. However, it can provide no strong grounds for dating the work later than those composed for the pope, and for the moment I would only maintain that it was written neither much before nor much after 1267.

The manuscripts indicate a division of the surviving part of the *De laudibus* into eight sections, which we shall consider serially. First, however, I would like to make some general observations. Sections 1, 2, 7, and 8 are very close to the *Opus maius*, often with identical wording, although they contain some additional material. On the other hand, sections 3–6 are independent; they may be seen as elaborating with variations on a single page of the *Opus maius* (1: 246). The main (but not the only) concern of the treatise is with the branch of mathematics that we would call astrology. Bacon of course was a firm believer in the subject, but a belief in astrology has two aspects—ontological and epistemological. It is one thing to assert that stellar configurations have important and fine-grained effects on happenings here below, but quite another to claim that precise prediction of these effects is possible. *Very* roughly we may say that the parts of the *De laudibus* that correspond to the *Opus maius* are mainly concerned with the ontological side of astrology, and how this relates to religion and theology, while sections 3–6 concentrate on the epistemological side, and especially the limitations and difficulties in the making of astrological judgments. With these schematic remarks made, let us look more closely at the treatise.

Section 1 starts with a comparison of the relative merits of experimental science and mathematics with respect to theology, both absolutely and rela-

[8] Little, "Roger Bacon's Works," p. 394.

[9] On this see Molland, "Roger Bacon's Appropriation of Past Mathematics," (n. 5) pp. 352–57.

tively (an oft-repeated Baconian theme), as they contribute to the ordering of the Church and of the commonwealth, the conversion of infidels, and the driving back of those who cannot be converted:

> Mathematics has more universal experiences, and so is prior in the order of learning, because, as became evident earlier, it is necessary to use mathematical considerations in the experience of the rainbow and similarly in many other cases, as was mentioned. But, although mathematics is prior in the order of learning, yet it is posterior in dignity and power, since the certitude of a particular experience of singulars is more perfect and greater than that of a universal experience. And yet mathematics quantitatively and sensibly avails more for theology than experimental science, although the latter is more useful in virtue [*virtute*].[10] Therefore in a certain way they exceed each other and are exceeded in utility with respect to theology.[11]

After alluding to the ninefold division of mathematics already mentioned, the text of the remainder of this section is much the same as that of Part IV of the *Opus maius* (1: 175–80). Its main concern is with the legitimation of mathematics vis-à-vis theology, and it starts with one of Bacon's favorite themes—the extreme antiquity of the subject:

> The patriarchs and prophets before and after the Flood discovered it and taught it to other men, first the Chaldeans and secondly the Egyptians, and from the Egyptians it descended to the Greeks, and it is not so clearly written that they labored thus in other sciences. But since the divine law was given to us by these men, and they were most holy, they occupied themselves only with sciences most useful to divine law. Therefore mathematics is most consonant with divine law.[12]

---

[10] The Latin term *virtus* and its cognates occur frequently in the treatise, usually in contexts more clearcut than the present one. It is often tempting to translate the term as "power," but I think that this is hazardous. *Virtus* and *potentia*, like modern "virtue" and "power," had a range of meanings that sometimes, but only sometimes, overlapped. Bacon himself was very conscious of this sort of thing—as may be seen, for instance, from D. C. Lindberg, *Roger Bacon's Philosophy of Nature: A Critical Edition, with English Translation, Introduction, and Notes, of "De multiplicatione specierum" and "De speculis comburentibus,"* (Oxford: Clarendon Press, 1983), p. 2; and Roger Bacon, *Communia naturalium*, ed. R. Steele, fasc. 2 of *Opera hactenus inedita Rogeri Baconi* (Oxford: Clarendon Press, n.d.), p. 80. (For a fuller discussion see MS London, British Library, Royal 7 F.viii, fols. 6r–7r.) Accordingly, at the risk of some unnaturalness—after all, we are dealing with a different culture—I have in general translated *virtus* as "virtue."

[11] "Mathematica habet experientias magis universales, et ideo prior est in ordine discipline, quoniam, sicut prius patuit, in experientia iridis oportet uti considerationibus mathematicis, et similiter in multis aliis, ut tactum est. Sed, licet mathematica sit prior in ordine discipline, tamen posterior est dignitate et potestate, eo quod perfectior et maior est certitudo experientie particularis in singulis quam universalis. Et tamen mathematica quantitative et sensibiliter plus valet theologie quam scientia experimentalis, licet illa plus virtute sit utilis, unde quodammodo se excedunt [extendunt J] et exceduntur in utilitate respectu theologie": *J*, fol. 72ra; *Q*, fol. 98ra.

[12] "Patriarche enim et prophete ante diluvium et post invenerunt ipsam et docuerunt ceteros homines, primo Caldeos, secundo Egiptios, et ab Egiptis ad Grecos descendit, et non ita evidenter scribitur quod sic laboraverunt in aliis scientiis. Sed cum per istos viros data est nobis lex divina, et fuerunt sanctissimi, non occupaverunt se nisi in scientia que maxime est

Among the authorities adduced to support this history are Josephus, Jerome, Cassiodorus, and Albumasar. The remainder of the section is mainly taken up by patristic testimonies to the value of mathematics, especially ones from Augustine and Cassiodorus.

Whereas section 1 tends to be blandly straightforward in its praises of mathematics, section 2 adopts a more defensive stance, by trying to free "true" mathematics from any taint of magic:

> Nor does anything hinder the praises of mathematics, except that in these times many are ignorant of the difference between true mathematics and false mathematics (which is by another name especially the magical art), and, since they do not know the conditions of either extreme, they do not understand how the saints attack magical but not true mathematics. Whatever they say has place only against false mathematics, and in no way against true. And certainly the saints do excellently reject magical mathematics, which the prophets also condemn.[13]

After this short introduction the text rejoins a later part of *Opus maius* IV (1: 239–45), and gives the faulty etymological argument from the Greek noted above to separate the two kinds of mathematics.

Soon afterward there occurs a significant deviation from the *Opus maius*. Both texts refer to false mathematics' superstitious use of characters, charms, conjurations, and sacrifices, its deploying of various frauds, and its appeal to celestial necessitarianism. The *Opus maius* merely tells us that these things are expressly asserted "in magical books"; the *De laudibus* is more expansive and says that they occur "in a magical book, whose title is infamous, namely *Theory of the Magical Art*, which several have changed into *Book of Rays*, because the book's author puts forward many excellent things about the multiplication of rays, which are philosophical and true, so that he may the better draw the readers' minds to the poison of falsity that is his principal intent."[14] The reference is unmistakably to al-Kindi's *De radiis stellicis*, which at an uncertain date had been translated into Latin and was widely diffused. It clearly had a great influence on Bacon's own doctrine of the multiplication of species, in which substances and qualities send

---

utilis legi divine. Ergo mathematica est maxime consona legi divine": ibid. On the place of ancient wisdom in Bacon's thought see George Molland, "Roger Bacon and the Hermetic Tradition in Medieval Science," *Vivarium* 31 (1993): 140–60.

[13] "Nec obstat aliquid contra laudes mathematice, nisi quod multi hiis temporibus ignorantes differentiam inter veram mathematicam et falsam, que est alio nomine ars magica precipue, cum utriusque extremi ignorant conditiones, non intelligunt quomodo sancti irruunt in mathematicam magicam, et non in eam que vera, quoniam quicquid dicunt locum habet solum contra falsam mathematicam, et nullo modo contra veram; optime enim reprobant sancti mathematicam magicam, quam etiam propheti dampnant": *J*, fol. 72va–b; *Q*, fol. 98vb.

[14] "In libro magico, cuius tetulus infamis est, scilicet Theorica Artis Magice, quem plures mutaverunt in Librum de Radiis, quia auctor illius libri multa preclara premittit de radiorum multiplicatione, que phisica sunt et vera, ut magis alliciat animos legentium ad venenum falsitatis quod principaliter intendit": ibid.

out likenesses of themselves in all directions, which are then transmitted through the medium in accordance with the laws of optics, until they impinge and act upon a suitable patient. Hence Bacon's reference to the "excellent things" that the book contained about the multiplication of rays, which for both writers provided theoretical underpinning for the ontological aspect of astrology. But al-Kindi's work did indeed also contain the suspect doctrines that Bacon alludes to, and, although he must have known of the work before 1267, it would be understandable if he did not refer to it in a work intended for the eyes of the pope; and this in fact removes one argument against dating the *De laudibus* before the *Opus maius*.[15]

Bacon quickly marshals philosophers as well as theologians to counter these errors of false mathematics; he notes that

> the reasoning of the saints was especially directed to the fact that from the beginning such mathematicians hindered the entry of the faith into the world, because, being imbued with this fantasy, they not only erred in faith and fastened their morals to celestial things (as if by celestial and other decrees they were of necessity rendered irascible or gentle, chaste or wanton, and so on), but they also ascribed the miraculous works proving the Christian faith to mathematics, saying that Christians were mathematicians and magi seducing the populace. For just as they themselves could by demonic skill perform many feats before a rude people, by which they led them into error and dominated them, so they charged that the apostles, martyrs, and other preachers of the faith did not perform true miracles on God's behalf, but instead [made marvels] through that magical art which they used. Therefore the saints and the Church thoroughly condemned this mathematics and its proponents.[16]

After his general attack upon false mathematics Bacon concentrates mainly on epistemological difficulties associated with astrology, making particular use of Ptolemy's *Tetrabiblos*, the pseudo-Ptolemaic *Centiloquium* (in two different translations) together with the commentaries of Haly,[17] and a work

[15] On Bacon and al-Kindi see also Molland, "Roger Bacon and the Hermetic Tradition" (n. 12), pp. 154–58.

[16] "Sed precipue ratio sanctorum fuit ad hoc, quod tales mathematici impediverunt a principio ingressum fidei in hunc mundum, quia, non solum imbuti hac fantasia, erraverunt in fide et fixerunt mores suos in celestibus, tanquam per celestia et cetera dicta de necessitate fierent iracundi vel mansueti, casti vel luxuriosi, et sic de aliis; sed opera miraculosa fidem Christi probantia ascripserunt isti mathematice, dicentes Christianos esse mathematicos et magos populum seducentes. Sicut enim ipsi per hanc demoniacam calliditatem potuerunt multa coram populo rudi facere, per que detinebant eos ad errores et dominabantur eis, sic [sicut *in MSS*] imposuerunt apostolis et martiribus et ceteris predicatoribus fidei quod non ex parte Dei fecerunt vera miracula, sed per artem magicam qua ipsimet usi sunt, et ideo hanc mathematicam et eius professores sancti et Ecclesia penitus dampnaverunt": *J*, fol. 73ra; *Q*, fol. 99ra.

[17] Haly is usually identified with Ali ibn Ridwan, but arguments have been put forward for identifying the commentator and perhaps even the author of the text itself with Aḥmad ibn Yūsuf. See F. J. Carmody, *Arabic Astronomical and Astrological Sciences in Latin Translation: A Critical Bibliography* (Berkeley: University of California Press, 1956), p. 16; Lynn Thorndike, *Latin Treatises on Comets between 1238 and 1368 A.D.* (Chicago: University of Chicago Press, 1950), p. v; R. Lemay, "Origin and Success of the Kitāb Tha-

that he refers to as Ptolemy's *De dispositione spere* or *Introductorius in Almagestum*, which can be identified with the Latin translation of the *Introduction to the Phenomena* of Geminus of Rhodes. He also appeals quite often to Avicenna, who was for him one of the greatest philosophers and who, he avers, "completed Ptolemy's works as he himself relates in the prologue to the *Liber sufficientie*."[18] Bacon uses his sources to emphasize the contingency of astrological judgments: one of his favorite quotations is from the beginning of the *Centiloquium*, where "Ptolemy" says that "the judgments which I deliver to you are between the necessary and the impossible."[19] He also notes that radically different outcomes could be astrologically very similar:

> On this account [Ptolemy] says in *Centiloquium* 59 according to one translation, "When you inquire of someone absent, you should not judge death of him until you rule out drunkenness, nor a wound until you rule out bloodletting from him, nor acquired wealth until you rule out money committed to him, for in all these there is the same [astrological] judgment."[20]

The remainder of section 2 gives more examples of the fallibility of astrological judgments, and emphasizes that it is easier for the astrologer to pronounce concerning universals than concerning singulars. Bacon also makes a special use of the *De dispositione spere*, consideration of which may conveniently be deferred to a later stage.

Sections 3–6, as noted above, depart widely from the text of the *Opus maius*, and substitute a much fuller discussion of the vulnerability of astrological judgments than that work contained. Little suggested that this part of the text may have been connected—identity is not in question—with a work mentioned in the *Opus tertium*, where, in the context of another defence of "true mathematics," Bacon writes to the pope: "If you wish to see more abundantly, you should command John to have written in a fair hand a fuller treatise that he has for you, so that you may see that the philosophers are most sober in judgments of astronomy, and that in nothing do

---

mara of Abū Ja'far Aḥmad ibn Yūsuf ibn Ibrāhīm from the Tenth Century to the Seventeenth Century in the World of Islam and the Latin West," *Proceedings of the First International Symposium for the History of Arabic Science, 5–12 April 1976* (Aleppo: University of Aleppo Institute for the History of Arabic Science, 1978), pp. 91–107.

[18] "Avicenna qui opera Ptholomei complevit, ut ipsemet recitat in prologo Avicenne Libri Sufficientie": *J*, fol. 73va; *Q*, fol. 99va. See A. Birkenmajer, *Études d'histoire des sciences et de la philosophie du Moyen Age*, Studia Copernicana, vol. 1 (Wroclaw: Ossolineum, 1970), p. 100.

[19] See, e.g. *J*, fol. 73ra; *Q*, fol. 99rb; and cf. *Liber Quadripartiti Ptholomei. Centiloquium eiusdem. Centiloquium Hermetis* ... (Venice: Bonetus Locatellus, 1493), fol. 107r.

[20] "Propter quod 59 [159 *in MSS*] Centilogii secundum unam translationem dicit, 'Cum interrogatus fueris de absente non iudices de eo mortem donec removeas ab eo ebrietatem, nec vulnus donec tollas ab eo sanguinis minutionem, nec substantiam acquisitam donec pecuniam sibi commissam removeas, est enim in omnibus idem iudicium'": *J*, fol. 73rb; *Q*, fol. 99rb. Cf. *Liber Quadripartiti Ptholomei* (n. 19), fol. 112r.

they presume or err."[21] Little also suggested with some plausibility that this text is referred to in a text included in Bridges's edition of the *Opus maius* (1: 393):

> I show in the treatise that I am sending that the soul can be induced to good or evil, both public and private, by the heaven's virtues without compulsion, as by way of example we see men change their wills [or in some MSS, "delights"] through masters, friends, and colleagues, and through the presence of new things, and in infinite ways without compulsion. I explain this sufficiently in the chapter on the power of judging concerning human affairs by the heaven, in which chapter there is the rationale of all the things that are mentioned about this matter.[22]

As Little himself pointed out, this quotation comes from a section on astrology and related matters that was originally part of the *Opus minus*, and was intended to fill a gap in the *Opus maius* that Bacon had left because of pressure of time.[23] In addition to being included in some manuscripts of the *Opus maius*, the section is also found, but with some rearrangement, in MS Vatican, Reg. Lat. 1317, fols. 97r–103r, a codex that also contains other parts of the *Opus minus*, including unpublished material. The version there preserved, unlike that in Bridges's edition, also has other references to a treatise that Bacon is sending, although possibly not all of them are to the same work: "This then is the intent of the treatise about the stars' virtues and actions, which I proposed to produce in the work that I am sending."[24] "And in the treatise that I am sending the possibility of this judgment is shown, and how it is certified by the saints, and especially by the patriarchs from the beginning of the world, from whom philosophers received the power of judging."[25] "I say this because [the Latins] do not yet have a complete science about these matters, because they have nothing of value

[21] Little, "Roger Bacon's Works" (n. 1), p. 394; J. S. Brewer, ed., *Fr. Rogeri Bacon Opera Quaedam Hactenus Inedita, Vol. 1, I. Opus tertium; II. Opus minus; III. Compendium philosophiae* (London: HMSO, 1859; reprint, Kraus, 1965), p. 270.

[22] "Ostendo enim in tractatu quem mitto quod induci potest animus ad bonum et malum, tam publicum quam privatum, per virtutes celorum sine coactione; sicut exemplariter videmus quod homines mutant suas voluntates/voluptates per dominos et amicos et socios, et rerum presentiam novarum, et infinitis modis sine coactione, hoc enim satis explano in capitulo de potestate iudicandi de rebus humanis per celum, in quo capitulo est ratio omnium que tanguntur de hac materia" (I base my text on MSS Vatican, Reg. Lat. 1317, fol. 106r; London, BL Royal 7.F.vii, fol. 70va, and Escorial, g.III 17, fol. 77v).

[23] A. G. Little, ed., *Part of the Opus Tertium of Roger Bacon, Including a Fragment now Printed for the First Time* (Aberdeen: University Press, 1912), pp. 18, 77; Little, "Roger Bacon's Works" (n. 1), p. 389.

[24] "Hec igitur est intentio tractatus quem in hoc opere quod mitto facere proposui de virtutibus et actionibus stellarum": MS Vatican, Reg. lat. 1317, fol. 97v (cf. Bridges, 1: 379).

[25] "Et in tractatu quem mitto ostensa est possibilitas huius iudicii, et quomodo certificatur hoc per sanctos, et precipue per patriarchas a principio mundi, a quibus habuerunt philosophi potestatem iudicandi": ibid., fol. 100v (cf. Bridges, 1: 388).

except from other languages, and translators are few and bad, as I make clear in the first part of the treatise that I am sending."[26]

However this may be, let us return to the text of the *De laudibus*. Early in section 3, in a passage reminiscent of Laplace, Bacon appeals to Avicenna to the effect that, "If it were possible for a man to know all things that are in heaven and on earth, and their natures, he would know what things and of what kind were to come."[27] But this immediately highlights epistemic difficulties, for it points to the necessity of precise knowledge of both sublunary and supralunary things. Accordingly, Bacon goes on to emphasize the paucity of our knowledge in both these areas. Even in celestial observations we could not achieve the necessary level of accuracy—and the relative motion of the fixed stars with respect to the equinoxes (later called the precession of the equinoxes) and the additional periodic disturbances (trepidation) proposed by Thabit ibn Qurra provide further complications. And these difficulties pale into insignificance when compared with those of setting up correspondences between celestial events and those here below, and of considering what disturbances in the effects of the stars may be produced by the air and other things in the elementary regions. It is therefore unsurprising that Bacon was inclined to follow the author and commentator of the *Centiloquium* in ascribing more immediate prognosticational significance to lower-level meteorological phenomena (e.g. comets, rainbows, haloes), which were generally known as the stars' seconds (*secundae stellarum*).

Section 4 continues the emphasis on the great obstacles to making good astrological judgments, again with frequent reference to the *Centiloquium*. The astrologer must not confine himself to celestial matters, but must also have an eye toward natural philosophy, for otherwise he will be like someone trying to read an unknown language. He must always be aware of the provenance (*origo*) of his object of consideration, for otherwise he may think that he is prognosticating for a man, when in fact he has been offered the nativity of a beast. Moreover, just as the power of a magnet to move iron by attraction drops off with distance or the size of the iron, so also may celestial influences be diminished. Again this means that the astrologer must pay special attention to the stars' seconds, and indeed it seems that he should be skilled not only in the mathematical sciences but in all branches of knowledge:

> By celestial things or [natural] impressions judgment is made about other natural things, but for judgments about man as regards the body medicine

[26] "Hoc dico quia nondum habent completam scientiam de hiis, quia Latini nihil quod valet habent nisi ab aliis linguis, et pauci fuerunt interpretes et mali, ut in prima parte huius tractatus quem mitto declaro": ibid., fol. 101r (cf. Bridges, 1: 389)

[27] "Si possibile esset alicui homini scire omnia ea que sunt in celo et in terra, et naturas eorum, sciret utique que et qualiter sunt futura": *J*, fol. 73va; *Q*, fol. 99va.

> must be known, and as regards the soul moral philosophy, to judge as much of morals as possible with freewill being preserved. On this account Ptolemy said, "No one can arrive at perfection in judging unless he is wise in morals." But physiognomy is necessary for judgments of both body and soul, because by his physiognomic complexion a man is inclined and disposed to different passions of soul and body. But alchemy, as is certain, makes known all inanimate things that are in water and earth, and certifies natural philosophy and medicine, wherefore an astrologer must know this science. And physical agriculture teaches in particular and by experience all natures and properties of animate things. And experimental science and natural astronomy complete all. On this account one cannot be an adequate judge unless one knows the other sciences, and this is rarely found and with difficulty.[28]

Section 5 moves the focus even more to the character of the astrologer himself, for in order to be perfect in the art he must be astronomically destined (*astronomice natus*), or otherwise he will have to depend on divine illumination and inspiration:

> A man will know that he is astronomically destined if from youth he has had an intense and effectual will toward astronomy, and if, when he has studied in it a little, he advances in it notably compared with other people, other things being equal. But he can know this by turning the time to the day and hour of his nativity and conception, if this time is known to him, for by knowing the hour of conception and birth he can know the heaven's disposition at those hours....
>
> But there are other useful and necessary conditions, even if they are not required with as much necessity, namely bodily goodness in composition and complexion, so that he be physiognomically destined. For a solid composition disposes toward longevity, and this is necessary to an astronomer, because his occupation demands experience and time, but a good complexion accompanies and facilitates skill. Therefore he must have a good complexion and composition, and size and situation of parts, and other such characteristics for engaging and continuing in these matters: he must have a virile heart, a strong and subtle brain, an open and well-disposed face, a free and lofty brow, as a sign of great and good imagination, because the instrument of the imagination is the anterior part of the brain; and he should have clear, strong, and healthy eyes for the multiple intuition of rays by instruments in subtle considerations, and many other things that signify wise dispositions, as is taught in books of physiognomy.

---

[28] "Per celestia vel impressiones fit iudicium de ceteris rebus naturalibus, sed propter iudicia de homine a parte corporis oportet medicinam cognosci, et a parte anime moralem philosophiam quantumlibet de moribus salva libertate arbitrii iudicare, propter quod dixit Ptolomeus, Non potest quis pervenire ad perfectionem iudicandi nisi sapiens morum. Phisionomiam vero oportet propter iudicia tam corporis quam anime, quia secundum phisinomicam complexionem homo inclinatur et disponitur ad passiones varias anime et corporis. Sed alkimia, ut certum est, facit cognoscere omnia inanimata que sunt in aqua et terra, et certificat naturalem philosophiam et medicinam, quapropter oportet astrologum scire hanc scientiam. Et agricultura phisica docet in particulari et per experientiam omnes naturas et proprietates rerum animatarum. Et experimentalis scientia et astronomia naturalis complent omnia. Quapropter non potest esse sufficiens iudicator, nisi sciat alias scientias, et hoc raro invenitur et cum difficultate": *J*, fol. 74va; *Q*, fol. 100vb.

> But on the soul's part it is very needful that he naturally have virtues of the soul appropriate to his office, namely a strong and long-lived intellectual virtue, on account of the magnitude of experience that is required, and that he have an ingenious and penetrating discretionary virtue, and a will strongly moved to the desire of this science. And thirdly it is necessary that he be of good morals, and above all he must be chaste and sober, gentle, poor, humble, and just, for will is most hostile to intellect, as was mentioned before.[29]

All these moral qualities are necessary because sin makes a soul like a rusty mirror, incapable of receiving species of things or their truth. This may be compared with the emphasis often found on the necessity that the alchemist (or, for that matter, the magician) be in the right spiritual state for the success of his operations. And, perhaps tendentiously, one might go further and cite both alchemy and astrology as illustrations of the private character of medieval science (as opposed to philosophy), when compared with the public science preached by Francis Bacon and other avatars of "modern" science.

Continuing the same theme, Roger Bacon emphasizes that the soul must be at peace with itself and free from the cares of this world. (Thus, as Aristotle says, idiots and simpletons sometimes judge better than sages.)[30] Even seemingly good characteristics, such as love and magnanimity, distort an astrologer's judgment:

---

[29] "Cognoscet autem homo an sit astronomice natus si a iuventute habuerit ad astronomiam voluntatem intensam et effectuosam. Et quando in ea aliquantulum studuerit si notabiliter profecerit respectu aliorum ceteris paribus. Potest autem hoc scire revoluendo tempus ad diem et horam et nativitatis et conceptionis sue si hoc tempus fuerit ei notum per partes. Cognita enim hora conceptionis et nativitatis potest scire dispositionem celi ad illas horas. Et hee due radices, scilicet completa cognitio per inventionem et doctrinam quantum pertinet ad officium astronomii et astronomica nativitas sunt omnino necessaria, sine quibus non est possibile astronomi iudicare, ut sufficit secundum artem. Et hee sunt conditiones principales que tamen raro inveniuntur in uno homine. Sed et alie sunt conditiones utiles et necessarie, licet non exigantur tanta necessitate, videlicet bonitas coporalis secundum compositionem et complexionem, ut sit phisionomice natus. Compositio enim solida disposuit ad longevitatem, et hec necessaria est astronomo, quia sua occupatio indiget experimento et tempore. Complexio vero bona attendit ingenium et habilitat, unde oportet quod habeat bonam complexionem et compositionem, et quantitatem et situm partium, et huiusmodi alia ad perseverandum et negotiandum in ea, videlicet quod habeat cor virile, cerebrum forte et subtile, faciem apertam et bene dispositam, et frontem liberam et eminentem in signum bone et magne imaginationis, quia imaginationis instrumentum est anterior pars cerebri habeatque claros oculos et fortes et sanos propter intuitum multiplicem radiorum per instrumenta in considerationibus subtilibus, multaque alia que significant dispositiones insipiales, ut in libris phisionomie docetur. Sed a parte anime expedit nimis quod virtutes anime habeat naturaliter ad officium eius convenientes utpote virtutem intellectualem validam et longevam, propter magnitudinem experientie que requiritur, et quod habeat virtutem discretivam, ingeniosam et perspicacem [sperpicatem *in MSS*] ac voluntatem fortiter motam ad desiderium istius scientie. Et tertio necessarium est ut sit bene morigeratus, et pre aliis maxime oportet ut sit castus et sobrius, mansuetus, pauper, humilis et iustus, voluntas enim est maxime rationi inimica, ut tactum est prius": *J*, fol. 74vb; *Q*, fols. 100vb–101ra.

[30] Aristotle, *De divinatione per somnum* 2, 464a20–27.

> Love will mean that an astronomer will have excessive regard for favorable signs in his friends, and place too much faith in them, and in judging he will easily be deceived and advance beyond what is just; and hate will excessively diminish this with respect to enemies, and will hinder justice. Similarly he will either distort a bad sign for the sake of love, or promote it for the sake of hate. "'Magnanimous,'" as Haly says, "signifies boldness and perfection of nature, but the person who is thus will undervalue great things ... and not discuss them, but pass them by on account of his greatness of soul. 'Pusillanimous' signifies a feeble soul which is turned towards trifles, so that long consideration hinders it from greater things.... Everything valued in each science is a mean between the two natures."[31]

Even with all precautions taken there must be an element of uncertainty in judgments concerning singulars—and ironically, this uncertainty proves to be a benefit, for it permits one to guard against the consequences of what the astrologer predicts, as when one stocks up with extra fuel and warm clothing against the expectation of a hard winter.

Section 6 of the *De laudibus* makes a significant appeal to what Bacon calls Ptolemy's *De dispositione spere*, a work that, as noted above, can confidently be identified with Geminus's *Introduction to the Phenomena*, in a Latin translation probably due to Gerard of Cremona.[32] At various points of his corpus Bacon spoke as if this work had some special relation to his doctrine of experimental science, and a notable reference that has often puzzled scholars occurs in his prefatory treatise to the pseudo-Aristotelian *Secretum secretorum*:

> According to Ptolemy in the book *De dispositione spere*, which by another name is called *Introductorius in Almagesti*, the part of philosophy concerned with getting to know future things according to possibility is twofold.... One science is natural astronomy, which judges of natural things in this inferior world, as has been said—namely, in the middle way between necessary and impossible, as Ptolemy teaches in the *Centiloquium* and the *Quadripartitum*, such that the astronomers impose no necessity on free will or contingent things.... But there is another science that concerns future alterations of this inferior world, which is called experimental science [*scientia experimentalis*] by Ptolemy in the aforesaid book, on which, as Ptolemy says, the noble Aris-

[31] "Amor enim faciet quod nimis collaudabit astronomus signa prosperata amicorum, et nimis sic in ea confidet, et in iudicando de facili decipietur et ultra quod iustum est promovebit, et odium nimis adnihilabit eam respectu inimicorum, et impediet contra iustitiam. Similiter signa malum aut pervertet propter amorem, aut promovebit propter odium. 'Magnanimus,' ut dicit Haly, 'audacium [iudicium *in MSS*] et nature perfectionem significat. Qui vero talis est grandia parvipendet ... et ea non discuterit sed preteriit propter suam animositatem. Pusillanimus vero animum inbecillum significat et versatur iuxta minima, ita ut longa eius consideratio eum a maioribus impediat.... Omnis [omnia *in MSS*] autem commendatus in omni scientia medius est inter utrasque naturas'": *J*, fol. 75ra–b; *Q*, fol. 101rb. Cf. *Liber Quadripartiti Ptholomei* (n. 19), fol. 108r.

[32] I use the version found in MS Paris, Bibliothèque Nationale, Lat. 16198, fols. 166r–177v. A Greek text with French translation may be found in Geminos, *Introduction aux Phénomènes*, ed. Germaine Aujac (Paris: Les Belles Lettres, 1975); and with German translation in *Gemini Elementa Astronomiae*, ed. K. Manitius (Leipzig: Teubner, 1898).

> totle relied, and a large assembly of the masters of judgments in astronomy, and which the philosopher Aratus much increased.[33]

It would be a matter of some interest if Ptolemy or Geminus had given a science the name "experimental science" (or a Greek equivalent), but we may say with some assurance that Bacon was using his powers of recall somewhat laxly. In the *De laudibus* he speaks, with reference to the *De dispositione*, of a variety of atmospheric phenomena as *nubila*, a term that clearly had a much wider meaning for him than just "clouds." We might think that these, taking place in the upper elementary regions, would be especially susceptible to astrological prediction, but Ptolemy:

> manifestly determines there that the astrologer cannot certify concerning these, nor must he be blamed for this. He only has to certify concerning those things that happen in the heaven, and if he errs there he is to be blamed, and so he says that we should not blame those who frame judgments from the stars when their judgments are not certified in atmospheric phenomena [*in nubibus*]. Thus, if one of them judges of an eclipse or the rise of a star and errs, blame and disgrace follow, because the course by which it should be judged follows a technical geometrical path [*est secundum viam geometricam artificialem*], and it follows that judgment should not err on it nor deceive at all. But atmospheric phenomena [*nubila*] do not follow the technical path, and so he says that the philosophers of the paths dismissed this path as insufficient, and sought another by the path of art together with experience.[34]

This passage (with overtones of modern weather forecasting) squares well with Geminus, where the author says that investigators of such meteorological phenomena do not follow a *via artificialis* or an *ingenium determinatum*, but proceed by means of a congruence seen by trial (*cum temptati-*

---

[33] "Nam secundum Ptholomeum in libro de Dispositione Spere qui alio nomine vocatur Introductorius in Almagesti, duplex est pars philosophie de futuris cognoscendis secundum possibilitatem, ut dictum est. Una est Astronomia naturalis, quia de rebus naturalibus in hoc mundo inferiori judicat, ut dictum est, scilicet medio modo inter necessarium et impossibile, sicut Ptolomeus docet in Centilogio et in Quadripartito, ita quod nullam necessitatem imponunt astronomi libero arbitrio, neque rebus contingentibus, ut prius expositum est. Sed est alia sciencia que considerat futuras alteraciones hujus mundi inferioris, que vocatur Sciencia Experimentalis a Ptolomeo in libro predicto, super quam, ut ait Ptolomeus, Aristotiles nobilis fundatus est, et multa turba dominorum judiciorum astronomie, quam Aratus philosophus multum ampliavit." Steele, *Secretum secretorum* (n. 5), pp. 9–10.

[34] "Sed manifeste ibi determinat [determinant *in MSS*] quod astrologus non potest certificare de hiis, nec propter hoc vituperari debet, sed solum certificare habet de hiis que in celo contingunt, et si ibi erret vituperandus est. Unde dicit nos non vituperamus opifices iudiciorum astrorum cum ipsorum iudicia non certificantur in nubibus. Quod, si aliquis eorum iudicat de eclipsi aut de ortu stelle et errat, consequitur vituperatio et dehonestatio, quoniam cursus est quo iudicaretur est secundum viam artificialem geometricam, et est consequens ut non erret in ea iudicium neque mentiatur penitus. Nubila vero non currunt secundum viam artificialem, et ideo dicit quod philosophi viarum dimiserunt hanc viam pro insufficienti, et quesierunt aliam per viam artis et experientie": *J*, fol. 75va; *Q*, fol. 100vb.

one) to follow from nearer matters.[35] Difficulties of precise interpretation may remain, and the phrase *scientia experimentalis* may be absent, but *temptatio* (*peira* in Greek) provides us with at least as active a concept of empirical testing. It is therefore easy to see how Bacon's study of the text (with the aid of a somewhat blurred literal memory) could make him recognize "Ptolemy" as an important ally in his insistence on the efficacy of experimental science.

But despite strong affinities, Bacon's conception of experimental science had major differences from the modern one, and it is not surprising to see him turn to the kind of experience that comes from divine revelation or inspiration:

> However much a man be destined for astronomy, informed by discovery and teaching, of good composition, complexion, and morals, and (the time of a hindering constellation having declined a bit) takes counsel in preparing for good significations and mitigating bad ones, as astrologer he will never, as they say, attain the certitude of his office, and certainly and perfectly know things to come, together with their remedies and aids, unless he is inspired and specially illuminated by God.[36]

The general revelation accorded to Christians makes them more receptive to philosophical wisdom and truth than infidels; but as regards special revelation, the *Centiloquium* can provide valuable support, for concerning this:

> Ptolemy speaks as if stammering, and similarly his commentator, because neither received the faith of Christ. They posited, however, that God illuminates the minds of those men he wishes, and when he wishes, but they did not understand that God does this most of all when the faith of Christ is supposed, in those prophesying both before and after Christ. And Haly says that when someone has this knowledge by divine inspiration he can give many judgments for which we find neither significations nor reasons, neither in the operator nor in what is operated upon, but what they say flows from their hearts, and from these we judge many things; and since this path is pure it is called divine by prophets. And so Ptolemy says that some have knowledge of things below the Moon's circle by teaching, some by inspiration, just as we see many hermits who predict things to come, and some epileptics when in a fit predict things to come, because then they do not use corporeal senses but only

[35] "Affirmaverunt illud non secundum viam artificialem neque ingenium determinatum a quo sumerent illud signum, sed assumpserunt illud per convenientiam cadentem cum temptatione secundum rem propinquiorem": MS Paris, BN Lat. 16198, fol. 176r.

[36] "Quantumcumque igitur sit homo natus astronomice, per inventionem et per doctrinam informatus, et fuerit compositionis et complexionis bone morigeratusque tempusque constellationis impedientis modicum declinaverit, consilium apponat in preparandis bonis significationibus et mitigandis malis, nunquam ut dicunt pertinget astrologus ad certitudinem sui officii et certitudinaliter et perfecte cognoscat futura et eorum remedia et adiutoria, nisi sit a Deo inspiratus et specialiter illuminatus": *J*, fol. 75vb; *Q*, fol. 102ra.

> the soul's power, and the signs of things that we see in dreams happen from this.[37]

After this, the text (section 7) immediately rejoins that of the *Opus maius* (1: 246–53), and we feel on more familiar ground. Bacon reiterates his assertions that true mathematicians do not claim infallible judgments, and that the saints attacked only false ones. Trouble then arose with the advent of certain theologians who did not know the true power of philosophy. The rational soul cannot be forced, but it can be influenced by celestial virtue, as it also can be by the species of sublunary things. Celestial influences do much to explain the differences between the various regions of the world and the characteristics of their inhabitants. It is easier to judge astrologically of communities and of public persons, such as princes and their counsellors, than of private individuals, and this means that a skilled astrologer can often be of value in affairs of state. The fragmentary section 8 announces a return to the usefulness of mathematics to theology, both absolutely and in Bacon's four favorite relative ways. The text is in close accord with an earlier passage of the *Opus maius* (1: 180–81), but, after some brief remarks about the value of contemplating celestial things and the use of astronomy in scriptural exegesis, it comes to an abrupt end.

As with so many of Bacon's writings, it is difficult to know whether he stopped writing at this point, or whether the vicissitudes of time and transmission have deprived us of a more complete version of the work. Further, Bacon is notorious for having continually employed the same writings (though selected and woven together, and augmented with new material) in different contexts. We may only speculate about the letter to which the fragmentary treatise that we have been considering was intended to belong, and wonder to whom it was putatively addressed. It seems clear that it was not directed to the pope. As noted, it contains much duplication of the *Opus maius* and also the potentially dangerous reference to al-Kindi. Moreover, it is perhaps significant that the short passage of the *Opus maius* (1: 246) that sections 3–6 replace contains the assertion that astrologers know that

---

[37] "Est autem revelatio specialis per aliquod bonum intentus vel sapientie vel alicuis raptus vel huiusmodi, de qua tanquam balbutiens loquitur hic Ptholomeus, similiter eiusque commentator quia fidem Christi neuter recepit. Posuerunt tamen quod Deus illuminat mentes hominum quando vult et quorum vult, sed non intellexerunt quod hic maxime facit Deus supposita fide Christi in prophetantibus ante Christum vel post. Et de hoc dicit Haly quod, quando aliquis per divinam inspirationem habuerit hanc scientiam, multa poterit dare iudicia, quorum nec significationes nec rationes invenimus, nec in operatore nec in operato, sed quod dicunt ex eorum cordibus manat (monalis [?] *in MSS*), et ex hiis multos iudicamus, et hec via cum pura fuerit a prophetis divina dicitur. Et ideo dicit Ptholomeus quod scientiam eorum sub circulo lune, quidam habent per doctrinam, quidam per inspirationem, sicut vidimus heremitarum plures qui futura predicunt, et epilentorum quidam dum epilentia laborant futura predicunt, quia tunc corporalibus sensibus non utuntur sed sola vi anime, et signa rerum quas videmus in sompnis ex illa contingunt": *J*, fol. 75vb; *Q*, fol. 102ra–b. Cf. *Liber Quadripartiti Ptholomei* (n. 19), fol. 107r.

"divine ordinance can change all things according to His will, and on this account they always add at the end of their judgments the expression that 'So it will be if God should wish it.'"[38] Sections 3–6 are more naturalistic in orientation, and I am inclined to see the "letter" as directed toward an aspiring "scientist." It cannot have been his pupil John, whom Bacon himself instructed orally, following, he said, a special method of his own. Perhaps it was directed to no one particular person, but to any aspirant to what Bacon saw as the meagre scientific community of his time.

In the sixteenth century John Leland, commenting on the chaotic state of Bacon's manuscript legacy, famously said that it would be easier to collect the leaves of the Sibyl than the names of the books that Bacon wrote.[39] Since then the labors of many scholars have helped to improve the situation, but much work remains to be done, and it is to this form of underlaboring that the present paper is principally directed—although in the course of it I hope that we have gained greater insight into the epistemological limitations and difficulties that Bacon saw as facing any genuine practitioner of the astrological art.

[38] "Et iterum sciunt, quod ordinatio divina potest omnia mutare secundum sui voluntatem, et propter hoc adjungunt semper in suis sententiis in fine hoc verbum, quod 'sic erit, si Deus voluerit.'"

[39] John Leland, *Commentarii de Scriptoribus Britannicis*, ed. A. Hall (Oxford, 1709), p. 258.

WHAT REALLY HAPPENED ON 7 MARCH 1277? BISHOP TEMPIER'S CONDEMNATION AND ITS INSTITUTIONAL CONTEXT

J. M. M. H. THIJSSEN*

On 7 March 1277 Bishop Stephen Tempier condemned 219 propositions as erroneous and prohibited their dissemination.[1] Tempier's condemnation has been widely considered to be the most dramatic and significant doctrinal censure in the history of the University of Paris. Since the beginning of the twentieth century, the syllabus of 219 prohibited propositions has also played a crucial role in the historiography of medieval science—and of medieval philosophy, for that matter. It borders on the commonplace to recall the momentous role that Pierre Duhem assigned to Tempier's action in the development of fourteenth-century natural philosophy, and the effects that his thesis had in founding the historiography of medieval science as a discipline. It has often been remarked that Duhem's identification of 1277 as the birth date of modern science directed the way for subsequent historians of late medieval science in determining the topics of their research agendas. The story of the lasting impression of Duhem's thesis on the historiography of late medieval science has been meticulously told by John Murdoch in his "Pierre Duhem and the History of Late Medieval Science and Philosophy in the Latin West."[2]

It was also Murdoch who emphasized that Duhem's work should be placed in the wider context of the studies of Anneliese Maier. It is, therefore, not unreasonable to suggest that John Murdoch might feel great sympathy for the procedure recommended by Maier in writing the history of late medieval science—an approach that is characterized by "putting things into their proper medieval context," and that aims at telling the story of

---

* Research for this article was carried out at the Netherlands Institute for Advanced Study in the Humanities and Social Sciences in Wassenaar, and was supported by a grant from the Netherlands Organization for Scientific Research (NWO). I wish to thank Chris Coppens for elucidating some canonistic points and Katherine Leigh-Van Vliet for correcting my English. This article is one of two studies that suggest a revisionist approach to the current interpretation of the condemnations of Bishop Stephen Tempier in 1277. The related article will appear as "1277 Revisited: A New Interpretation of the Doctrinal Investigations of Thomas Aquinas and Giles of Rome," *Vivarium* 35 (1997), forthcoming.

[1] H. Denifle and E. Chatelain, *Chartularium Universitatis Parisiensis*, 4 vols. (Paris: Delalain, 1889–91), henceforth cited as *CUP*, 1: 543–55.

[2] John E. Murdoch, "Pierre Duhem and the History of Late Medieval Science and Philosophy in the Latin West," in *Gli studi di filosofia medievale fra otto e novecento*, ed. Ruedi Imbach and Alfonso Maierù (Roma: Edizioni di storia e letteratura, 1991), pp. 253–302. See also H. Floris Cohen, *The Scientific Revolution. A Historiographical Inquiry* (Chicago and London: University of Chicago Press, 1994), pp. 45–53, 260–68.

medieval science "as it really was." Against this background it seems only natural to honor John with a contribution that tries to be especially attentive to the medieval context of a text that has had such an impact on the history of medieval science.

In the historiography of science, Tempier's condemnation has generally been perceived as a symptom of the conflict between science and theology, though a conflict that, for once, did not lead to the subordination of science but, on the contrary, helped it to take a new direction, away from the prevailing Aristotelian tradition. In particular, the emphasis on God's absolute power in the condemned propositions has seemed to have opened the way to speculations that were contrary to the principles of Aristotle's natural philosophy.[3] The impact of Tempier's action in thirteenth- and fourteenth-century natural philosophy is evident from the many texts that make reference to propositions that occur in Tempier's syllabus. While historians have long recognized the importance of these explicit and implicit allusions, most of their attention has focused on the content of Tempier's censure, rather than on its functional significance. In this study I will take a different perspective and make central what hitherto has been regarded as of secondary importance in the historiography of medieval science and philosophy—namely, the institutional and juridical context of the document issued by Bishop Tempier on 7 March 1277.

THE COLLECTION OF PARISIAN ARTICLES

Tempier's condemnation is only one of the approximately sixteen lists of censured theses that were issued at the University of Paris during the thirteenth and fourteenth centuries. Most of these lists of propositions were combined into systematic collections of prohibited articles. One of the earliest of these collections originated toward the end of the thirteenth century under the name of *Collectio errorum in anglia et parisius condempnatorum*. Initially this collection consisted of twenty-two chapters, presenting the errors condemned in 1277 by Archbishop Robert Kilwardby at Oxford, in 1241 by Bishop William of Alvernia in Paris, and in 1270 and 1277 by Bishop Stephen Tempier in Paris (in that order).[4] Heinrich Denifle and

[3] See Edward Grant, "The Condemnation of 1277, God's Absolute Power and Physical Thought in the Late Middle Ages," *Viator* 10 (1979): 211–44. The role of arguing *de potentia absoluta* in late medieval (natural) philosophy is controversial. See William J. Courtenay, *Capacity and Volition: A History of the Distinction of Absolute and Ordained Power* (Bergamo: Pierluigi Lubrina, 1990), for a fundamental study of the genesis and historical context of this idea.

[4] See Luca Bianchi, *Il vescovo e i filosofi: La condanna parigina del 1277 e l'evoluzione dell'aristotelismo scolastico* (Bergamo: Pierluigi Lubrina Editore, 1990), pp. 25–27; and William J. Courtenay, "The Preservation and Dissemination of Academic Condemnations at the University of Paris in the Middle Ages," in *Les philosophes morales et politiques aux*

Émile Chatelain conjectured that the collection had been composed by some unknown person, without doubt in England, on his own initiative, very much as Giles of Rome had done in the thirteenth century with his collection of "Errors of the philosophers" (*De erroribus philosophorum*).[5] The latter work, incidentally, was not concerned with academic condemnations, but with the allegedly erroneous views of Aristotle, Averroes, Avicenna, Algazel, Alkindi, and Maimonides.[6]

During the fourteenth century, the *Collectio errorum in anglia et parisius condempnatorum* grew significantly, so that, in fact, a second, augmented *Collectio errorum* came into existence. All the propositions that were added to the first compilation had been condemned in the fourteenth century at the University of Paris.[7] It is not certain, therefore, whether the *Collectio errorum* was the result of private initiative, as Denifle and Chatelain suggested, or whether it was the result of a deliberate attempt of the University of Paris to convey information regarding erroneous teaching. The *Collectio* is comparable to other collections of court decisions that circulated during the Middle Ages and the early modern period. It can be seen as a collection of verdicts that emanated from the University of Paris. The opening of each set of verdicts contains information about the parties involved and the date, but motivations for the verdicts are not given, nor does the collection provide an account (*narratio*) of the events leading to the verdicts.

In any case, the collection of Parisian Articles must have had some kind of official status, and must have circulated among medieval scholars, because bachelors in theology were required by oath not to maintain anything "in favor of articles that have been condemned at the Roman curia or in

---

Moyen Age, ed. C. Bazàn, E. Andújar, and L. Sbrocchi, 3 vols. (New York, Ottawa, and Toronto: Legas, 1995), 3: 1659–67, for a discussion of this *Collectio errorum*. There are also other collections of condemned articles—for instance, the one transmitted in MSS Munich, Bayerische Staatsbibliothek, Clm. 311 and 9558, which presents the errors derived from the *Evangelium aeternum* (1254), together with heresies of the Jews, and errors of the Poor of Lyons. Another compilation can be found in MS Paris, Bibliothèque de l'Arsenal 532, containing the articles condemned by Bishop William of Alvernia in 1241, and articles stating which views of Peter Lombard are not to be followed. These latter articles also occur in the early *printed* tradition of the *Collectio errorum* (see Appendix 1). An excellent description of this manuscript is given in *Aegidii Romani Opera Omnia: Catalogo dei Manoscritti* 1/3**, a cura di Concetta Luna (Florence: Olschki, 1988), p. 27.

[5] *CUP* 1: 556. There is no evidence, however, that the first compilation was made in England.

[6] Giles of Rome, *Errores Philosophorum*, critical text with notes and introduction by Josef Koch, English translation by J. O. Riedl (Milwaukee, Wis.: Marquette University Press, 1944). Giles's collection and the *Collectio errorum* are often to be found joined together in the manuscripts; see, for instance, Vatican, Borgh 360; Erfurt, B. Amploniana Q 151; Munich, Bayerische Staatsbibliothek, Clm. 28126; Paris, BN lat. 16533; Schlägl, Prämonstratenser Abtei, Cpl. [819] 132; Utrecht, Universiteitsbibliotheek 286 (283 kk).

[7] This makes it highly likely that the second, enlarged compilation had its origin in Paris. See also n. 5.

Paris."[8] Moreover, as I mentioned above, many medieval texts contain references to and quotations from the Parisian Articles.[9]

If one compares Tempier's 1277 syllabus with the other lists of errors assembled in the compilation of Parisian Articles, two features stand out: its anonymity, and its promulgation by a bishop.[10] Tempier does not specify the persons behind the false views, but merely states that the errors were disseminated by "certain scholars at the faculty of arts" (*nonnulli Parisius studentes in artibus*).[11] The other condemnations, however, all concern specific scholars whose names are explicitly mentioned in the *Collectio errorum* (see Appendix 1). In addition, Tempier's is one of the few censures in which a bishop was involved.

The anonymity of the 1277 condemnation and the episcopal intervention might seem small details, but, in fact, they are highly significant for what they can tell us about the procedure that Tempier followed in issuing his syllabus. The significance of the anonymity can be appreciated only in the larger context of how scholars at the University of Paris suspect of disseminating false teaching were censured.

University cases of suspect teaching—that is, cases in which the allegations of false teaching arose in the context of typical university activities, such as lecturing or disputing—were in the first instance reviewed by the chancellor and the (regent) masters of theology.[12] The body composed of the chancellor and theologians constituted the lowest level of jurisdiction. The suspects were held accountable on the grounds of their membership in the university, even if they belonged to a religious order. The type of juris-

[8] *CUP* 3: 120–21: "jurat quod...non dicet, tenebit, aut dogmatizabit aliquid quod sit...in favorem articulorum in Romana curia vel Parisius condempnatorum...sed sanam doctrinam tenebit et dogmatizabit." The exact wording of the oath does not appear among the edited statutes of the theological faculty of Paris. Jean Gerson, *Oeuvres complètes*, introduction and notes by Mgr. P. Glorieux, 10 vols. (Paris: Desclée, 1962–73), vol. 5, p. 430, also refers to such an oath: "et jurant baccalaurei priusquam legant Sententias in manu cancellarii Parisiensis quod si quid audierint dici in favorem articulorum Parisius condemnatorum, revelabunt infra octo dies episcopo vel cancellario Parisiensi qui erunt pro tempore."

[9] In the past, references to the "articuli Parisienses" have almost always been considered allusions to Tempier's condemnation. They may, however, concern any of the articles in the different versions of the *Collectio errorum*.

[10] The only other condemnation in the *Collectio errorum* that is anonymous was also issued by Tempier, on 10 December 1270: see *CUP* 1: 486–87 (#432). Much scholarly literature has been devoted to the interpretation of this censure. The judicial proceedings are studied in William J. Courtenay, "Dominicans and Suspect Opinion in the Thirteenth Century: The Cases of Stephen of Venizy, Peter of Tarentaise, and the Articles of 1270 and 1271," *Vivarium* 32 (1994): 191–92. Note that the so-called Ockhamist statute of 1340, which appears in the printed tradition of the *Collectio errorum* (see Appendix 1), is also anonymous; see *CUP* 1: 505–7 (#1042).

[11] Note that the 1277 condemnation is one of the very few censures that concerned the faculty of arts. This aspect could be considered the third distinctive feature of Tempier's syllabus.

[12] What follows is more fully discussed and documented in J. M. M. H. Thijssen, *Censure and Heresy at the University of Paris, 1200–1278* (forthcoming), chap. 1.

diction exercised by the consistory of chancellor and masters was disciplinary. Strictly speaking, this disciplinary council did not *adjudicate* cases of false teaching, that is decide a complaint of false teaching with judicial powers. Their procedure is probably best characterized as extrajudicial: it was a pretrial review that hinged on the idea of fraternal, rather than strictly judicial, correction. Only if the disciplinary procedure at the university level failed, either because the suspect bluntly refused to recant, or because he appealed, was the case transferred to the episcopal or papal courts—that is, to courts with criminal jurisdiction.

The disciplinary character of university censures is reflected, for instance, in the recantation (*revocatio*), which is clearly penance, rather than punishment. When a scholar was found guilty of disseminating the charged errors, he was "corrected" in a public ceremony in which he solemnly withdrew one by one the condemned propositions, which were typically summed up in a list of articles (*articuli*). During the recantation, the culprit demonstrated before the academic community that he was not pertinaciously defending his errors and, consequently, was not a heretic. Moreover, the ceremony informed the academic community about which views were condemned, views that they, from then on, were prohibited from disseminating in their own writings and lectures. Interestingly, the lists of errors compiled in the *Collectio* are in the form of recantations. The *Collectio* in one of its fourteenth-century varieties includes the recanted errors of the following scholars at the University of Paris (see Appendix 1): Stephen of Venizy (1241), Nicholas of Autrecourt (1346), John of Mirecourt (1347), John Guyon (1348), Simon (1351), Guido (Giles of Medonta?) (1354), Lewis of Padua (1362), John of Calore (1363), and Denis of Foullechat (1369).[13]

The disciplinary character of university censures is also reflected in the documentary evidence—or, rather, in the lack thereof. The disciplinary correction of allegedly erring scholars at the university did not leave a paper trail. Of most university condemnations, only the final lists of censured views gathered in the *Collectio errorum*—the verdicts, so to speak—were preserved (see Appendix 1). Documents concerning the proceedings themselves were only generated at the moment when a case was transferred from the disciplinary council of chancellor and masters to the episcopal or papal courts. Only then did it become necessary to recount the events that had led to the transfer of the case to another jurisdiction.

This way of proceeding in cases of suspect teaching at the University of Paris finds confirmation in a little-studied source from the sixteenth cen-

[13] See also Courtenay, "Preservation" (n. 4). The only lists that are not recantations are those of 1270 and 1277 issued by Bishop Tempier, and the condemnation of 1277 by Bishop Kilwardby.

tury. The document is a register of pronouncements and judgments by the faculty of theology, which was kept in the faculty archives. In its meeting of 15 October 1523, the faculty of theology had decided to charge its syndic, the theologian Noël Beda (c. 1470–1536), to compile a survey of (medieval) censures that concerned the University of Paris.[14] Beda's activities resulted in the production of two Registers: the *Liber primus registri [determinationum] facultatis theologie schole Parisiensis in materia fidei et morum incipiens ab anno domini 1284*, and the *Liber secundus registri determinationum ... ab anno domini 1524*.[15]

The first register, which is the only one relevant here, covers the period 1210–1523.[16] If one were to look in this register at the documents concerning suspect teaching from the thirteenth and fourteenth centuries, one would find that it only contains some records concerning the censures of John of Brescain (1247), Denis of Foullechat (1369), and John of Monzón (1389) (see Appendix 2). The register does not reproduce any documents of the censures of Stephen of Venizy (1241), Nicholas of Autrecourt (1346), John of Mirecourt (1347), John Guyon (1348), Simon (1351), Guido (Giles of Medonta?) (1354), Lewis of Padua (1362), or John of Calore (1363). Nor does it contain Bishop Tempier's condemnations of 10 December 1270 and 7 March 1277.

What may one infer from these omissions? First, when Beda drew up his register, he ignored the *Collectio errorum*, even though this collection too concerns the suppression of suspect teaching by the faculty of theology. Apparently, he relied on other documents. Study of the documents that have been preserved in Beda's register shows that they were all original records. Beda probably found them in the faculty's archives. Most of these original records are still extant today.[17]

Second, there was a juridical reason why the cases of John of Brescain, Denis of Foullechat, and John of Monzón generated original records that

---

[14] See James K. Farge, *Biographical Register of Paris Doctors of Theology, 1500–1536* (Toronto: Pontifical Institute of Mediaeval Studies, 1980), pp. 31–36, for a convenient biography of Noël Beda.

[15] The second book covers the period 1524–32. The registers are currently preserved in MSS Paris, BN nouv. acq. lat. 1826, and lat. 3381B. They were consulted by Charles Du Plessis d'Argentré for his *Collectio judiciorum de novis erroribus*, 3 vols.; vol. 1 (Paris: Coffin, 1724); Vols. 2–3 (Paris: A. Cailleau, 1728–36); who indicates on several occasions that he derived a certain document from "the larger register of censures" (*in majori registro censurarum*), or from the "large volume by Noël Beda" (*magnum volumen N. Bedae*). Jules-Alexandre Clerval, *Registre des procès-verbaux de la faculté de théologie de Paris*, vol. 1: 1505–1523 (Paris: Lecoffre and Gabalda, 1917), pp. xi–xv, identified Noël Beda's register as one of the sources used by Du Plessis. I owe this reference to Zénon Kaluza. See also Farge, *Biographical Register*, pp. 453–55.

[16] Although the title indicates otherwise, the first entry is from 1210, not from 1284.

[17] Some of the documents reproduced in Beda's register have been critically edited in *CUP*. The accompanying critical apparatus made me aware that Noël Beda must have relied on the same original charters that the editors of *CUP* used for their edition.

Beda could retrieve from the archives of the Sorbonne, whereas the other cases did not leave a paper trail in the archives. The common characteristic of these cases is that they were all problematic.[18] John of Brescain was cited to appear before the court of the papal legate Odo of Chateauroux because he had failed to recant and oppose his errors, although he had been ordered to and had promised to do so in a consistory of the chancellor, the masters, and the bishop.[19] Denis of Foullechat appealed on the day that he was expected to pronounce his rehearsed recantation. The chancellor and masters of theology perceived his refusal to recant as an act of contempt and stubbornness, and they called in the help of the episcopal court and the inquisitor of heretical depravity to enforce their decisions.[20] John of Monzón also refused to surrender to the correction of the chancellor and masters of theology.[21] As a consequence, his case too was transferred to the bishop of Paris, "the ordinary judge in this location"; the faculty and university handed the bishop the dossier of the Monzón case and begged him to proceed judicially against the defendant "as was the custom in similar cases."[22] In short, all three cases involved a transfer from the "private" proceedings at the university level, within the self-regulating community of masters and scholars, to the "outside" jurisdictions of papal legate, bishop, and pope.

Beda's register does not report any records pertaining to the inquiry that led to the 1277 condemnation, nor have any such records been found elsewhere, for that matter. From this omission I would conclude that a transfer of jurisdiction never occurred in this investigation. A scenario involving a shift from the university level to the bishop's jurisdiction would have generated records, which would surely have been included in Beda's register. It would thus appear that the 1277 investigation did not start at the level of the university, but at the episcopal level, and that Bishop Tempier was involved

---

[18] See Thijssen, *Censure* (n. 12), chap. 1, for a fuller discussion of these censures.

[19] *CUP* 1: 206–7 (#176). This document also concerns the censure of a Master Raymond, who was condemned to return to prison because he had only feigned recanting his errors: upon his release he had continued to disseminate his false teachings.

[20] *CUP* 3: 121–22 (#1299): "Et adhuc in majorem contemptum contra prohibitionem nostram et promissum suum veniens."

[21] *CUP* 3: 503 (#1564): "Qui ita se facturum intra terciam diem promisit; sed ipse promissum suum in hoc, sicut et in omnibus aliis, violavit."

[22] *CUP* 3: #1559, which records the sentence of the episcopal court and reproduces the record of Monzón's condemnation by the consistory of theologians; see p. 495: "Super quibus proposicionibus sic in forma per Universitatem Parisiensem ad requestam dicte facultatis theologie reverendo in Christo patri episcopo Parisiensi, ordinario judici in hac parte, judicialiter exhibitis, factoque super hiis processu." See further *CUP* 3: 503 (#1564): "7a est, quod postquam Universitas et facultas antedicte quod potuerant et debuerant, quantum in ipsis erat, perfecerant, postea nunciaverunt hec omnia reverendo in Christo patri domino episcopo Parisiensi, judici ordinario in hac parte, et presentata eidem cedula facultatis predicte, sibi prout in similibus casibus fieri solitum est, requirendo supplicaverunt, quatenus super hiis vellet judicialiter procedere." This passage is taken from the brief that Pierre d'Ailly prepared when presenting the university's case at the papal court.

in the inquiry right from the start. As will become clear below, this episcopal involvement is an important factor in explaining another distinguishing feature of the 1277 condemnation, namely its anonymity.

Why did Bishop Tempier pronounce an anonymous censure? Why were those condemned of disseminating false teaching not named explicitly, as was the case with the proponents of the other lists of censured errors? This anonymity becomes even more perplexing if one realizes that there existed a medieval tradition, however shallow, that linked the 1277 condemnation to the names of Siger (of Brabant) and Boethius (of Dacia).[23] The reliability of this tradition is partly confirmed in the study of Roland Hissette, who examined the proximate background of the 219 condemned theses. He established that thirty condemned articles did, indeed, seem to be aimed at Siger of Brabant directly, whereas thirteen seemed to have been derived from the works of Boethius of Dacia. In an effort to cast some light on why the name of Siger of Brabant was suppressed from Tempier's condemnation, I shall now return to the University of Paris in the years 1276 and 1277 and propose a new way of understanding some old facts.

THE EVENTS LEADING UP TO 7 MARCH 1277

The traditional picture of the events leading to Tempier's condemnation looks something like this.[24] On 18 January 1277, Pope John XXI (1276–77) informed Stephen Tempier, Bishop of Paris, that he had heard rumors of heresy, and he charged Tempier with the task of examining (*facias inspici vel inquiri*) where and by whom these errors had been disseminated.[25] On 7 March 1277, Tempier published his list of 219 articles and of some books that were condemned. Anyone teaching or listening to the listed errors would be excommunicated, unless they turned themselves in to the bishop

[23] Only two medieval manuscripts bear a rubric identifying the holders of the condemned errors. MS Paris, BN lat. 4391, fol. 68, presents the syllabus of 219 errors under the rubric "Contra Segerum et Boetium hereticos." MS Paris, BN lat. 16533, fol. 60, mentions "Principalis assertor istorum articulorum fuit quidam clericus boetius appellatus." See Pierre Mandonnet, *Siger de Brabant et l'Averroïsme Latin au XIIIe siècle*, 2 vols. (Louvain: Institut Supérieur de Philosophie, 1908–11), 1: 220.

[24] This traditional picture goes back to Mandonnet, *Siger*, 1: 214–86 and has been codified in Fernand van Steenberghen, *Maître Siger de Brabant* (Louvain: Publications Universitaires, 1977), pp. 139–49 and 159–65, though with one important correction—namely, the date of the inquisitor's citation of Siger of Brabant (see n. 35 below). All the subsequent literature has basically accepted Mandonnet's and van Steenberghen's portrayal of the events. See, for instance, Bianchi, *Vescovo* (n. 4), pp. 17–18; John F. Wippel, "Thomas Aquinas and the Condemnation of 1277," *Modern Schoolman* 72 (1995): 237; and François-Xavier Putallaz, *Insolente Liberté: Controverses et condemnations au XIIIe siècle* (Paris: Éditions du CERF, 1995), pp. 51–55, to mention a few of the more recent studies.

[25] *CUP* 1: 541.

or the chancellor within seven days, in which case the bishop would inflict proportionate penalties.[26]

Since this papal letter precedes Tempier's condemnation, it has been generally assumed that Tempier acted on papal initiative. Two claims have been built on this alleged sequence of events, which recur in almost all the literature that has been written about Tempier's condemnation. First, since the pope merely ordered Tempier to investigate rumors of erroneous teaching and then report back to him, Tempier's pronouncement of the 1277 syllabus was the action of an overzealous bishop. Tempier went far beyond his mandate when he issued his decree, which, moreover, he did not previously submit to the Holy See. For this reason, some scholars have characterized the condemnation of 1277 as a proof of the competition between pope and bishop.[27] Second, since only about six weeks elapsed between the papal instructions and Tempier's condemnation, the latter's inquiry was hasty. Further proof of Tempier's haste is found in the repetitions, contradictions, and general disorderliness of the list of 219 censured propositions.

The two claims are considered established conclusions, but they are corollaries from a rather unsophisticated and unfounded *post quem propter quem* argument. The instructions in Pope John's letter are rather vague. Moreover, nowhere does the bishop mention that he was acting on papal orders; nor did he need a papal mandate to investigate allegations of false teaching at the University of Paris. In his introductory letter to the list of condemned articles Tempier merely indicates that he has received information from important people (*magnarum et gravium personarum crebra zeloque fidei accensa insinuavit relatio*). Who these "important people" may have been is a question that shall be addressed below.[28]

The evidence suggests that Tempier acted independently from the pope and that when he received the papal letter of 18 January 1277 he was already in the process of preparing his condemnation. If Tempier received this papal letter at all before 7 March 1277, it must only have encouraged him to continue what he had already been doing, namely preparing his con-

---

[26] *CUP* 1: 543–55.

[27] Jürgen Miethke, "Papst, Ortsbischof und Universität in den Pariser Theologenprozessen des 13. Jahrhunderts," in *Die Auseinandersetzungen an der Pariser Universität im XIII. Jahrhundert*, ed. Albert Zimmermann (Berlin: De Gruyter, 1976), pp. 86–87. See also Roland Hissette, "Étienne Tempier et ses condemnations," *Recherches de théologie ancienne et médiévale* 47 (1980): 239–42, for a discussion of this thesis.

[28] But see Bianchi, *Vescovo* (n. 4), p. 206, who draws attention to similar formulas in other university documents, and suggests that it may be a standard phrase, or a topos.

demnation.[29] Historians have simply been misled by the near contemporaneity of Pope John's letter and Tempier's prohibition.

The absence of any coordination between bishop and pope becomes more apparent in a second letter—*Flumen aquae vivae*—from John XXI to Bishop Tempier.[30] This letter is dated 28 April 1277—that is, more than forty days after Tempier had promulgated his list of condemned articles. Curiously enough, this letter gives no indication whatsoever that the pope knew about Tempier's action. On the contrary, the pope grants a mandate to Tempier to notify him, the pope, of new errors, and to inform him about the names of the propagators of these errors, about their followers, and about their writings. John's second letter has generally been understood as a *new* mandate, now aimed not only at persons of the arts faculty, but also at theologians: the second letter is supposed to have been induced by new (heterodox) doctrinal developments that were not covered by Tempier's condemnation of 7 March 1277.

It seems more plausible, however, to consider John's second letter as a further specification of his first.[31] In both letters Tempier is requested to give information about errors that have been newly disseminated. In the first letter this is phrased as "certain errors to the disadvantage of faith are said to have come forth anew" (*quidam errores in prejudicium eijusdem fidei de novo pullulasse dicuntur*). In the second letter the errors are described as "errors that have been newly invented or taken up again or renewed" (*errores qui de novo inventi vel resumpti seu renovati sunt*). In the first letter Tempier receives a mandate to inquire "by which persons and in which locations" in Paris these errors have been disseminated (*a quibus personis et in quibus locis*). In the second letter, the pope himself already indicates the perpetrators of the errors, namely "some scholars of arts and in the faculty of theology at Paris" (*nonnulli tam in artibus quam in theologica facultate studentes Parisius*). The second letter, finally, is more specific in stating the purpose of the bishop's investigation. The pope will use the dossier that he has requested from the bishop to establish—with the help of an advisory committee—the nature of the errors, and to decide whether they will have to be recanted, or condemned, and whether the University of Paris will need to be reformed.[32]

---

[29] According to Robert Wielockx in *Aegidii Romani Opera Omnia*, vol. 3, pt. 1, *Apologia*; ed. and comm. Robert Wielockx (Florence: Olschki, 1985), p. 92, n. 65, correspondence from the papal court to Paris took about a month to arrive.

[30] This letter has been edited in A. Callebaut, "Jean Pecham et l'Augustinisme: aperçus historiques," *Archivum Franciscanum Historicum* 18 (1925): 459–60.

[31] This suggestion is also made by Miethke, "Papst" (n. 27), p. 85.

[32] Callebaut, "Jean Pecham," p. 460: "ut receptis eisdem ad discussionem, determinacionem seu reprobationem errorum ipsorum vel etiam ad ordinacionem ... , nec non et statu eiusdem studii reformando in premissis viderimus faciendam, de fratrum nostrorum consilio procedamus."

In short, the second papal letter combines the theme of the first letter with rumors of false teaching at the faculty of theology. These rumors may have originated from the inquiry against the theologian Giles of Rome,[33] an investigation that took place shortly after the condemnation of 7 March 1277.[34] This papal letter *Flumen aquae vivae* must have crossed the letter in which Tempier announced the condemnation.

The problems raised by the traditional picture, which was based on the *post quem propter quem* argument, become less serious if one links Tempier's action, not to the papal letter of 18 January 1277 but to events that occurred on 23 November 1276. On that date Simon du Val, the Inquisitor of France, cited Siger of Brabant together with Bernier of Nivelles and Goswin of Chapelle to appear before his court. The citation, which was published in 1947 by Antoine Dondaine, was preserved in a manual for inquisitors.[35] The purpose of the manual was to provide examples for drawing up official documents; the example of how to compose a citation to appear before the court of an inquisitor happened to pertain to Siger of Brabant and his two fellow masters. It is surprising that this document has not been linked before to the events of 7 March 1277.[36]

In the wake of the studies of Pierre Mandonnet and Fernand van Steenberghen, it is generally thought that Siger of Brabant and Goswin of Chapelle had already fled from Paris to Italy when this citation was issued. The purpose of their flight was to appeal to the papal court. The latter conclusion is based on a few lines from a thirteenth-century poem that place

[33] Wielockx, *Aegidii Romani* (n. 29), p. 102, also suggests that *Flumen aquae vivae* may contain echos of the inquiry against Giles of Rome. At the same time, one should keep in mind that such vague letters expressing concern over the orthodoxy of teaching at the university had a topical character. Compare, for instance, the opening of *Flumen aquae vivae* with the well-known letter of Pope Clemens VI of 20 May 1346, addressed to the masters and scholars in Paris and edited in *CUP* 2: 587–90 (#1125), which also refers to the theme of *Flumen aquae vivae*.

[34] Wielockx, *Aegidii Romani*, pp. 77–88, suggests a date after 7 March and before 28 March 1277. In my view, Wielockx's portrayal of the inquiry against Giles of Rome is not very plausible. A different scenario is proposed in J. M. M. H. Thijssen, "1277 Revisited: A New Interpretation of the Doctrinal Investigations of Thomas Aquinas and Giles of Rome," *Vivarium* 35 (1997), forthcoming.

[35] Antoine Dondaine, "Le manuel de l'Inquisiteur," *Archivum Fratrum Praedicatorum* 17 (1947): 186–92. Dondaine was the first to reestablish the correct date of the summons, namely 23 November 1276, not 24 November 1277 as Mandonnet, *Siger* (n. 23), 1: 255, n. 1, thought. The significance of the correct date is that it places the summons before, not after, the condemnation of 7 March 1277. Dondaine's corrected edition of the inquisitor's citation has been accepted by all scholars, unanimously, who, otherwise, have remained faithful to Mandonnet's portrayal of the events leading to the censure of 7 March 1277. See also n. 24.

[36] Dondaine, "Manuel," p. 187, merely observed that, according to customary inquisitorial proceedings, it was the inquisitor's citation that opened the process. Unfortunately, he did not apply this insight to a reinterpretation of Tempier's censure.

Siger in Orvieto, saying that he was stabbed at the papal court there.[37] The Brabant continuator of Martin of Troppau's chronicle (*Chronicon pontificum et imperatorum*) adds that Siger was stabbed by his own cleric, who had gone mad;[38] supposedly, this cleric was Goswin of Chapelle.[39] Siger's death at the papal court must be situated after 22 February 1281, for it was on that date that Pope Martin IV moved the papal court to Orvieto. Siger must have died before 10 November 1284, for on that date the theologian John Pecham wrote a letter containing a passage that has generally been understood as a reference to the already-dead Siger of Brabant: Pecham reports the rumor (*dicuntur*) that the two main defenders of the theory of the unicity of form ended their lives in a wretched manner in countries beyond the Alps, although they were not born there.[40]

The evidence on which this picture is built is extremely tenuous, however. There are, for instance, no records that show that Siger *fled* to Orvieto, nor that he lodged an appeal there. Yet René Gauthier was the first and perhaps the only scholar to challenge seriously the traditional picture of Siger of Brabant's whereabouts after 1275.[41] Before suggesting an alternative scenario of the events leading up to the 1277 condemnation, let me review what is wrong with the traditional picture.

Apart from the fact that there is no documentary evidence that Siger fled to the papal court—which, incidentally, was not yet in Orvieto in 1277, but in Viterbo—the suggestion that he ever lodged an appeal there is highly implausible.[42] First, the concept of appellate jurisdiction implies the review of a sentence rendered in a lower court, or, more generally, at a lower level

---

[37] "Mastro Sighier non andò guari lieto. / A ghiado il fe' morire a gran dolore, / Nella corte di Roma, ad Orbivieto." These lines are from the poem *Il Fiore*, and are reproduced in Mandonnet, *Siger*, 1: 264; an extensive discussion is provided on pp. 264–77.

[38] The passage is quoted in Mandonnet, *Siger*, 1: 277 n. 4: "Sygerus, natione Brabantinus, eo quod quasdam opiniones contra fidem tenuerat, Parisius subsistere non valens, Romanam curiam adiit ibique post parvum tempus a clerico suo quasi dementi perfossus periit." See also ibid., pp. 259–61, for a discussion.

[39] But see note 53.

[40] Mandonnet, *Siger*, 1: 258: "Nec eam [opinionem] credimus a religiosis personis, sed saecularibus quibusdam duxisse originem, cuius duo praecipui defensores vel forsitan inventores miserabiliter dicuntur conclusisse dies suos in partibus transalpinis, cum tamen non essent de illis partibus oriundi." The entire letter has been edited by Franz Ehrle, *John Pecham über den Kampf des Augustinismus und Aristotelismus in der zweiten Hälfte des 13. Jahrhunderts*, reprinted in Franz Pelster, *Gesammelte Aufsätze zur Englischen Scholastik* (Rome: Edizione di Storia e Letteratura, 1970), p. 63. The passage is dicussed in Mandonnet, *Siger*, 1: 258–59 and 278–79, and further in van Steenberghen, *Maître Siger* (n. 24), pp. 163–65.

[41] René Gauthier, "Notes sur Siger de Brabant. II. Siger en 1272–1275. Aubry de Reims et la scission des Normands," *Revue des sciences philosophiques et théologiques* 68 (1984): 26–28.

[42] The itineraries of the papal court in the years between 1276 and 1284 are summed up in Mandonnet, *Siger*, 1: 276, and van Steenberghen, *Maître Siger*, p. 163. If we assume that Siger joined the papal court in 1277, we must also assume that he followed its itineraries to Viterbo, Rome, and Orvieto.

of jurisdiction, such as an episcopal court, or a disciplinary tribunal of chancellor and masters.[43] But against which sentence would Siger have appealed?[44] Since he is supposed to have fled, the citation could not even have been served, and Siger and his fellow masters would have failed to appear in court. Consequently, his supposed flight would have resulted in an arrest warrant, or in a condemnation for contempt of court (*contumacia*), rather than for disseminating false teaching.[45] Furthermore, it is highly unlikely that, pending Siger's appeal, the bishop would have moved against him. An appeal to the papal court would have put a stop to the judicial actions at the inquisitorial and episcopal levels. Following papal appeal, Siger of Brabant (and the two other masters) would have obtained relief from the bishop's jurisdiction. Yet, it is generally accepted that Tempier's condemnation of 7 March 1277 includes thirty propositions that were held by Siger of Brabant. In conclusion, the known facts about the events in 1277 contradict the suggestion that Siger of Brabant ever fled to the papal court to lodge an appeal.

This does not mean, however, that Siger of Brabant and the other two masters, Goswin of Chapelle and Bernier Nivelles, were still in Paris when the summons to appear before the inquisitor was issued. On the contrary, the document contains many indications that the three masters had already left Paris.[46] At the beginning of the summons, for instance, the inquisitor

[43] This is demonstrated, for instance, in the cases against the theologians Denis of Foullechat and John of Monzón, who both appealed against a sentence. See Thijssen, *Censure* (n. 12), ch. 1.

[44] Theoretically, Siger could have turned to the pope and asked him, as judge over all major causes, to decide his case. In that scenario, the papal court would not have been an appellate court, but a court of first instance. There is no documentary evidence, however, that Siger took this course, nor is there any trace of a papal judgment in an inquiry against him.

[45] In essence, contumacy was considered to be disobedience to an ecclesiastical court. See F. D. Logan, *Excommunication and the Secular Arm in Medieval England* (Toronto: Pontifical Institute of Mediaeval Studies, 1968), pp. 44–49, for a discussion of contumacy in canon law.

[46] Dondaine, "Manuel" (n. 35), pp. 191–92. The text is presented under the medieval rubric "Forma citandi eos qui regnum exierunt et in regno commiserunt." Since the text of the citation is not easily accessible, and since it is so crucial for the interpretation suggested here, I shall quote it in its entirety: "Frater S[ymon] de Valle ordinis fratrum predicatorum, inquisitor heretice pravitatis auctoritate apostolica in regno Francie deputatus, religiosis viris [priori] predicatorum fratrum et Gardiano fratrum minorum et eorum vices gerentibus salutem in auctore et consummatore fidei iesu Christo. Cum in autentico apostolico nobis inquisitoribus misso contineatur expresse, quod nos citationes, examinationes testium, denunciationes sentenciarum fratribus ordinis Predicatorum et Minorum possimus committere, non obstante quocumque privilegio a Sede apostolica sibi dato. Necnon et quod nos contra eos, qui in crimine heresis in regno Francie predicaverunt et se ad partes alias transtulerunt, libere procedamus. Qua fungimus [!] auctoritate districte vobis et cuilibet vestrum precipiendo mandamus, quatenus [magistrum] Sugerum de brebancio canonicum sancti Pauli Leodiensis et Magistrum Gossioynum de Capella canonicum sancti Martini Leodiensis et Magistrum Bernerum de Vuulla [*leg.* Nivilla] concanonicum eiusdem, de crimine heresis probabiliter et vehementer suspectos, et qui in regno Francie dicuntur in tali crimine commisisse, citetis peremptorie presentibus testibus fide dignis, ut die Domenica post octavam Epiphanie, apud sanctum quintinum in Viromandia noviomensis diocesis, compareant personaliter coram

recalls that he also has jurisdiction to pursue suspects of heresy outside the realm of France, if they have propagated their heresy in France.[47] But where did the three masters go, if not to the papal court in Italy?

Gauthier has suggested that the three masters were simply staying in Liège when the citation was served.[48] The citation itself contains two clues in support of this suggestion. First, the inquisitor, Simon de Val, indicates that the three suspects had to appear before his tribunal in Saint Quentin (*apud sanctum quintinum*), a town that lies halfway between Liège and Paris; yet it is certain that he was still in Paris when he issued his order (*datum parisius*). Second, he spells out in detail the ecclesiastical titles of the three suspects, who were all canons in Liège: Siger of Brabant at Saint Paul's, and Goswin of Chapelle and Bernier of Nivelles at Saint Martin's. As a corollary, Gauthier concludes that the Prior of the Dominicans and a Guardian of the Franciscans to whom the citation is addressed (*religiosis viris [priori] predicatorum fratrum et gardiano fratrum minorum et eorum vices ferentibus*), and to whom the responsibility was delegated for seeing that the three masters would really appear before the inquisitor on 18 January 1277, were the Prior and the Guardian in Liège. Since the citation stems from a formulary, the names of the destinaries have been suppressed in the document itself.[49]

In sum, then, it appears that the inquisitor in Paris knew that the three masters were in Liège. According to Gauthier, they had probably returned there at the end of the academic year of 1275 (or, in any case, no later than 1276) to resume their ecclesiastical offices.[50] Siger's departure from Paris may well have been due to the aftermath of an administrative conflict in which he had been involved, which had divided the arts faculty since 1272. The cause of the dissension had been that a minority party, led by Siger,

---

nobis. Responsuri de fide, et dicturi tam de se quam de aliis vivis et mortuis super crimine heresis et super contingentibus ipsum crimen puram et plenariam veritatem. In signum vero et testimonium recepti et executi mandati presentibus sigilla vestra faciatis apponi. Datum parisius, anno domini m cc lxxvi die lune in festo beati Clementis."

[47] The "autenticus apostolicus" that grants this jurisdiction to the inquisitor is identified and quoted in Mandonnet, *Siger* (n. 23), 1: 254, nn. 2 and 3; and van Steenberghen, *Maître Siger* (n. 24), p. 143 n. 8. The jurisdiction to delegate certain tasks to the Dominicans and Franciscans and to pursue suspects of heresy outside France was granted by Pope Gregory X in a papal letter of 20 April 1273, which is edited in E. Martène and U. Durand, *Thesaurus novus anecdotorum*, 5 vols. (Paris, 1717), 5: 1819. Mandonnet and van Steenberghen interpret this reference as evidence that the masters had fled; note, however, that the inquisitor's citation allows the conclusion only that they had left France, not that they had fled.

[48] Gauthier, "Notes" (n. 41), p. 26.

[49] Dondaine, "Manuel" (n. 35), p. 191 n. 59, observes that this was normal practice. Mandonnet, *Siger*, 1: 254, believed that the citation was addressed to all the priors and guardians, although in a footnote he also considered that the names of the addressees might have been suppressed. Van Steenberghen, *Maître Siger*, p. 143, believed that the citation was addressed to the Prior and Guardian in Paris, and that in this way the Dominicans and Franciscans were mobilized to be on the lookout for the three fugitives.

[50] Gauthier, "Notes," p. 26.

had rejected the legitimacy of the election of Alberic of Reims as new rector of the arts faculty. On 7 May 1275 Simon of Brion, the papal legate, settled the dispute to the disadvantage of Siger's party.[51] This administrative conflict, together with the doctrinal controversies that had been raging since the beginning of the 1270s, may have induced Siger to return to his country of origin in 1275–76, never to set foot in Paris again. This suggestion is based on conjecture, but it is consistent with what we know of Siger's academic career. The *Quaestiones super librum de causis*, Siger's last known work, was written in 1275–76.[52]

But if Siger and the other suspects received the summons in Liège, what was their response? Did they obey the citation and really go to Saint Quentin? Gauthier thinks that they appeared before the inquisitor's tribunal on the stipulated day, but that they were acquitted. He bases this conclusion mainly on the fact that there is no documentary evidence about any conviction for heresy. After the reference in the inquisitor's document, Goswin of Chapelle completely disappears from the picture.[53] He probably remained in Liège. The names of Bernier of Nivelles and Siger of Brabant only surface again in the 1280s. Bernier of Nivelles reappears in the documents as a theologian and member of the Sorbonne College, to which he left a legacy of twenty-five books. In 1286 he copied Thomas Aquinas's commentary on the *Sentences*, and he bore the ecclesiastical title of curator of the Church of Saint Martin in Liège.[54] He was the only one of the three masters who resumed his studies in Paris, and proceeded to the faculty of theology. Siger of Brabant probably remained in Liège, like Goswin of Chapelle, until the time the Italian poem places him at the papal court in Orvieto.[55]

This alternative picture of the events resulting from the inquisitor's decree raises two new questions, not discussed by Gauthier. First, for what reason had the three masters been summoned in the first place? Second, what is the relation between the inquisitor's decree and Bishop Tempier's condemnation of 7 March 1277?

[51] *CUP* 1: 521–30 (#460). See Gauthier, "Notes," pp. 22 and 24: he has convincingly argued that this was a purely administrative conflict, not a doctrinal one.

[52] Van Steenberghen, *Maître Siger*, pp. 133, 218, and 221.

[53] Louis-Jacques Bataillon, "Bulletin d'histoire des doctrines médiévales: La treizième siècle (fin)," *Revue des sciences philosophiques et théologiques* 65 (1981): 107, has convincingly argued that it is very unlikely that the mad *clericus* who reportedly stabbed Siger of Brabant is identical to Goswin of Chapelle: his ecclesiastical rank was too high to be Siger's servant.

[54] Van Steenberghen, *Maître Siger* (n. 24), p. 144; Gauthier, "Notes" (n. 41), p. 26.

[55] Bataillon, "Bulletin," p. 107, suggested that Siger of Brabant went to Orvieto for matters that concerned his chapter. This suggestion was followed by Gauthier, "Notes," p. 27.

We do not know in any detail the charges against the three masters.[56] Yet, it seems natural to assume that they were accused of disseminating false teaching at the University of Paris. The inquisitor's document clearly states that the three masters were "in probability and gravely suspect of the crime of heresy" (*de crimine heresis probabiliter et vehementer suspectos*). Apparently, the complaints of disseminating false teaching were lodged with the inquisitor, who, ex officio, decided to start an inquiry. Since there is some evidence that the suspects resumed their academic and ecclesiastical careers, and since there is no evidence that they were condemned or had recanted their errors, it is likely that they were acquitted.[57] This scenario finds far more support in the documentary evidence than the older scenario in which Siger of Brabant fled to the papal court to appeal.

WHO WERE CONDEMNED ON 7 MARCH 1277?

The second question, the relation between the inquisitor's summons and Tempier's condemnation, brings me to the disputed issue of the targets of the episcopal action of 7 March 1277. The introductory letter to Tempier's condemnation contains some interesting clues. It is uncontested that the targets of the condemnation are unspecified members of the arts faculty in Paris: *nonnulli Parisius studentes in artibus*.[58] The rather vague "nonnulli studentes in artibus," some people engaged in the arts, instead of the more precise "magistri in artibus," even suggests that not all the propagators of false views were full-fledged masters. But of what precisely were these members of the arts faculty accused? Tempier rebukes them for disseminating (*tractare et disputare*) manifest and damned errors (*manifesti et exsecrabiles errores*). The errors are specified in the rotulus or leaves connected to the introductory letter (*in rotulo seu cedulis, praesentibus hiis annexo seu annexis*). They are the 219 censured propositions. Tempier does not state, however, that the members of the arts faculty are the *authors* of these errors. I think that this point has not been duly recognized in the scholarly

[56] Van Steenberghen, *Maître Siger*, p. 144, observes: "On ne sait rien des accusations qui avaient été formulées contre les trois maîtres dénoncés à Simon du Val, ni, dès lors, des griefs qui justifiaient leur citation devant le tribunal de l'inquisiteur."

[57] There are no documents to suggest why the suspects might have been acquitted. Possibly the evidence was inconclusive, or perhaps the Bishop of Liège came to the rescue of his canons and coerced the inquisitor to acquit them.

[58] Here follows the beginning of Tempier's introductory letter (*CUP* 1: 542): "Universis praesentes litteras inspecturis Stephanus, permissione divina Parisiensis ecclesiae minister indignus, salutem in filio virginis gloriosae. Magnarum et gravium personarum crebra zeloque fidei accensa insinuavit relatio, quod nonnulli Parisius studentes in artibus proprie facultatis limites excedentes quosdam manifestos et execrabiles errores, immo potius vanitates et insanias falsas in rotulo seu cedulis, presentibus hiis annexo seu annexis contentos quasi dubitabiles in scolis tractare et disputare presumunt."

literature. In his introductory letter Tempier separates the 219 censured errors from their propagators. Only the propagators have to be sought in the arts faculty in Paris: on pain of excommunication, they are prohibited to disseminate in any way (*dogmatizare, aut defendere seu sustinere quoquo modo*) the propositions collected by Tempier. The origin of these propositions, however, is not stated in the introductory letter.

The suggestion that those *artistae* who were castigated for disseminating false teaching were not necessarily disseminating their *own* views finds support in the admirable study by Roland Hissette.[59] From Hissette's summary of the results of his careful examination it appears that surprisingly few of the censured propositions could be identified with any degree of certainty in the known works of thirteenth-century *artistae*. Of the 219 propositions, only 79 are identified, with various degrees of probability, in the works of Siger of Brabant, Boethius of Dacia, or the three anonymous writings from the arts faculty that are accessible in a modern edition. Of 72 propositions the attribution is uncertain, whereas 68 propositions could not be identified at all.[60] Moreover, many censured propositions that seem to have been derived from the examined works that were generated at the arts faculty in the thirteenth century do not really represent the author's own view, but rather appear to be quotations or paraphrases from Aristotle, from Arabic philosophers, or from "the philosophers," as Hissette indicates.[61]

In light of this evidence one can only conclude that research into the proximate background of the censured propositions has to be broadened. The directions that such research should take are indicated, either implicitly or explicitly, in Hissette's study and in subsequent studies, such as those by John Wippel and Calvin Normore. It is generally agreed today that a considerable number of the 219 censured propositions have a bearing on the reintroduction of pagan philosophy into the arts faculty, and on the ensuing crisis over the relations of faith and reason.[62] Consequently, Greek or Arabic sources may prove to be at the origin of a number of censured propositions.[63] Other propositions may well have been derived from the teaching of

[59] Roland Hissette, *Enquête sur les 219 articles condamnés à Paris le 7 mars 1277* (Louvain: Peeters, 1977).

[60] See ibid., pp. 314–17. Note that in the actual discussion of these articles, Hissette is more careful about claiming whether it is "beyond doubt" that an article is derived from the works of Siger of Brabant or Boethius of Dacia. Moreover, there are hardly any literal quotations or accurate paraphrases among the identified propositions. In some cases, the almost literal similarity goes back to a text of Aristotle or Averroes; see, for example, articles 13, 117, 123, and 129 in Hissette's numbering.

[61] Ibid., p. 317.

[62] See especially John F. Wippel, *Mediaeval Reactions to the Encounter between Faith and Reason* (Milwaukee, Wis.: Marquette University Press, 1995), and the literature cited there.

[63] Most recently, Calvin G. Normore, "Who Was Condemned in 1277?" *Modern Schoolman* 72 (1995): 273–81, explores sources from late antiquity and the Islamic world.

theologians, such as Thomas Aquinas.[64] In this respect, the often-quoted statement from Tempier's introductory letter that members of the arts faculty were transgressing the limits of their own faculty (*propriae facultatis limites excedentes*) acquires new meaning. Some members of the arts faculty were rebuked not only for teaching suspect philosophical views, but also for teaching suspect theological views.

Only a limited number of the 219 censured propositions represent erroneous teaching that was authored by *artistae*. Among the authors were Siger of Brabant and Boethius of Dacia—but they were clearly not the only members of the arts faculty who were criticized for propagating false views, or whose own views were condemned on 7 March 1277. Siger and Boethius, however, appear to have been the most prominent targets, or, in any case, among the most easily identifiable for modern historians. They may have been the "heresiarchs," so to speak, of the crisis over the encounter between faith and reason that became manifest in Tempier's condemnation. Yet, their names appear nowhere in the syllabus.

In Siger of Brabant's case the surviving evidence suggests a specific juridical reason why his name was omitted, and why his teaching was included in a rather general anonymous censure. If he had already been acquitted by the inquisitor toward the end of 1276, his views could not be censured *nominatim* by the bishop. According to the juridical principle that one cannot be tried twice for the same crime (*ne bis in idem crimen judicetur*), the bishop could not start a new inquiry against Siger for disseminating false teaching at the University of Paris.[65] Yet, it seems very likely that Tempier based his own examination of Siger's views on the inquisitor's inquiry. The inquisitor of France was also located in Paris, and the summons for Siger of Brabant and the two other masters was actually issued from there. The bishop of Paris, who had jurisdiction over the University of Paris, would surely have been informed about the outcome of an

---

[64] Wippel, "Thomas Aquinas" (n. 24), argues that Aquinas's views were targeted in Tempier's condemnation. Roland Hissette, *Saint Thomas et l'intervention épiscopale du 7 mars 1277*, in *Studi*, ed. D. Lorenz, O.P., and S. Serafini, O.P. (Rome: Istituto San Tommaso, 1995), pp. 204–58, is more cautious: he believes that Tempier's syllabus was only indirectly aimed against Aquinas, and that the adherents of the censured views have to be sought primarily in the arts faculty, as Tempier's introductory letter indicates. Note that their disagreement can be resolved if one follows my suggestion and acknowledges that Tempier's introductory letter clearly distinguishes between the propagators of the censured views, who are members of the arts faculty, and the censured views themselves, whose origin is not stated.

[65] This juridical principle can be found in Gratian's *Decretum*, C.2 q.1 c.14 par. 1. John the Teuton offers the following comment in his *Glossa ordinaria* C.2 q.1 c.14 par. 1, "Non potest": "Sive enim quis sit condemnatus, sive absolutus, tamen super eodem crimine saepius agi non potest, ut extra De acusa. De his." He was the first to phrase this general principle, which applied in both accusatorial and inquisitorial proceedings. See Peter Landau, "Ursprünge und Entwickelung des Verbotes doppelter Strafverfolgung wegen desselben Verbrechens in der Geschichte des kanonischen Rechts," *Zeitschrift der Savigny-Stiftung für Rechtsgeschichte*, Kan. Abt. 87 (1970): 124–56, esp. 138–52.

inquiry against some of its members. He and the theologians who were charged with examining Siger's views may have disagreed with the outcome of the inquisitor's process, and, consequently, had no other option than to censure those views anonymously. They simply included them in the longer list of false propositions that had already been prepared.

THE DISCIPLINARY PROCEDURES

The way of proceeding outlined above finds some support in Tempier's introductory letter. There he indicates that he has received information from important people (*magnarum et gravium personarum crebra zeloque fidei accensa insinuavit relatio*). This means that Tempier did not take the initiative, as has been generally assumed in the scholarly literature, but that he reacted to allegations of suspect teaching, allegations that may have been derived from the inquisitor's dossier. The general nature of these allegations is also made more explicit in the episcopal letter than in the inquisitor's citation. The important persons on whose information Tempier acted had informed him precisely that "some scholars of arts at Paris" (*nonnulli Parisius studentes in artibus*) had been transgressing the limits of their own faculty (*proprie facultatis limites excedentes*). Such a complaint must surely have come from theological circles, though probably not directly from members of the faculty of theology: in Tempier's prefatory letter the theologians are clearly distinguished, as a group, from the "important persons" who denounced the suspects of false teaching.[66]

In his introductory letter Tempier reports that he sought the advice "not only of the doctors of Sacred Scripture, but also of other wise men" (*tam doctorum sacrae Scripturae, quam aliorum prudentium virorum communicato consilio*). From other cases of suspect teaching we know that the task of the theologians was to examine certain works and draw up a list of errors. In cases where a list of alleged errors already existed, the theologians were charged to assess the degree of error of the listed propositions. The theologian John of Pouilly reports that sixteen masters of theology were Tempier's assessors for the condemnation.[67] One of the members of this

[66] Tempier distinguishes the important persons who reported the allegations of false teaching (*magnarum et gravium personarum crebra zeloque fidei accensa insinuavit relatio*) from the theologians and other wise men who gave him advice in this matter (*tam doctorum sacrae Scripturae, quam aliorum prudentium virorum communicato consilio*). See *CUP* 1: 542.

[67] "Iidem magistri fuerunt assessores episcopi Stephani in condendo articulos et in concedendo praedictam propositionem. Et ideo cum praedicta magistralis propositio interimat articulum praedicto modo intellectum, si praedicto modo deberet articulus intelligi, illi magistri sibi ipsis contradixissent, omnes etiam XVI magistri qui illam propositionem concesserunt excommunicationis sententiam incurrerent, quae omnia non sunt dicenda." The text is quoted in Wielockx, *Aegidii Romani* (n. 29), p. 98 n. 6.

commission was Henry of Ghent, as he himself testifies in his Quodlibet II.[68] It is unknown when these masters met, but it must have been after Henry of Ghent had become a regent master in theology, a position that he obtained in 1276. That there were some tensions between Tempier and the theologians is attested by the theologian Giles of Rome, a contemporary witness of the events of 1277: he claimed that some articles were condemned, not on the basis of the advice of the masters, but rather due to the "stubborness of a few."[69] This observation has been taken to concern Tempier, but it might also include some of the "wise men" who had assisted him.

The identity of these other wise men is unknown. Since, however, they are so clearly distinguished from the theologians, they have to be sought among the prelates. Of these, only the involvement of the chancellor, John of Alleux is directly substantiated by textual evidence: the introductory letter to Tempier's condemnation stipulated that offenders had to report themselves either to the bishop himself, or to the chancellor. Other likely candidates are Simon of Brion, the papal legate, and Ranulphe of Houblonnière, Tempier's future successor as bishop of Paris.

In the present state of documentary evidence it is not possible to establish which method Tempier and his advisors had used to draw up their syllabus of 219 errors. This can only be established when we have identified with more precision the suspect works that Tempier perused. As I mentioned above, it would be too narrow to confine research to works produced at the faculty of arts in the thirteenth century. Tempier and his advisors relied on more sources, written or oral, than those used by Hissette. Possibly, their list even incorporates older lists of suspect views.

In light of the above, the generally accepted conclusion that Tempier's syllabus of condemned propositions is not very well organized and is "broad in scope to the point of confusion" appears somewhat gratuitous.[70] The lack of doctrinal cohesion is also present in other lists of the *Collectio*

---

[68] Henry of Ghent, *Quodlibet II*, ed. R. Wielockx (Louvain: Leuven University Press, 1983), p. 67: "In hoc enim concordabant omnes magistri theologiae congregati super hoc, quorum ego eram unus, unanimiter concedentes quod substantia angeli non est ratio angelum esse in loco secundum substantiam." See Miethke, "Papst" (n. 27), p. 86; van Steenberghen, *Maître Siger* (n. 24), pp. 146–47; Hissette, "Étienne Tempier" (n. 27), pp. 234–36.

[69] E. Hocedez, "La condamnation de Gilles de Rome," *Recherches de théologie ancienne et médiévale* 4 (1932): 56: "Nam nos ipsi tunc eramus Parisiis, et tamquam de re palpata testimonium perhibemus, quod plures de illis articulis transierunt non consilio magistrorum, sed capitositate quorundam paucorum." Hissette, "Étienne Tempier," p. 238, observes that the "quorundam" refers to the bishop and not to some masters of theology, as some other scholars believed.

[70] See, for the latter assessment, Richard Knapwell, *Quaestio disputata de unitate formae*, edited with an introduction and notes by Francis E. Kelley (Paris: Vrin, 1982), p. 12. An exception to the unfavorable views concerning the consistency of the condemnation is Kurt Flasch, *Aufklärung im Mittelalter? Die Verurteilung von 1277* (Frankfurt: Suhrkamp, 1989), p. 56.

errorum, simply because the order in which the errors appeared on the *rotulus* was determined by other factors—such as, for instance, the order in which they appeared in the examined work. Shortly after 1277 the extremely long list of 219 prohibited views was reorganized, possibly to facilitate its use in the academic community.[71] Similarly, the theologian Hugolin of Orvieto reorganized the list of errors recanted by John of Mirecourt.[72] At the beginning of this century, Pierre Mandonnet once again put Tempier's articles into a new order.[73]

CONCLUSION

The title of the present contribution seems to make the claim that the portrayal of Tempier's condemnation suggested here truly represents what really happened in the years 1276–77. This is not the case, however. Without entering into the discussion of whether or not it is at all possible to give an historical account of "how it really was," I believe that many aspects of Tempier's 1277 condemnation cannot be reconstructed, because the evidence that has survived is simply too scanty.

Rather, I have suggested a revisionist approach to Tempier's syllabus by emphasizing some previously unperceived connections. This new interpretation resulted from exploring the historical context in which Tempier's syllabus originated—that is, the context of teaching authority and censure at the University of Paris. In interpreting the document that Bishop Tempier issued on 7 March 1277 I have tried to be especially attentive to this context.

What new perspective has been gained by placing Tempier's syllabus in its proper medieval context? There is no documentary evidence that the 1277 censure followed a pattern similar to that of other university condemnations in Paris: the investigation did not start at the faculty of theology only to be moved to the episcopal court at a later stage, after the procedure

---

[71] This new medieval edition is part of the *Collectio errorum in Anglia et Parisiis condemnatorum* and has been printed in Du Plessis, *Collectio* (n. 15), 1,1: 188–200. A modern edition is now available in H. Anzulewicz, "Eine weitere Überlieferung der *Collectio errorum in Anglia et Parisius condemnatorum* im *Ms. lat. fol. 456* der Staatsbibliothek Preussischer Kulturbesitz zu Berlin," *Franziskanische Studien* 74 (1992): 375–99. Tempier's original list is edited in *CUP* 1: 543–55, and reprinted with a German translation and a discussion in Flasch, *Aufklärung*, pp. 99–261. The transmission of Tempier's list of errors is discussed in *CUP* 1: 556–57, and in Roland Hissette, "Une *Tabula super articulis Parisiensibus*," *Recherches de théologie ancienne et médiévale* 52 (1985): 171–72.

[72] *CUP* 2: 610–13 (#1147).

[73] Mandonnet, *Siger* (n. 23), 1:175-91. Hissette, *Enquête* (n. 59), follows Mandonnet's order and provides on pp. 319-21 a useful concordance of the three different editions of Tempier's articles. Other thematic discussions of the articles are given in John F. Wippel, "The Condemnations of 1270 and 1277 at Paris," *Journal of Medieval and Renaissance Studies* 7 (1977): 187-94; and in Wippel, *Medieval Reactions* (n. 62), pp. 19-27.

of fraternal correction had failed. Nor did Tempier himself initiate the investigation of Siger of Brabant, or the inquiry that resulted in his syllabus of censured propositions: the condemnation was the result of an episcopal reaction to denunciations that came from outside the University of Paris. However, the first papal letter, written on 18 January 1277, did not play a role in that reaction. The picture of Tempier as an overzealous bishop is simply untrue.

More likely, Tempier was disturbed by the charges that were raised against Siger of Brabant's teaching toward the end of 1276. Probably, the bishop used the dossier collected against Siger and the other two masters in drawing up the censure of 1277. Since, however, the inquisitor's proceedings against these men had been dropped, Tempier could not single them out by name when he issued his condemnation. One can only speculate whether it was the inquisitor's dossier that triggered the entire inquiry against the arts faculty. Interestingly, Henry of Ghent's testimony does not contradict the possibility that the deliberations of the theologians go back as early as 1276.

Admittedly, this alternative account of the events that led to Tempier's censure is hypothetical, and only partly explains why this condemnation is anomalous when compared to other censures of suspect teaching at the University of Paris. There were other members of the arts faculty whose teaching was condemned in 1277, and they too are unnamed. But in Siger's case, there is enough documentary evidence to suggest a reason why he was unnamed in Tempier's condemnation, and, moreover, to destroy the story that he ever lodged an appeal at the papal court against this condemnation. Despite the absence of conclusive evidence, the interpretation here offered may provide an account that comes closer to what really happened in 1277, in that it is more attentive to the historical context of Tempier's condemnation and, in this way, has gained in plausibility.

APPENDIX 1:

THE TRANSMISSION OF THE PARISIAN ARTICLES

As William Courtenay has pointed out, the collection of Parisian Articles was only one of the documentary vehicles by which lists of errors were preserved and circulated in the academic community in Paris.[74] Another procedure for disseminating lists of censured views was to attach them to commentaries on the *Sentences*. The lists of condemned errors of Durand of St. Pourçain and John of Mirecourt, for example, were attached to copies of their commentaries on the *Sentences*.[75] Finally, some lists have survived as original records in the archives of the faculty of theology; a fine example of this is the list of errors attributed to Denis of Foullechat. As will be demonstrated in Appendix 2, only records of this type found their way into Noël Beda's register.

The *Collectio errorum in anglia et parisius condempnatorum* has survived in many manuscript copies. In addition, it was often included in early modern editions of the *Sentences*. Below, I will give two samples of lists of Parisian Articles, one as preserved in a manuscript from the late-fourteenth century, the other representing the early printed tradition. The tradition of transmission of the *Collectio errorum* is quite complicated and will not be explored here; instead, let the following few observations suffice. Not all compilations of Parisian Articles contain the same lists of errors, nor, consequently, the same division into chapters.[76] Not all manuscripts include the prefatory letter to Tempier's condemnation of 1277. Sometimes Tempier's letter is at the head of the entire *Collectio*, and sometimes it precedes the 1277 condemnation. Furthermore, there are significant differences in the numbering and wording of the articles between the printed tradition, on the one hand, and the manuscript tradition, on the other hand, and also between the different manuscript copies. When, in the sample lists of Parisian Articles that follow, discrepancies between the date in the text and in the heading occur, it is the heading that contains the correct date.

The short *Collectio errorum in anglia et parisius condempnatorum*—that is, the collection that does not go beyond Tempier's articles of 1277—

[74] Courtenay, "Preservation" (n. 4).

[75] See F. Stegmüller, "Die zwei Apologien des Jean de Mirecourt," *Recherches de théologie ancienne et médiévale* 5 (1933): 43–45; and Josef Koch, *Kleine Schriften*, 2 vols. (Rome: Edizione di Storia e Letteratura, 1972), 2: 31–32.

[76] See Anzulewicz, "Eine weitere Überlieferung" (n. 71), pp. 377–80, for the division in chapters.

has been transmitted in more than thirty medieval manuscripts.[77] The oldest of these seems to be Paris, BN lat. 15661. The number of copies of the longer *Collectio errorum*—that is, the compilation of condemned articles that also includes those of the fourteenth and fifteenth centuries—is far smaller. The following manuscripts are known: Auxerre, Bibliothèque municipale 243; Erfurt, Bibl. Amploniana F 179 and Q 151; Munich, Bayerische Staatsbibliothek, Clm. 3798 and Clm. 28126; Paris, BN lat. 16533; Rouen, Bibliothèque municipale 587 (A. 263).[78]

Charles Du Plessis d'Argentré and the editors of the *Chartularium* not only relied on manuscripts, but also used the early printed tradition of the *Collectio errorum*, of which, according to Du Plessis, there were as many editions as there were of Peter Lombard's *Sentences*.[79] An important procedure for disseminating the lists of errors, indeed, was attaching them to the *Sentences*. According to Du Plessis the following printed editions of Lombard's *Sentences* also contained the *Collectio errorum in anglia et parisius condempnatorum*: Venice: Vendelinum de Spira, 1477; Venice, 1480; Venice, 1507 (*cum Nic. de Orbellis interpretatione*). Likewise, the following editions of the *Sentences* together with the commentary by Henry of Gorichem (Gorkum): Nuremberg, 1478, 1499, 1528; Basel, 1492 [Hain 10197], 1498 [Hain 10198], 1502, 1507, and 1513; Rouen, 1653; Venice: Simon de Luere, 1506, 1570; Paris: Jean Roigny, 1536, 1550; and Paris: Jean Petit, 1536. To these latter editions can be added Basel, 1487 [Hain 10192], 1488 [Hain 10195], and 1489 [Hain 10196]. Moreover, the following separate editions of articles can be mentioned: *Articuli Parisius condemnati*: Padua: Mathaeus Cerdonis, c.1485 [nr. 2709]; *Articuli in Anglia et*

---

[77] See *CUP* 1: 556–57, and Bianchi, *Vescovo* (n. 4), pp. 207–8 n. 7, for a survey of manuscripts not listed in *CUP*, to which can be added the Berlin manuscript that Anzulewicz ("Eine weitere Überlieferung") used for his new edition of the short version of the *Collectio errorum*.

[78] Most of these manuscripts are also mentioned in Courtenay, "Preservation" (n. 4). MS Paris, BN lat. 16533 contains the condemnations of 1241 and 1277, but also those of Autrecourt and Mirecourt, the *Evangelium aeternum*, and the "Errors of the Philosophers" by Giles of Rome. See *Aegidii Romani Opera Omnia: Catalogo dei Manoscritti* 1/3** (n. 4), pp. 258–60, for an extensive description of this manuscript. In the Rouen manuscript, the initial brief *Collectio errorum* has been extended with articles from the fifteenth century (Jean le Marchand and Jean Laillier), but those of the fourteenth century have been omitted. In MS Munich, Bayerische Staatsbibliothek, Clm. 3798, articles of the Amalricians, John of Calore, John of Mirecourt, John of Guyon, Lewis of Padua, and Giles of Rome's *De erroribus* have been added to the short *Collectio errorum*. MS Munich, Bayerische Staatsbibliothek, Clm. 28126 contains the short *Collectio errorum*, together with the fifteenth-century commentary on Tempier's condemnation of 1277, the articles recanted by John of Monzón, and a copy of Giles of Rome's *De erroribus philosophorum*. The other fourteenth-century articles, however, have been omitted. See *Aegidii Romani Opera Omnia*: *Catalogo dei Manoscritti* 1/5*, a cura di B. Faes de Mottoni (Florence: Olschki, 1990), pp. 37–39; 222–24.

[79] Du Plessis, *Collectio* (above, n. 15), 1,1: 212.

Parisius condemnati: Paris: Antoine Caillant, c. 1483 [nr. 2710]; Cologne: Heinrich Quentell, c. 1488, and c. 1490 [nrs. 2711 and 2712].[80]

The List of Parisian Articles in MS Auxerre, Bibl. municipale 243

[1] (fol. 78rb) *Ten articles condemned by Bishop William of Alvernia on January 13, 1241* (*Edition*: *CUP* 1: #128). Isti sunt errores detestabiles contra veritatem catholicam reperti in quibusdam scriptis. Quos quicumque dogmatizaverit sub domino Guillelmo parisiensis episcopo convocato consilio omnium magistrorum theologie parisius regentium cum vinculo anathematis est innodatus. Ideo eos cavere debent summopere omnes professores fidei orthodoxe. Primus error est quod divina essentia in se nec ab angelo nec ab homine videtur vel videbitur....

[2] (fols. 78va–82rb) *Condemnation by Bishop Stephen Tempier on 7 March 1277, including the prefatory letter* (*Edition*: *CUP* 1: #473). Universis presentes litteris inspecturis stephanus permissione divina ecclesia parisiensis.... Primus articulus condempnatus contradicit quod deus non sit trinus et unus, quoniam trinitas non stat cum summa simplicitate. Ubi enim est pluralitas realis, ibi est additio et compositio; exemplum de acervo lapidum. Error....

[3] (fols. 82rb–84ra) *Condemned articles of Nicholas of Autrecourt, 1346* (*Edition*: *CUP* 2: #1124, pp. 580–84). Anno domini 1348 post festum omnium sanctorum hii sequentes articuli fuerunt in romana curia condempnati et per magistrum nycolaum de ultrecuria publice in universitate parisiensi revocati. Sequitur primus articulus. Dixi quod multa venerunt ad animam meam ex quibus perpendi seu bonum esse simpliciter scribere istum tractatum....

[4] (fol. 84ra) *Recantation of John Guyon OFM on 12 October 1348* (*Edition*: *CUP* 2: #1158). Isti sunt articuli revocati anno domini 1348 secunda die octobris per fratrem Johannem goerum ordinis fratrum minorum in domo predicatorum. Primus est quod generare secundum suam rationem formalem non sit in patre....

[5] (fol. 84ra–va) *Recantation of John of Mirecourt in 1347* (*Edition*: *CUP* 2: #1147, though in the order given by Hugolin of Orvieto). Articuli condempnati parisius et revocati per fratrem Johannem de mirecuria ordinis cisterciensum anno 1349, quod satis erit possibile quod per volitionem creatam christus aliquid voluit quod nunquam sic voluit evenire....

[6] (fols. 84va–85rb) *Recantation of Lewis of Padua in 1362* (*Edition*: *CUP* 3: #1270). Articuli ludovici revocati anno domini 1362 primo probandum prima partem suppositionis sic arguitur: ad quodlibet nomen velle vel nolle....

[7] (fol. 85rb–va) *Recantation of Guido OSA on 16 May 1354* (*Edition*: *CUP* 3: #1218). Revocatio facta parisius anno domini 1354. Primo dixi illam conclusionem: caritas que solum labitur vel deperditur nunquam fuit vera caritas, et eam probavi contra bachalarium beate marie de carmelo. Hanc revoco tanquam falsam et hereticam....

---

[80] The numbers refer to the *Gesamtkatalog der Wiegendrucke* (Leipzig, 1925–).

[8] (fols. 85va–86ra) *Recantation of Simon, bachelor of theology in 1351 (Edition: CUP* 3: #1201). Anno domini 1351 quia ego symon respondendo in vesperis declarando vel exponendo terminos questionis dixi quod hoc nomen "ihesus" dictum est de deo accidentaliter vel ex parte significat filium dei qui assumpsit naturam humanam....

[9] (fol. 86ra) *Repetition of the first three articles condemned by Bishop William of Alvernia on 13 January 1241* (*Edition*: *CUP* 1: #128). Isti sunt articuli dampnati parisius ab episcopo parisiensis et magistris regentibus parisius in facultate theologie anno domino 1340 in octavo epiphanie. Primus est quod sua divina in se nec ab homine nec ab angelo videbitur....

[10] (fols. 86ra–87rb) *Recantation of Denis of Foullechat OFM, 1364 and 1369* (*Edition*: *CUP* 3: #1298, pp. 117–19 and #1352, p. 185). Revocatio quorundam articulorum per fratrem dionisum foulechat ordinis fratrum minorum etc. Reverendi magistri mei et domini, quia aliqua dixi in principio meo sententiarum, que in auribus nonnullorum male sonarunt et ex causa, ego veritate coactus et ex ordinatione reverendi patri domini cancellarii ac facultatis theologie magistrorum tamquam ecclesie et dicte facultatis filius humilis ipsa dicta in prefato principio meo male posita corrigo sub hanc formam.... /fol. 87ra/ Item, cum debui componere dicta mea in curia romana coram reverendissimis patribus dominis cardinalibus nemansensi et vabrensi, addidi propositiones que secuntur, videlicet quod christus in morte sua omnia simpliciter abdicavit. Istam rationam tamquam falsam, erroneam, et hereticam....

The Early Modern Printed Tradition

Textus Sententiarum Cum conclusionibus magistri Henrici Gorichem et concordantiis Biblie ac Canonum necnon in principio singularum distinctionum utilimis summariis diligentissime iam primum appositis. Item errores quidam Parisius revocati et articuli in quibus Magister communiter non tenetur. Item registrum totius libri. Basel, 1498. [Hain 10198]

[1] *Prefatory letter of Tempier's condemnation in 1277* (*CUP* 1: #473). Sequuntur varii articuli erronei omnium pene facultatum in anglia et parisius studiose et auctoritative condemnati cum revocationibus eorundem. Prefatio. Universis presentes litteras inspecturis Stephanus permissione divina Parisiensis ecclesie minister indignus Salutem in filio virginis gloriose.... Collectio errorum in anglia et parisius condemnatorum que sic per capitula distinguuntur. Et primo de erroribus condemnatis in anglia.... [there follows an index of 35 chapters]

[2] *Condemnation by Archbishop Robert Kilwardby at Oxford in 1276* (*CUP* 1: #474). Isti sunt errores condemnati a fratre Roberto kilimardbi archiepiscopo Cantuariensi de consensu omnium magistrorum tam regentium quam non regentium apud Oxoniam die iovis xx ante festum sancti Cunberti in quadragesimo Anno domini Mcclxxvi. Capitulum i. Errores in grammatica. Ego currit, tu currit. Currit et curro eque sunt perfecte et congrue.... Capitulum ii. Errores in logica. Quod contraria simpliciter possunt esse vera in aliqua materia.... Capitulum iii. Errores in naturali philosophia. Quotquot sunt composita, tot sunt omnino principia prima....

[3] *Ten articles condemned by Bishop William of Alvernia on 13 January 1241* (*CUP* 1: #128). Capitulum iv. Isti sunt errores detestabiles contra catholicam veritatem reperti in quibusdam scriptis. Quos quique dogmatizaverit vel defenderit a venerabili patre guilermo parisiensi episcopo convocato concilio omnium magistrorum tunc parisius degentium vinculo anathematis est innodatus. Et ideo eos cavere debent omnes professores fidei orthodoxe. Primus est quod divina essentia in se nec ab angelo nec ab homine videatur vel videbitur....

[4] *Condemnation of thirteen theses by Stephen Tempier on 10 September 1270* (*CUP* 1: #432). Capitulum v. Isti sunt errores condemnati et excommunicati cum omnibus qui eos scienter docuerint vel assuerint a domino stephano parisiensis episcopo Anno domini Mcclxx Die mercurii ante festum beati Nicolai hiemalis. Primus articulus est quod intellectus omnium hominum est unus et idem numero....

[5] *Condemnation by Bishop Stephen Tempier on 7 March 1277* (*CUP* 1: #473). Capitulum vi. Isti articuli que sequuntur condemnati sunt a domino stephano parisiensis episcopo de concilio magistrorum theologie Anno domini Mccxxvi die domenica qua cantantur Letare hierusalem in ecclesia parisiensis. Ubi excommunicaverit in scriptis omnes illos qui eos docuerint vel defenderint. Et primo ordinantur illi qui sunt de deo. Primus quod deus non est trinus et unus.... Capitulum vii. Errores de angelo vel intelligentia. Item quod omnia separata coeterna sunt primo principio.... Capitulum viii. Errores de anima et intellectu. Item quod intellectus non est forma corporis.... Capitulum ix. Errores de voluntate sive libero arbitrio. Item quod de sui natura non est determinandum ad esse vel non esse.... Capitulum x. Error de toto coniuncto, id est de toto composito naturali perfecto sive de homine. Item quod homo pro tanto dicitur intelligere, pro quanto celum dicitur ex se movere vel intelligere vel vivere.... Capitulum xi. Errores de mundo et mundi eternitate. Item quod nihil est eternum a parte finis quod non sit eternum a parte principii.... Capitulum xii. Errores de celo et stellis. Item quod corpora celestia moventur a principio intrinseco quod est anima.... Capitulum xiii. Errores de natura generabilium et corruptibilium. Item quod forme non recipiunt diversionem, nisi secundum divisionem materie.... Capitulum xiiii. Errores de necessitate eventus rerum. Item quod nihil a casu, sed omnia a necessitate eveniunt.... Capitulum xv. Errores de accidente. Item quod cum deus non comparatur ad entia in ratione cause materialis vel formalis, non facit accidens esse sine subiecto.... Capitulum xvi. Errores de scientia sive philosophia. Item quod omnes scientie sunt necessarie preter phisicas disciplinas.... Capitulum xvii. Errores de sacra scriptura. Item quod sermones theologice fundati sunt in fabulis.... Capitulum xviii. Errores de fide et sacris. Item quod non est curandum de fide si dicatur aliquid esse hereticum.... Capitulum xix. Errores de raptu. Item quod raptus et visiones non habent fieri, nisi per naturam.... Capitulum xx. Errores de vitiis et virtutibus. Item quod peccata contra naturam utpote abusus in coitu, licet sint contra naturam speciei, non tamen contra naturam individui.... Capitulum xxi. Errores de resurrectione. Item quod non contingit corpus corruptum redire idem numero.... Capitulum xxii. Errores de felicitate seu beatitudine. Item felicitas non potest a deo infundi immediate....

[6] *Prohibition of six "Ockhamist errors" by the faculty of arts on 29 December 1340* (*CUP* 2: #1042). Capitulum xxiii. Sequuntur articuli condemnati parisius. Isti articuli que sequuntur fuerunt condemnati parisius in facultate artium Mcccxliiii in festo nativitatis domini. Quod nullus magister, baccalarius vel scolaris in artium facultate legens parisius audeat propositionem famosam illius autoris cuius librum legit dicere simpliciter esse falsam de virtute sermonis....

[7] *Recantation of John Guyon OFM on 12 October 1348* (*CUP* 2: #1158). Capitulum xxiiii. Isti sunt articuli revocati anno domini Mcccxviii ii die octobris parisius per fratrem iohannem Guion ordinis minorum in domo predicatorum. Item quod generare secundum suam rationem formalem non sit in patre....

[8] *Recantation of John of Mirecourt in 1347* (*CUP* 2: #1147). Capitulum xxv. Errores iohannis de mercuria ordinis cisterciensis. Articuli condemnati per magistros parisiensis anno domini Mcccxlvii et prohibiti omnibus baccalaris legentibus qui legunt vel legerunt sententias sub pena privationis ab omni honore facultatis quorum articulorum aliqui reputantur erronei aliqui suspecti aut male sonantes in fide. Item quod satis erat possibile quod per voluntatem aut volitionem creatam christus aliquid voluit quod nunquam debuit evenire....

[9] *Condemned articles of Nicholas of Autrecourt, 1346* (*CUP* 2: #1124). Capitulum xxvi. Articuli contra magistrum Nicolaum de ultricuria anno domini mcccxlviii post festum omnium sanctorum. Isti sequentes articuli fuerunt in romana curia condemnati per magistrum Nicolaum supradictum publice in universitate parisiensi revocati. Dixi quod multa venerunt ad animam ex quibus perpendi seu iudicavi bonum esse simpliciter scribere istum tractatum....

[10] *Repetition of the articles condemned by Bishop William of Alvernia on 13 January 1241* (*CUP* 1: #128). Capitulum xxvii. Isti articuli que sequuntur sunt condemnati ab episcopo Parisiensi et magistris theologie regentibus Parisius Anno domini mcccxl in octavo Epiphanie domini. Item quod divina essentia in se nec ab homine nec ab angelo videbitur....

[11] *Recantation by Denis of Foullechat OFM, 1364 and 1369* (*CUP* 3: #1298; #1352) Capitulum xxviii. Copia cedule revocationis quorundam articulorum fratris Dionisii foulechat ordinis fratrum minorum facte per eundem Parisius et in curia romana. Reverendi magistri mei et domini dixi aliqua in principio meo sententiarum que in nonnullorum auribus male sonuerunt et ex causa.... Item cum debui exponere dicta mea in curia romana coram reverendissimis patribus dominis cardinalibus nemansensi et vabrensi addidi propositiones que sequuntur....

[12] *Recantation of John of Calore, 1363* (*CUP* 3: #1288). Capitulum xxix. Revocatio magistri Johannis de calore Anno domini mccclxiii. Antequam descendam ad conclusiones habeo aliqua dicere, quia in primo articulo mearum vesperiarum posui duas propositiones.... Fuerunt parisius quidam discipuli

cuiusdam Almarici nomine studentes parisius qui dixerunt mortuo eo quod illud quod alias est peccatum mortale, ut stuprum factum in charitate, non est peccatum impunitatem peccatorum promittentes, deum bonum et non iustum dicentes. Qui propter huiusmodi condemnati a rege francie extra portas sunt combusti. Heresiarcha vero eorum Almaricus a generali concilio rome condemnatus est et excommunicatus et ex civitate exhumatus, et cinis et ossa per sterquilinum sunt dispersa et merito....

[13] *Recantation of Lewis of Padua in 1362* (*CUP* 3: #1270). Capitulum xxx. Articuli Ludovici revocati Anno domini Mccclxii. Primo probando primam partem prime suppositionis arguitur sic: Ad quodlibet nomen velle vel nolle respectu cuiuslibet producibilis ad extra sequitur necessario mutatio in voluntate perfecta....

[14] *Recantation of Guido, OSA, on 16 May 1354* (*CUP* 3: #1218). Capitulum xxxi. Forma et modus revocationis facta Parisius per fratrem Guidonem ordinis heremitarum sancti Augustini actu legentis Parisius sententias in scholis dicti ordinis Anno domini Mcccliiii xvi die mensis Maij. Et hoc anno legendo ambulavi in manis et mirabilibus super me, verbum veritatis non recte tractando, sed verbis contendendo prophanis et vaniloquis....

[15] *Recantation of Simon, bachelor of theology, in 1351* (*CUP* 3: #1201). Capitulum xxxii. Revocatio facta Anno domini Mcccli. Quia ego symon in respondendo in vesperiis exponendo vel declarando terminos questionis quod hoc nomen Jesus dictum est de deo accidentaliter et ex parte significat filium dei qui assumpsit naturam humanam....

[16] *Theses from Peter Lombard's Sentences that are not to be upheld by the Parisian masters* (Du Plessis, *Collectio* 1,1: 118; and another, shorter list of theses in *CUP* 1: #194). Capitulum xxxiii. Isti sunt articuli in quibus magister sententiarum non tenetur communiter ab omnibus, et primo primi libri. Primo quod charitas qua diligimus deum et proximum est spiritus sanctus....

[17] *Decree of the faculty of theology issued in 1398, against superstition and sorcery* (Du Plessis, *Collectio* 1,2: 154-57; and *CUP* 4: #1749) [Capitulum xxxiiii] Determinatio Parisius facta per almam facultatem theologicam Anno domini Mccexcviii super quibusdam superstitionibus noviter exortis. Prefatio. [U]niversis orthodoxe fidei relatoribus Cancellarius ecclesie Parisiensis et facultas theologie in alma universitate Parisiensi matre nostra cum integro divini cultus honore spem habere in domino ac in vanitates et insanias falsas non respicere....

[18] *Excerpts from Peter of Ailly, Treatise against John of Monzón OP, written in 1388* (Du Plessis, *Collectio* 1,2: 75–129). [Capitulum xxxv] Sequuntur excerpta principalium articulorum tractatus cuiusdam contra errores fratris Johannis de montesone ordinis predicatorum Parisius condemnatos. Prefatio. [A]postolicis verbis edocti parati semper ad satisfactionem omni poscenti nos rationem de ea que in nobis est fide ad tuitionem eius quam profitemur orthodoxe....

APPENDIX 2:

AN INVENTORY OF NOËL BEDA'S REGISTER

On 15 October 1523, the faculty of theology approved Beda's suggestion to compile a survey of its pronouncements and judgments. The first resultant volume was the *Liber primus registri [determinationum] facultatis theologie schole Parisiensis in materia fidei et morum incipiens ab anno domini 1284*, which has been preserved in MS Paris, BN nouv. acq. lat. 1826. Below follows an inventory of those documents reproduced in Beda's register which concern the thirteenth and fourteenth centuries. As I have already mentioned, the first item is not from 1284, as the register's title indicates, but from 1210.

fols. 1r–3r: *Heading*: Instrumentum continens modum eligendi per facultatem theologie aliquem vicarium seu vicesgerentem decani eiusdem facultatis qui habeat portare onera officii decani. *Incipit*: In nomine domini amen. Per hoc presens publicum instrumentum cunctis pateat evidenter quod anno eiusdem domini millesimo trecentesimo octuagesimo quarto, inditione octava, mensis novembris die septima pontificatus sanctissimi in Christo patris et domini nostri domini Clementis. *Edition*: *CUP* 3: #1494, based on an original record that is preserved in the university archives.

fols. 3r–4r: *Heading*: Sequntur errores Albigensium qui fuerunt tempore Innocentii tertii circa annum domini millesimum CC^{m} decimum. *Incipit*: Primus error est quod ipsi constituebant duos creatores: invisibilium scilicet quem vocabant benignum deum et visibilium quem malignum deum nuncupabant.

fols. 4r–5v: *Heading*: Condamnatio errorum Johannis de Brescain. *Incipit*: Odo miseratione divina Tusculanus episcopus apostolice sedis legatus universis magistris et scolaribus Parisius studentibus. *Edition*: Du Plessis, *Collectio* 1,1: 158–59; *CUP* 1: #176, based on Noël Beda's register. The original record has not survived.

fols 5v–11v: *Heading*: Errores magistrorum Marsilii de Padua et Johannis de Janduno. *Incipit*: In nomine domini amen. Universis et singulis presens publicum instrumentum inspecturis pateat evidenter quod anno a Nativitate eiusdem millesimo ccc^{mo} lxxv, indictione tertiadecima more Romana curie. *Edition*: Du Plessis, *Collectio* 1,1: 397–400; *CUP* 3: #1406, based on an original record that is preserved in the university archives.

fols. 11v–14r: *Incipit*: Universis presentes litteras inspecturis officialis Parisiensis salutem in domino. Item, notum facimus notarios nostros subsignatos curie nostre iuratos.... Serenissimo principi ac domino Phillippo dei gratia regi francorum illustri sui devoti cappellani eiusdem miseratione Petrus. *Edition*: Du Plessis, *Collectio* 1,1: 316–17; *CUP* 2: #981, based on an original record that is preserved in the university archives.

fols. 14r–17r: *Incipit*: Urbanus episcopus servis servorum dei dilecto filio Johanni sancte Romane ecclesie presbitero cardinali olim episcopo Belvaccensi salutem. *Edition*: Du Plessis, *Collectio* 1,1: 382–86; *CUP* 3: #1350, based on an original record that is preserved in the university archives.

fols. 17r–20v: *Heading*: Littera continens determinationem factam per facultatem theologie atque condemnationem domini episcopi Parisiensis de propositionibus et erroribus dictis per fratrem Johannem de monte sono. *Incipit*: Anno domini millesimo trecentesimo octogesimo septimo die sexta mensis Jullii, post primam Sancti Jacobi. *Edition*: Du Plessis, *Collectio* 1,2: 149–51; *CUP* 3: #1559, based on Noël Beda's register. The original record is probably no longer extant.

fols. 20v–23v: *Heading*: Littera qualiter universitas Parisienis approbavit ordinationem et condemnationem errorum per fratrem Joannem de monte sono. *Incipit*: Universis Christi fidelibus rector et universitas magistrorum et scolarium Parisius studentium salutem in domino. *Edition*: *CUP* 3: #1560, based on an original charter that is preserved in the university archives.

fols. 23v–63v contain copies of the recantations of those scholars who had sided with John of Monzón (*fautores*) in the debate over the Immaculate Conception. They have all been edited in *CUP* from the original records, still extant in the archives of the University of Paris. They have been reproduced in Noël Beda's register in the following order:

fols. 23v–25v: Recantation of William of Volan; *CUP* 3: #1571.
fols. 25v–30v: Recantation of John Thomae OP; *CUP* 3: #1572.
fols. 30v–33r: Recantation of Adam of Suessione OP; *CUP* 3: #1574.
fols. 33r–38r: *Heading*: Sequitur condemnatio quorundam errorum fratris Richardi Marie presbiteri ordinis fratrum predicatorum concernentium conceptionem gloriosissime virginis Marie. *Incipit*: Universis presentes litteras inspecturis Matheus Anquetil decretorum doctor, rector ecclesie parrochialis de Sassetot la maucondivyt Rothomagensis diocesis iudex seu commissarius in hac parte auctoritate apostolica specialiter deputatus, salutem in domino. *Edition*: Du Plessis, *Collectio* 1,2: 135–38, on the basis of Noël Beda's register. *CUP* did not include this recantation, probably because the original record was no longer extant.
fols. 38r–43r: Recantation of Godfrey of Sancto Martino OP; *CUP* 3: #1576.
fols. 43r–52r: Recantation of John Adae OP; *CUP* 3: #1577.
fols. 52r–56r: Recantation of Peter of Chanceyo OP; *CUP* 3: #1578.
fols. 56r–63r: Recantation of John Nicolai OP; *CUP* 3: #1579.

ARMENGAUD BLAISE AS A TRANSLATOR OF GALEN

MICHAEL R. MCVAUGH*

Armengaud Blaise was overshadowed in almost every aspect of his career by his famous uncle, Arnau de Vilanova. The uncle was the most renowned master on the medical faculty of Montpellier in the 1290s; the nephew had barely squeezed out his degree from the local bishop in 1289, over the objections of the same faculty.[1] The uncle wrote sophisticated treatises of medical theory and practice, many of which soon found a wide audience; the nephew composed one or two short, simple medical handbooks that circulated hardly at all. When Count-King Jaume II of Catalunya-Aragò fell ill in 1305, the nephew admitted he was helpless to treat the royal illness, and the uncle had to be called in to bring about a cure.[2]

But in one respect, at least, the nephew surpassed the uncle: as a translator of medical works into Latin. Arnau did complete a few such translations (one of Galen's *De rigore* [*Peri tromou*] in 1282; another of Avicenna's *De viribus cordis*, probably in the same decade), but Armengaud produced a dozen or so—starting with Avicenna's *Cantica* (and Averroes' commentary on it) in 1285, before he had been awarded his degree, and concluding with a series of Maimonides' short treatises (*Contra venena*, *De emorroidis*, *De asmate*) in the first years of the fourteenth century, while he was practicing medicine in Languedoc and Catalunya.[3] Armengaud's career paralleled Arnau's so closely that it is somehow not surprising to find that he, like Arnau, chose to include Galen among the authors whom he translated, and it is this aspect of his work as a translator that will be explored here.

Armengaud translated two works ascribed to Galen while at Montpellier about 1300, one titled *Yconomia*, the other *De cognitione propriorum defectuum et viciorum*. In 1904 Moritz Steinschneider's listing of translations from the Arabic recorded them as existing only in the first volume of a splendidly illuminated two-volume collection of Galen's works in Dresden (MS Db 92–93), where they are found next to each other at fols. 16–17v

*A version of this paper was presented to the XIXth International Congress of History of Science (Zaragoza) in August 1993. I am particularly grateful to Vivian Nutton for his comments on that version, which have corrected a number of errors and have led me to articulate my ideas more fully.

[1] The controversy is recorded in *Cartulaire de l'université de Montpellier*, vol. 1 (Montpellier, 1890), pp. 213–15.

[2] A certain amount of information about Armengaud's career is collected in Michael R. McVaugh, *Medicine Before the Plague* (Cambridge: Cambridge University Press, 1993), pp. 8–15.

[3] Armengaud's translations are listed conveniently in Ernest Wickersheimer, *Dictionnaire biographique des médecins en France au moyen âge* (Paris: E. Droz, 1936), p. 41.

and 17v–19, respectively.[4] In a 1921 Leipzig dissertation directed by Karl Sudhoff, Theodor Trotz briefly analyzed the contents of the Dresden MS and transcribed the text of the *Yconomia*, singling it out as a Galenic work—on household management—that had never yet been printed.[5] Seven years later, however, Martin Plessner concluded that the Greek original of Armengaud's translation was not a work of Galen's at all.[6] In a careful study of the fragments of that original and of its later renderings into Arabic and Hebrew as well as Latin, Plessner decided that it should be attributed instead to a figure roughly contemporary with Galen, the neo-Pythagorean writer Bryson, and he supplied texts of the various translations, reediting Armengaud's Latin version from the Dresden MS in order to correct errors in Trotz's earlier transcription.

Because Plessner had depended upon Steinschneider's catalogue for information about manuscripts of the *Yconomia*, while Trotz made no attempt to identify other manuscripts, neither of their editions made use of a second copy of the work that was described, four years after Steinschneider's work had appeared, among the manuscripts of the Académie de Médecine at Paris. The Académie's MSS 51–56 make up, again, a multivolume collection of Galenic works, and the *Yconomia* is found at fols. 223–226v of the first of these codices, followed there, as at Dresden, by *De cognitione*.[7] The importance of this Parisian manuscript (which I will call *A*) is now the greater because in World War II the Dresden codex (*D*) suffered severe water damage and is today in large part illegible. In 1977 Vivian Nutton attempted to assess the relation of *D* to other medieval collections of Galen's work (including *A*), collating what he could of another of the texts it contained, *De prognosticatione*, and he concluded in part that *A* and *D* were closely connected:

> Where D is now missing, A fills the gap well. Its errors are on the whole trivial compared with those of D and, like D, it provides a sound supplement and corrective to the errors and omissions of B [another Galenic collection, MS Paris BN lat. 6865, which does not include either of Armengaud's translations]. Coming as it does from the same area and possibly the same scriptorium as D—for the hand is similar, although not identical, in the two manu-

[4] Moritz Steinschneider, *Die europäischen Übersetzungen aus dem Arabischen bis Mitte des 17. Jahrhunderts* (Sitzungsberichte der Kaiserlichen Akademie der Wissenschaften in Wien, phil.-hist. Klasse, 149, pt. 4 [1904]; rpt. Graz: Akademische Druck- u. Verlagsanstalt, 1956), p. 6.

[5] Theodor Trotz, *Der Inhalt der Dresdener lateinischen Galenhandschrift aus dem Anfänge des 15. Jahrhunderts* (inaug.-diss.; Dresden: Gärtner, 1921).

[6] Martin Plessner, *Der OIKONOMIKOC des Neupythagoreers "Bryson" und sein Einfluss auf die islamische Wissenschaft*, Orient und Antike, 5 (Heidelberg: Carl Winter, 1928).

[7] *Catalogue général des manuscrits des bibliothèques publiques de France; Paris*: vol. 1 (Paris: Plon-Nourrit, 1909), pp. 367–68.

> scripts—A is a more than adequate replacement for the textual loss suffered by the near destruction of Db 92.[8]

My collation of *A*'s text of the *Yconomia* with that transcribed from *D* by Plessner fully supports Nutton's assessment of *A*'s relative merits.

The relationship that has thus been established between *A* and *D* has even greater interest for a study of Armengaud's translation of *De cognitione*, which has never been edited. Again, *A* and *D* are the only manuscripts known of the translation, and *D* is today very difficult to use; the codex is in such poor condition that it cannot be reproduced photographically, although the particular folios containing *De cognitione*, while damaged (stained and smeared), are reportedly still legible.[9] But, as we have seen, the previous examples of the *Yconomia* and *De pronosticatione* suggest that a complete collation of *D* might in any case have little to contribute to a text of *De cognitione* prepared from *A*.

As a matter of fact, a few very brief excerpts of *D*'s version of *De cognitione* happen to have been published before the codex was damaged, and these excerpts tend to confirm that *D* has little to add to *A*. Unlike the *Yconomia, De cognitione* is based on a genuine work by Galen (*Peri psyches pathon*); when Johann Marquardt edited the Greek text among Galen's *Scripta minora* in 1884, he recognized this fact and used the Latin version he found in *D* to propose emendations to the Greek.[10] In Marquardt's apparatus criticus, eight short passages from *D* are preserved, comprising in all about 5 percent of the entire text, and all but one coincide exactly with the corresponding passages in *A*. Then in 1910 there was published a facsimile

---

[8] Vivian Nutton, "A Forgotten Manuscript of Galenus Latinus," *Studia codicologica*, Texte und Untersuchungen zur Geschichte der Altchristlichen Literatur, 124 (Berlin: Akademie-Verlag, 1977), p. 340.

[9] I owe this account of the state of the manuscript to the kindness of Mr. Perk Loesch of the Handschriftenabteilung, Sächsische Landesbibliothek, Dresden (letter of 30 September 1996).

[10] Ioannis Marquardt, ed., *Claudii Galeni Pergameni scripta minora*, vol. 1 (Leipzig, 1884), p. x; the text of *Peri psyches pathon* (with its quotations from *D*) is printed on pp. 1–44. Marquardt had made no use of *D* in idem, *Observationes criticae in Cl. Galeni librum peri psyches pathon kai amartematon* (inaug.-diss.; Leipzig, 1870). The criticisms of Marquardt's edition by A. Nauck ("Zu Claudii Galeni Pergameni Scripta minora, vol. 1," *Mélanges gréco-romains tirés du Bulletin de l'Académie impériale des sciences de St. Petersbourg*, 5 [1888]: 253–78 [repr. from *Bulletin* 31 (1886): 396–414]); and by Johannes Ilberg (*Berliner philologische Wochenschrift* 5 [1885], cols. 469–71), do not add further material from *D*.

Subsequent editors have acknowledged that *D* is relevant but have been sceptical of its utility, and they too have added no more passages: Otto Hennicke, *Observationes criticae in Cl. Galeni Pergameni commentarios ΠΕΡΙ ΨΥΧΗΣ ΠΑΘΩΝ ΚΑΙ ΑΜΑΡΤΗΜΑΤΩΝ* (inaug.-diss.; Berlin: Ebering, 1902), pp. 3, 50; Wilko de Boer, *In Galeni Pergameni libros ΠΕΡΙ ΨΥΧΗΣ ΠΑΘΩΝ ΚΑΙ ΑΜΑΡΤΗΜΑΤΩΝ observationes criticae* (inaug.-diss.; Marburg: Noske, 1911), pp. 10–12, 25–26; idem, *Galeni De propriorum animi cuiuslibet affectuum dignotione et curatione*, Corpus Medicorum Graecorum V 4, 1.1 (Leipzig and Berlin: B. G. Teubner, 1937), p. vi.

collection of the illuminated initials that begin each of the works in the Dresden codices, and the facsimile reproduction of the initial at the beginning of *De cognitione* includes another—different—4 percent of the text, in which only five words vary from the text given in *A*.[11] Thus on several counts it seems unlikely that a complete collation of *D*—were it possible—would have significance for the text of *De cognitione*, and I shall therefore speak about Armengaud's translation of this Galenic work with some confidence on the basis of a transcription of *A* alone. (The transcription will be given as an appendix to this study.)

Galen's brief *Peri pathon*—only fifty-seven pages in Kühn's edition—proves to be something a little surprising for Armengaud to have chosen, for it is not a medical work but a philosophical or moral treatise.[12] A young friend has asked Galen to discuss the subject of a recent book "On Controlling One's Passions," and Galen agrees. He begins by explaining that it is important to distinguish between passions (springing from an irrational power within us) and errors (springing from false opinion), and that he will reserve the latter for discussion in a second book. (That complementary book, *Peri psyches amartematon*, had apparently become detached from *Peri pathon* by Armengaud's day, since Armengaud does not translate it or even refer to it.) Our own self-love, Galen explains, makes it difficult for us to recognize our own failings, but he has a method for overcoming this difficulty: Find a truly disciplined, temperate, friend and beg him to call your attention to your immoderate outbursts of passion. Do not believe him if he has nothing to criticize—no one is without such faults, and there are obvious reasons why he might at first be hesitant to speak, so insist. Accept his judgments uncomplainingly and praise him for his candor when he has spoken.

After describing his method for achieving self-awareness, Galen proceeds to distinguish between two powers that can overwhelm the reason, and against which we must be on guard; somewhat repetitiously, he explains his method of detection in connection with each. The irascible power, first, once you have been made aware of it, can be controlled by reason: you can teach yourself never to strike out immediately in anger, but

[11] E. C. van Leersum and W. Martin, eds., *Miniaturen der lateinischen Galenos-Handschrift der Kgl. Öffentl. Bibliothek in Dresden Db 92–93 in phototypischer Reproduktion* (Leiden: A. W. Sijthoff, 1910), p. 2, nr. 6.

[12] An English translation by Paul W. Harkins is available as *Galen on the Passions and Errors of the Soul* (Columbus: Ohio State University Press, 1963); a portion was also translated by Arthur J. Brock, *Greek Medicine* (London: Dent; New York: Dutton, 1929), pp. 165–74. Its content is analyzed by Pierluigi Donini, "Tipologia degli errori e loro correzione secondo Galeno," in *Le opere psicologiche di Galeno*, ed. Paola Manuli and Mario Vegetti (Naples: Bibliopolis, 1988), pp. 67–116. An English summary is in Owsei Temkin, *Galenism: Rise and Decline of a Medical Philosophy* (Ithaca: Cornell University Press, 1973), pp. 37–38.

to check your outburst instead and respond later, more reflectively. The other power, however, the concupiscible power, cannot be controlled and must be suppressed; temperance, its opposite, must be put in its place as a goal to be pursued. This practice should be taught to the young so that it may become habitual—Galen's own upbringing, he explains, led him to check a display of passions, to accept calmly the ups and downs of life. Our irrational greed for food and possessions and honor arouses grief when we cannot have what we crave—yet, if we reflect, we will recognize that we *need* very little of these things.

When we compare this treatise with the translation made by Armengaud Blaise, it is surprising to find that the latter is hardly like a translation at all—it is certainly very different from Armengaud's quite literal renditions of Maimonides, for example.[13] Rather than presenting the full text of the original, Armengaud's version is a pastiche, a condensation created by entirely omitting about two-thirds of the work and translating and weaving together the scraps that are left. The result corresponds roughly to the second, third, and sixth chapters (of the original ten) in the Greek text, except that Armengaud has broken his composite, condensed version into five new chapters and has introduced his own chapter headings to indicate what now appears in each one:

> Chap. 1: That we often err through ignorance ... [one sentence from Greek chap. 1, half of Gk. chap. 2].
>
> Chap. 2: He shows how difficult it is to recognize our own failings ... [most of the rest of Gk. chap. 2].
>
> Chap. 3: He teaches how to avoid our failings and to find a faithful mentor ... [part of Gk. chap. 3].
>
> Chap. 4: He teaches how to behave towards your mentor and dominate the irascible and concupiscible powers [a fusion of part of the remainder of Gk. chap. 3 with the first part of chap. 6].
>
> Chap. 5: He provides special advice on maintaining virtue and avoiding faults [a part of Gk. chap. 6 with two sentences from chap. 9 tacked on, bringing the whole abruptly and incongruously to an end].

The cuts made by Armengaud have eliminated, first, much of Galen's repetitious instructions as to how one must choose a mentor and accept his criticism calmly. But they also leave out what we might call the *exempla*, as it were the case histories, that Galen employed to bring home to his young

---

[13] I owe much of my understanding of this material to an unpublished paper by Dr. Charles Burnett that he has generously shared with me: "The Latin Translations of the Medical Works of Maimonides: A Survey Followed by an Edition of John of Capua (?)'s translation of Maimonides' *De coitu*."

friend, concretely, the dangers of allowing the passions to rule: not even the story of how the emperor Hadrian put out a slave's eye with his stylus, and repented too late, has been preserved. Nor have Galen's anecdotes about his parents and his relations with them been included. It is through *Peri pathon* that we know so much today about Galen's mother, "so prone to anger that sometimes she bit her handmaids,"[14] and about his father—her temperamental opposite, calm and thoughtful—and the course of education he planned out for his son; but none of this autobiographical material was passed on to the Middle Ages in Armengaud's *De cognitione*.

The changes made by Armengaud in Galen's work are not simply matters of omission. He has regularly modified wording and meaning slightly when joining formerly distinct blocks of text, in order to make the transitions smoother. Moreover, many sections that have been translated entire have not been rendered literally, though sometimes they have simply been paraphrased in a way that may reflect a slight misunderstanding of Galen's text just as easily as a deliberate attempt to reformulate the original. Thus Galen's admonition that "you imagine that you are something other than a human being if you mislead yourself into believing that you have done nothing but good actions for a whole day, much less for a whole month," becomes ironic praise in Latin: "If you manage to do only good actions and not commit any errors, though for less than a month, even just one day, you can count yourself more than human."[15] But there is convincing evidence of other deliberate change. Armengaud has felt free to add material here and there to give rhetorical force: a vicious dog ("cano maligno") is added to the untamed horse that Galen uses to represent the bestial power of the passions that must be subdued; the gratuitous phrase "reflecting on their harmony and truth" ("contemplando eorum decorum et veritatem") now fills out Galen's own terse statement that every day he reads Pythagorean teachings, at first silently and then out loud.[16] One might with some justice characterize Armengaud's translation as a new creation: tighter and more straightforwardly didactic but at the same time drier and flatter, a succinct moral lecture on why you should allow a friend to point out your failings to you.

Yet was it really Armengaud who was its creator? It is after all possible that he simply translated a condensed version of *Peri pathon* that was already available in Arabic or Hebrew, or that had already existed in Greek; unfortunately, no Arabic (or Hebrew) text of *Peri pathon* seems to be ex-

---

[14] Harkins, *Galen on the Passions* (n. 12), p. 57.

[15] "Unde si possis sic operationes medias et temperatas exercere quod non committas errorem in eis in minori tempore quam sit mensis, nisi una die, scito te ipsum potuisse et fecisse aliud ab homine"; see the edition below, lines 125–28.

[16] Below, lines 179–82.

tant today with which comparison can be made.[17] Ḥunain ibn Isḥāq recorded revising the Syriac translation by Thomas of Edessa, adding that of the two treatises originally in Greek ("maqâlatain"; referring to *Peri pathon* and *Peri amartematon*?) he had seen only one, and that one imperfect (*nâqisa*).[18] This makes it seem as though the changes in the text might have been introduced before it came into Ḥunain's hands, or by Ḥunain himself in the act of translation. And it remains possible that a literal text prepared by Thomas and Ḥunain could have been epitomized later in Arabic, even though no such epitome is known.

Martin Plessner worried about exactly these difficulties when he addressed the striking differences between the *Yconomia* translated by Armengaud and the much fuller texts of the same work available in Arabic (and, fragmentarily, in Greek): "Whether he himself first shortened the text in the way we have it cannot be decided with certainty. We can recognize in R an Arabic and in H a Persian epitome [of the *Yconomia*], and it is perfectly possible that L [the Dresden text] was produced from one such."[19] Subsequently Plessner decided that Armengaud had not abbreviated the text himself, but must have worked from an epitomized Arabic translation;[20] he believed that he had found evidence for such an epitome in the commentary on Maimonides' *Ethics* composed in the thirteenth century by Shem Tob ibn Falaquera.[21]

However, what Plessner did not appreciate, not having studied Armengaud's other Galen-translation, was that the *Yconomia* and *De cognitione* had been treated in exactly the same way. In both works, whole sections were omitted in the Latin translation, except for isolated sentences included

[17] Manfred Ullmann, *Die Medizin im Islam* (Leiden: E. J. Brill, 1970), p. 51 no. 65, identifies no manuscripts of the work. It is not treated in Fuat Sezgin, *Geschichte des arabischen Schrifttums*, vol. 3 (Leiden: E. J. Brill, 1970).

[18] G. Bergstrasser, "Hunain ibn Ishaq über die syrischen und arabischen Galen-Übersetzungen," *Abhandlungen für die Kunde des Morgenlandes* 17 pt. 2 (1925); Arabic text pp. 48–49, German translation pp. 39–40 (no. 118).

[19] "Ob er den Text selbst erst so verkürzt hat, wie er uns vorliegt, ist nicht mit Sicherheit zu entscheiden. Wir werden in R eine arabische und in H eine persische B[ryson]-Epitome kennen lernen; und es ist durchaus möglich, dass L [the Dresden text] auf Grund einer solchen entstanden ist" (Plessner, *OIKONOMIKOC* [above, n. 6], p. 22).

[20] Martin Plessner, "Der Dresdner Pseudo-Galenus über Pädagogik und Kinderhygiene," *Monatsschrift für Kinderheilkunde* 114 (1966): 479.

[21] Martin Plessner, "Über die Bedeutung Schemtobs Ibn Falaquera für das Studium der Geschichte der Philosophie" (in Hebrew), *Homenaje a Millás Vallicrosa*, vol. 2 (Barcelona: CSIC, 1956), pp. 176-79. Here Plessner argued that the Latin translation and a second Hebrew translation had been made, independently, from an abridged Arabic text, and that each had been further abridged in the process. He provided parallel passages of Latin, Hebrew, and Arabic to support his claim that the Latin and Hebrew both proceeded from such an abridged Arabic version, but for the rest of his argument he referred the reader to his forthcoming study of Rabi Shem Tov, which unfortunately he never completed..

I am grateful to my colleagues Frederick Behrends and David Halperin for translating the Hebrew for me.

to tie together the remaining blocks of text, and new chapter headings were introduced to give shape and structure to the result.[22] It seems to me, therefore, just as likely that Armengaud employed the same technique of abridgement on these two "Galenic" works as that the manuscripts from which he worked had each been separately epitomized in the same way before they reached him. And there is a certain amount of evidence for this interpretation, too, although it is in no sense conclusive. Maimonides's disciple Ibn ʿAqnîn (d. 1226) has left us an aphoristic collection on "The Hygiene of the Soul" (*Ṭibb al-Nufûs*), in which at one point he quotes from Galen's *Peri pathon*. The passage is in Arabic (though written in Hebrew characters) and proves to be an almost exact rendering of the corresponding Greek text; it is quite possible that it is a fragment of the full Arabic translation of *Peri pathon* made by Ḥunain.[23] But in that case, it is significant that the passage in question (from chap. 8 of Galen's work) is *not* reproduced in Armengaud's Latin translation, for this is entirely consistent with the assumption that the reworking of *De cognitione* originated with Armengaud rather than with Ḥunain (or a later epitomizer).

Plessner pointed out certain stylistic peculiarities of the Latin *Yconomia* that he felt sure were features of Armengaud's translating style rather than of the original from which he worked, and this seems to be confirmed by their presence in *De cognitione* as well. One, already commented on above, was the translator's departure from a relatively strict translation when he expanded certain brief statements that he evidently thought of as elliptical and in need of amplification. Another was described by Plessner as "a piling up of expressions. In almost every paragraph we find two nouns, adjectives, or verbs linked together with an *et* or a *seu*, word pairs of which only one is found in the original."[24] This had not originally struck me about *De cognitione*, but once it was suggested to me by Plessner's account of the *Yconomia*, I could not help noticing with increasing amusement such doublets in, as he said, every paragraph, and even every other sentence. So in both large and small matters I think we can identify an approach to translation that is common to these two works and that we can presume originated with Armengaud.

Assuming this is so, why then might Armengaud have chosen to cut and rework these two "Galenic" texts so drastically? His approach to translating other works, like Avicenna's *Cantica* or Maimonides' *De venenis*, was a

[22] Plessner, *OIKONOMIKOC* (n. 6), pp. 23–24.

[23] "The passage is a faithful translation of the Greek original … (ed. Kühn), 40-41, except that it omits the comparison of his mother's treatment of her father to Xanthippe's treatment of Socrates" (A. S. Halkin, "Classical and Arabic Material in Ibn 'Aknin's 'Hygiene of the Soul,'" *Proceedings of the American Academy for Jewish Research* 14 (1944): 71).

[24] "Häufung des Ausdrucks. Fast in jedem Paragraphen finden sich zwei durch *et* oder *seu* verbundene Substantiva, Adjectiva oder Verba, von denen in den orientalischen Texten jeweils nur eines steht" (Plessner, *OIKONOMIKOC* [n. 6], p. 24).

much more literal one; those translations occasionally carried over into Latin minor features of Arabic style that could perfectly well have been paraphrased or even omitted. A close, word-for-word rendition of the Arabic had similarly marked his uncle Arnau's translation of *De rigore*, a translation that despite its transmission through at least one intermediate language can still provide useful emendations to the surviving Greek manuscripts—something that *De cognitione* just cannot do for *Peri pathon*.[25] Perhaps the obvious answer is correct, at least for *De cognitione*: that in translating this work Armengaud simply decided to eliminate repetition and other material he thought irrelevant in order to pare Galen's argument down to its essentials. We can understand why he might have been unwilling to abridge a medical text, to condense a recipe or eliminate a case history, for there anything might be relevant to his readers; but the subject matter of *De cognitione* (like that of the *Yconomia*) was far from medical.

Yet that answer only raises another question: Why should Armengaud have singled out *De cognitione* for translation when it had so little professional relevance? His translation of the *Yconomia*, which is paired with *De cognitione* in transmission and is so like it in style, might have been inspired by contemporary European interest in domestic economy,[26] but that argument leaves the much more moralizing *De cognitione* unexplained. Why should Armengaud have chosen to translate that didactic work when there were still—as Niccolò da Reggio was to begin to demonstrate ten years later—Galenic medical writings aplenty remaining to be translated? Again, there are of course obvious possible answers—that those medical writings were not available to Armengaud, or that they represented more work than he wanted to undertake—and they may well be the right ones. Nevertheless, I want to move away from the obvious and to step firmly into the realm of speculation, offering another possible explanation for Armengaud's choice of a text to translate.

I have not yet discussed the colophon to *De cognitione*, and I will turn to it now, because it bears directly on the question of Armengaud's motives. Armengaud ordinarily concluded his translations by recording the date and place where they were finished; *De cognitione* is unique, I believe, in con-

---

[25] Wilko de Boer wrote scornfully of Marquardt's attempt to use *De cognitione* in this way: "Codex igitur Dresdensis ad veras Galeni lectiones restituendas *perparvi momenti est* et cavendum, ne hac interpretatione adducti genuinas scripturas foedemus potius quam investigemus" (de Boer, *In Galeni* [above, n. 10], p. 12; emphasis in the original). He expressed the same opinion exactly a quarter-century later in *Galeni De propriorum* (above, n. 10), p. vi. For the utility of Arnau's Latin in establishing the original Greek of Galen's *Peri tromou*, see examples in David Sider and Michael McVaugh, "Galen on Tremor, Palpitation, Spasm, and Rigor," *Transactions and Studies of the College of Physicians of Philadelphia* 1 (1979): 183–210.

[26] This is the suggestion of Luis García Ballester, "Arnau de Vilanova (c. 1240–1311) y la reforma de los estudios médicos en Montpellier (1309)," *Dynamis* 2 (1982): 137–38.

cluding with a much longer, personal statement (though *De venenis* has a fairly long dedicatory preface to Pope Clement V). The colophon has apparently never been published in full—not by the cataloger of the Paris MS, nor by the editors of *Peri pathon* who described the Dresden MS—and it reads as follows:

> Here ends Galen's book "On an awareness of one's own defects and vices," translated from Arabic into Latin by master Armengaud Blaise at Montpellier on 14 August 1299. I translated it in honor of master Guillaume de Mazet, reverend and venerable and expert master of medicine. And I mean to work on the translation of many other very useful books in medicine, by Averroes and rabbi Moses [Maimonides], Avenzoar and many others, written shortly before the present day, in honor of other men—so long as there shall be an awareness of their own vices, otherwise not. For what is given to someone who does not appreciate it [*ingrato*] is rightly considered lost; and one regrets having given what is seen to have been lost.[27]

What are we to make of these cryptic remarks?

First of all, they force us at least to raise the question of whether Armengaud made his translations from Arabic or from Hebrew, and of whether he worked typically with a cotranslator, a Jew who could read the language from which the translation was to be made. These issues have long been debated, almost invariably strictly on the basis of the colophons to his works, not from a study of the texts themselves.[28] Plessner's judgment, ar-

[27] "Explicit liber Galieni de cognitione propriorum defectuum et viciorum translatus ab arabico in latinum a magistro Armengaudo Blazii apud Montempessulanum xix^{no} kl. septembris anno domini m^{o}.cc^{o}.xcix^{no}; transtuli autem ipsum in reverentiam magistri Guillermi de Mazeto, reverendi et venerabilis magistri in medicina periti. Ad reverentiam etiam aliorum laborare intendo in translatione plurium aliorum perutilium in medicina, ab Averroy et a Rabi Moyse Avensdare et pluribus aliis paulo ante hec tempora compilatorum—dum fuerint propriorum viciorum cognitiones, alias non; nam merito perditum reputatur quod ingrato datum est, aut dedisse penitet quod periisse videtur": below, lines 221–30.

In the Paris catalogue the colophon is transcribed, with one long ellipsis, only as far as "periti." The Dresden colophon was transcribed by Marquardt, but with some errors, and Karl Kalbfleisch transcribed it more carefully for de Boer in 1911: "Explicit liber Galieni de cognitione propriorum defectuum translatus ab arabico in latinum a magistro ermengando blazii apud montem pessulanum .xix. kl' Septembris anno domini m.^{o}cc [blank space enough for four letters follows]. Transtuli autem ipsum in reverentiam magistri guillermi de mezeto reverendi et venerabilis magister in medicina periti" (de Boer, *In Galeni* [above, n. 10], pp. 10–11). If the remainder of the Paris colophon followed in the Dresden MS, neither Marquardt nor Kalbfleisch nor de Boer (who had photographs of the MS but did not inspect it personally) felt that it was necessary to quote it.

[28] In manuscript colophons, Armengaud's translations of the Avicennan *Cantica*, the *Yconomia*, and *De cognitione* are described as made from the Arabic; his translation of Profatius's *Quadrans novus*, as made from the Hebrew "secundum vocem eiusdem [Prophatii]"; and his translation of Maimonides' *De asmate*, as made "ab arabico in latinum ... mediante fideli interprete." (This information has been usefully assembled in Catherine Ann Blue, "Ermengaud Blasi: Medieval Physician and Scholar" [M.A. thesis, University of North Carolina, 1978], pp. 37–41.) One way to explain these statements is to understand Armengaud as having been able to translate Arabic, but as having required assistance when Arabic words

rived at by comparing the Latin *Yconomia* systematically with the Hebrew and Arabic textual traditions, seems surprisingly to be little known: that Armengaud translated *that* work, at least, from an Arabic rather than a Hebrew original.[29] Lacking a Hebrew or an Arabic text of *De cognitione*, it has been impossible for me to make a similar test; all I can do is point out that here, too, Armengaud claims to be translating from Arabic. The subsidiary question of Armengaud's need for a cotranslator remains open; yet I think it is worth pointing out that the ability to create an intelligent abridgment of a work in another language requires on *someone*'s part probably greater ease in that language than would a purely word-for-word translation.

Let me now turn to the speculation I promised, which arises out of what seems to me to be Armengaud's curious attitude toward the Montpellier faculty, as expressed in the colophon. If I read that text correctly, he is honoring master Guillaume de Mazet and promising Guillaume's colleagues that he will do the same for them, too, if they are willing to acknowledge their faults as Galen advises. What can this mean? We know very little about Armengaud's life before about 1297, when he first makes an appearance in the enormously rich archives of the Crown of Aragon, and I know very well how risky it is to try to build a hypothesis on one of the isolated facts of his career that have survived. Still, I think it is suggestive that, just a decade before his translation of *De cognitione*, Armengaud (as I mentioned at the outset) had received his medical degree from the bishop of Maguelonne, over the violent objections of the Montpellier faculty, who proceeded to protest to the pope that the bishop had acted against their wishes; so far as we know, Armengaud never actually taught (and, perhaps embarrassed, rarely assumed the title *magister in medicina* in formal documents). You will probably not think it far-fetched to suppose that the faculty's hostility toward Armengaud was matched by his bitterness and resentment toward them, and that even ten years later he would have had few if any friends in that group besides his uncle—except, it would seem, for master Guillaume de Mazet.[30] I find myself speculating, therefore, that Armengaud chose *De cognitione* deliberately for translation, attracted by its theme that one ought to accept criticism and to admit one's mistakes—a theme that he focused and gave increased emphasis by his reworking of the text. His dedication to Guillaume might even suggest that Guillaume had

---

were written in Hebrew characters, as was Maimonides' practice (or that of Ibn ᶜAqnîn, alluded to in n. 22 above).

29 "Ebenso steht nunmehr fest, dass Armengandus Blazii wirklich Übersetzungen aus dem Arabischen angefertigt hat" (Plessner, *OIKONOMIKOC* [n. 6], p. 28).

30 Very little more is known about master Guillaume. He eventually became physician to Pope Clement V but is referred to as dead in papal bulls of 1309, which say of him merely that "longo tempore in Studio [Montepessulano] rexerat" (*Cartulaire* [above, n. 1], 1: 223, 225; Wickersheimer, *Dictionnaire* [above, n. 3], p. 255.

once been unsympathetic but had finally come round and made up with Armengaud, and Armengaud's colophon would thus be a taunt addressed to the rest of the faculty: if you too will accept your master Galen's advice and admit you were in the wrong, I will dedicate a translation to you as well. Perhaps it is not surprising that none of his later translations *was* dedicated to a Montpellier master; as I am imagining the situation, I cannot believe he thought any of his enemies would accept his offer.

Once more I have to say that this story about Armengaud's motives is *only* a story. It has the merit, perhaps, of making sense out of that unique and enigmatic colophon in the light of what we know of his history, but I am sure there are other possible stories that could do the same. I have developed my story, speculative though it is, for another reason besides mere conviction. When we examine the transmission of learning in the Middle Ages via translations, we concentrate on subject matter and on technique; motivation and personality are nothing we try very far to understand. We ignore motives and character because they are so difficult to get at, and if we think about them at all I suspect we usually fall back on the easy explanations: that translations were made either altruistically, for the good of mankind (Gerard of Cremona's biography provides our model here),[31] or else selfishly, out of the hope of patronage or professional advancement. My little story is meant to remind us that human psychology is complicated and sometimes perverse, and that other motives—passions, if you like—might sometimes play a part: like Armengaud's anger, if my speculation has any foundation; an anger that matured and eventually achieved thoughtful expression in *De cognitione*, just as that work might have advised. If I am right, Armengaud himself may have learned a lesson from Galen.

[31] An English translation is in Edward Grant, ed., *A Source Book in Medieval Science* (Cambridge, Mass.: Harvard University Press, 1974), p. 35.

APPENDIX

ARMENGAUD BLAISE'S TRANSLATION OF GALEN, *DE COGNITIONE*

This edition of Galen's *De cognitione* is based on MS Paris, Académie de Médecine 51, fols. 227–231 (= *A*), with modifications from MS Dresden, Landesbibliothek Db 92, fols. 17v–19 (= *D*), as reported in two published sources: E. C. van Leersum and W. Martin, eds., *Miniaturen der lateinischen Galenos-Handschrift der Kgl. Öffentl. Bibliothek in Dresden Db 92–93 in phototypischer Reproduktion* (Leiden: Sijthoff, 1910), p. 2, no. 6, which reproduces a portion of the text from *D* that is printed in bold face in lines 17–27 below; and Ioannis Marquardt, *Claudii Galeni Pergameni Scripta Minora*, vol. 1 (Leipzig, 1884), whose passages from *D* are printed in bold face in lines 46, 61, 68–69, 115–17, 123, 133–34, 146–47, 151, 217–25.

* * *

Liber Galieni de cognitione propriorum defectuum et viciorum translatus ab Armengaudo Blazii de Arabico in Latinum in Montepessulano continet quinque capitula:

Capitulum primum quod pluries erramus per ignorantiam et qui ex nobis errant plus et qui minus et que decent volentem esse perfecte bonum;

Capitulum secundum sub quadam similitudine ostendit difficultatem noscendi propria vicia et confirmat illam ratione et dat duos modos generales noscendi ea;

Capitulum tertium docet modum vitandi vicia et eligendi sibi virum[32] fidum qui doceat ipsum si sit opus;

Capitulum quartum, invento fido amico, docet modum se habendi ad eum et domandi virtutes irascibilem et concupiscibilem;

Capitulum quintum dat specialia documenta circa virtutes colendas et vicia vitanda.

---

[32] *ante* fidum *del. A* frigidum

Capitulum primum quod pluries[33] erramus per ignorantiam et qui ex nobis errant plus et qui minus et que decent volentem esse perfecte bonum.

Inquit Galienus: Quesivisti a me ut compilarem[34] tibi tractatum ad notificandum et manifestandum hominis vicia in eodem existentia. Volo ergo respondere tibi in eo quod postulasti, et dico quod pluries accidit nobis quod erremus,[35] nec tamen scimus vel cognoscimus nos errasse. Hoc autem probatur et manifestatur vobis[36] ex eo quod dicam, et est quoniam videmus fere totum populum opinantem se etiam non posse errasse. Et qui ex popularibus opinantur se errare,[37] videtur eis quod error eorum sit valde modicus secundum opinionem eorum, cum tamen apud alios sit valde magnus; et ego quidem vidi hoc pluries et probavi. Sum etiam expertus quod qui probat alium a se, et examinat omnes errores suos, et iudicat seipsum super illos, quod talis quidem errat minus multum eo qui hoc non facit. Verum qui opinantur quod sint boni et virtuosi,[38] (227v) iudicantes se secundum hoc, non autem secundum opinionem aliorum, ego quidem reperio tales multum errantes. Unde dico quod decet volentem esse perfecte bonum et omni vicio circumscriptum ut sciat et opinetur semper sibi non constare de omnibus viciis et delictis suis, ymo credat et opinetur sibi hoc esse occultum. Decetque ipsum esse attentum, primo ad sciendum errores proprios et ad inveniendum eos qualitercumque possit, et est possibile ut notificetur ei modus per quem possit hoc decenter apprehendere. Ego tamen nundum notificabo hic, et hoc quia scio possibile esse ut hic meus liber veniat in possessionem[39] alterius a te, priusquam eis sit promptum dare operam ad inveniendum modum quo sciant et cognoscant propria vicia et delicta. Et sicut feci hoc in te, scilicet quod dixi et interrogavi te ut notificares michi quid esset in te ex hoc, nec volui tibi dicere aliquid ex hoc priusquam intellexissem quid ex hoc in te esset, sic etiam facio nunc, querens ab illo qui inspiciet hunc librum **ne[40] separet ipsum a manu sua**, et quod investiget in mente sua modum quo sit ei promptum scire et cognoscere propria delicta.

Capitulum secundum sub quadam similitudine ostendit difficultatem noscendi propria vicia et confirmat illam ratione et dat duos modos generales noscendi ea.

---

[33] plures *D*
[34] copularem *D*
[35] erramus *D*
[36] nobis *D*
[37] errasse *D*
[38] *post* virtuosi *del. A* iudicantes
[39] possessionem *emendavi ex* pt'atem *A*
[40] ne *corr. A m. rec. ex* in

Inquit enim Liton quod ad collum cuiuslibet ex nobis suspenduntur due burse, una anterius et altera posterius. In anteriori autem sunt vicia aliorum in quam continuo intuemur; vicia autem et delicta propria sunt in posteriori bursa, quam neque parum neque multum inspicimus vel videmus. Et inde est quod aliorum vicia bene cognoscimus et propria ignoramus, nec invenimus quemquem non prebentem assensum huic sermone aut ipsum falsificantem. Quin ymo Plato cum hoc quod assensit ei notificavit ratione huius,[41] **dicens amicum in amato esse secum.** Sicut igitur homo magis diligit seipsum, sic absque dubio (228) est secum et non videns in seipso. Unde cum inspiciet omnia vicia et delicta propria sibi[42] nedum aliqua ex eis erit secum et non videns in conspectu eorum, et cum inspiciet eorum maliciam erunt post non ante ipsum. Sic igitur video quod simile quod Liton posuit[43] et ratio quam super ipsum Plato assignavit et notificavit faciunt nos desperare ne possimus errores proprios invenire, et hoc cum sit impossibile ne quis non diligat seipsum et **etiam[44] impossibile quin quilibet[45] sit secum et non videns in suo amato**. Nec quero ab eo qui inspiciet hunc meum librum ut sit attentus per seipsum invenire modum quo sciat et cognoscat propria vicia nisi propterea quia hoc est valde difficile et longinquum. Pro quanto volo ut ille qui intuebitur librum istum cogitet et attendat ad inveniendum per seipsum modum super hoc et consequenter speculetur in eo quod ego inveni ex hoc; et si forsitan invenerit modum alium ab illo quem recte inveni et super quem fundavi et compilavi hunc meum librum, erit hoc siquidem ei utilius—et hoc quia tunc[46] invenerit duas vias seu modos ad salutem suiipsius, scilicet illum quem noverit per sui cogitationem et investigationem et illum quem ex hoc libro sumpserit. Verum si quis sit attentus in inventione huius methodi quam retulimus nec possit eam invenire donec prius invenerit aliam per sui investigationem, nequaquam erit deterius.

Capitulum tertium docet modum vitandi vicia et eligendi sibi virum fidum qui doceat ipsum si sit opus.

Ex nunc autem dicemus methodum quam invenimus, et ponam in principio sermonis mei et dicam quod ratione erroris existentis in hominibus in falso et pravo consilio, viciis contingentibus ei<s>dem ex appetitu et concupiscentia bestiali, videtur michi expedire quod preponam modum quo possint

---

[41] *ante* huius *del. A* eius
[42] sibi *ins. A*
[43] *ante* posuit *del. A* possint
[44] est *D*
[45] quisque *D*
[46] tunc *ins. A*

vitari vicia, et homo ab eis evadere. Et si non possis eum per se scire vel cognoscere, decet quidem ut huius scientiam et cognitionem committas aliquibus (228v) qui nec amici aut inimici tui erunt. Unde si dicatur tibi quod aliquem ex eis qui sunt de loco aut villa tua, non amicus vel inimicus tuus, populi laus exaltet et gratum habeat preter quod ipse amorem adulativum gentibus pretendat, decet siquidem ut reddas te sibi familiarem, probans nichilominus ipsum posse tuo. Si ergo videas ipsum frequentare domos divitum et regum, scito non esse verum quod dictum fuit tibi ex eodem et quod non est possibile ut omnes sermones talis viri veridici sint et certi, et hoc quia appetitus ostendendi amorem et adulationem cogit eos ad prosecutionem falsitatis proculdubio. Verum si invenias quod ipse non curet de frequenti acceptione et salutatione divitum magnatuum et regum nec de frequenti visitatione domorum ipsorum aut habitatione cum eisdem, spero esse vera que tibi de eo dicta fuerunt. Pro quanto frequenta et continua esse secum, et quere ab eo ut inspiciat omnes mores tuos et quod notificet tibi qui ex eis non placeat eidem,[47] dicasque ei quod tu regraciaberis ei de hoc plus quam si quis curaret te ab aliquo magno morbo in tuo corpore existente—et hoc quia ex eo quod in te inde faciet fiet medicina morborum in anima tua existentium. Qui si promittat tibi hoc facere et quocienscumque videbit aliquid in te quod non placeat ei quod manifestabit et reserabit illud tibi, tunc si elapsis multis diebus sit preter quod aliquid notificaverit tibi inde, immo asseruerit tibi quod non viderit aliquid in te in tempore predicto quin placuerit eidem, nullo modo credas ei, ne credas quod subito et absque animadversione factus sis virtuosus. Ymo decet quod opineris de eo alterum duorum, scilicet vel quod ex sua negligentia non curaverit de moribus et negociis tuis, parvipendens te, negligens quod ab eo postulasti, **aut quod erubescit de correctione tui, timens quod haberes inde ipsum odio**; (229) scis enim totum populum ut plurimum habere illum odio qui manifestat ei veritatem suorum delictorum et qui corrigit eos ex eisdem. Verum si non[48] cesset ab hoc[49] propter aliquam ex predictis duabus rationibus, scito ipsum nolle[50] commodum et utilitatem tuam et quod propter hoc noluit tibi aliquod ex delictis tuis reserare. Quod tibi ex hoc constare potest, nam vides omnes homines omni die millesies errantes, committentes[51] multa turpia in operibus suis, **preter quod sciant hoc et cognoscant**. Constat autem tibi esse hominem sicut et ipsa sunt, pro quanto decet tibi manifestum esse quod tu etiam erres sicut ipsi. Unde si possis sic operationes medias et temperatas exercere quod non committas errorem in eis in minori

[47] *ante* eidem *del. A* eadem
[48] non *ins. A*
[49] *mg. A* scilicet a dicendo quod non invenit aliquid in te quin placeat ei
[50] *corr. A ex* molle
[51] *ante* commitentes *del. A* committantes

tempore quam sit mensis, nisi una die, scito te ipsum potuisse et fecisse aliud ab homine.

Capitulum quartum, invento fido amico, docet modum se habendi ad eum et domandi virtutes irascibilem et concupiscibilem.

Et scito quod primum quod decet te facere in hoc est quod sis usus et exercitatus pati et recipere correctiones. Quin ymmo decet quod regracieris et laudes illum et notificabit tibi omnia vicia tua et quod recedas contentus ab eo et laudans ipsum. Nec cures si fuerit falsitas aut malicia quod tibi dicet in tua correctione. Unde decet nos esse contentos si attendamus ad rectificandum nos ipsos et transferendum nos ad laudabilem dispositionem, quamvis desperemus nos posse consequi in hoc illos philosophos qui scilicet animadverterunt[52] super seipsos a puericia. Decet etiam quemlibet amplius ex propria confusione verecundari quam ex confusione[53] omnium aliorum. Cum autem quis hoc faciet, erit quidem possibile ipsum vincere et superare vim irascibilem in eo existentem, que scilicet bruto pravo et malicioso similatur, et quod mitiget eam, nec similabitur ex tunc equo aut cani maligno, ymo corrigetur et temperabitur in parvo tempore. Iam autem locuti fuimus (229v) sermone lato super hoc **in libro nostro de consuetudinibus,** declaravimusque ibidem qualiter possit fieri eorum correctio cum sint illaudabiles. Rursus reserabimus ex diversis causis predicti libri qualiter notificetur et corrigatur vis seu virtus concupiscibilis, quia est vis inducens nos ad delectationes et voluptates corporales et sensibiles sine premeditationibus, **nec comparabo vim istam sicut primam** equo et cani, sed potius yrco commoto et porco et feris silvestribus. Reprimitur autem vis predicta si non concedatur ei[54] prosecutio alicuius voluptatis sensibilis et corporalis. Consecutio enim voluptatis predicte auget et fortificat vim predictam. Cum autem vis predicta deprimitur, minuitur et debilitatur, ex cuius debilitate viget ratio et vis intellectiva. Fuit autem quidam ex antiquis qui Ciseleus dicebatur qui prosecutus est concupiscentias[55] anime sue, et hoc valde approbavit nec voluit eas reprimere, et ob hoc vis eius cognitiva non repressit vim eius concupiscibilem. Sunt enim in animabus nostris due vires bestiales, ex quibus una causat in nobis iram facitque tenere nos et servare odium. Ipsa[56] etiam[57] firmat in nobis longo tempore rancorem qui quidem morbus deterior et maius malum est quam ira; durat enim et perseverat diutius quam ira. Alia autem vis facit nos desiderare, trahitque nos ad voluptu-

[52] *ante* animadverterunt *del. A* ad
[53] verecundari ... confusione *mg. A*
[54] *ante* ei *del. A* e
[55] *ante* concupiscentias *del. A* ad
[56] *corr. A ex* ipsam
[57] *ante* etiam *del. A* enim firmat

osum et delectabile secundum apparentiam, priusquam sciamus et cogitemus an hoc sit bonum et utile aut malum et contrarium. Decet autem ut rumpatur vis et fortitudo[58] huius concupiscentie, priusquam augeatur et fortificetur in tantum quod non possit vinci et superari; nam cum erit in tali dispositione nequaquam poterit vinci et superari, quantumcumque velit quis et solicitus fuerit super hoc.

Capitulum quintum dat specialia documenta circa virtutes colendas et vicia vitanda.

Unde decet quod teneamus in mentibus nostris modos (230) laudabilium morum et correctionum, et quod omnibus diebus recordemur et mentionem faciamus eorum multotiens, si fuerit nobis oportunum. Quod si non fuerit hoc nobis bene oportunum, faciamus hoc tamen bis in die, scilicet semel de mane cum surgimus a sompno priusquam inchoemus opera exercere, et iterum alia vice in vespere ante introitum lecti. Nos autem usi fuimus studere documenta a Pictagora tradita bis omni die, et prima vice legimus solum et vidimus ea, secunda vero vice proferimus ea[59] oretenus contemplando eorum decorem et veritatem. Decet etiam ut contemplemur et attendamus omni die et continuo an expediat nos solum vivere subicientes nosmet ipsos nequam rebus illaudabilibus, aut quod pocius faciamus omnia negocia nostra cum discretione et ratione. Decet etiam ut vitemus et elongemus nos posse nostro a prosecutione voluptatis et inordinate concupiscentie, et quod sobrie vivamus nec crapulemur et inebriemur et rebelles efficamur, vitantes semper ocium et invidiam. Nec decet famelicum esurientem cibis repleri ad saturitatem, nec desiderantem et delectantem in cibis, quod repleatur cibariis dulcibus et delectabilibus, nec utatur eis frequenter, sicut faciunt canes vorantes cibum suum impetuose[60] et cum magno appetitu. Nec decet sitibundum bibere aquam frigidam magnis haustibus et continuis, sicut facit inflammatus cum febre. Ymo decet quidem nos multo minus ex cibo capere quam faciat aliquis ex eis qui presentes erunt nobiscum, intendentes semper in hoc quantitatem qua indigemus ad salutem et conservationem nostrorum corporum. Congruit etiam quod animadvertas super teipsum cum aliqua die presenti vel preterita magis retinueris vel compresseris appetitum tuum comedendi. Quod cum feceris, manifestabitur tibi singulis diebus facilitas et possibilitas huius rei in teipso. Debes autem valde letari si fuerit[61] ex eis qui diligunt anime nobilitatem et mores nobiles (230v) et laudabiles. Et scito quod amantes vinum ipsumque[62] multum desiderantes gaudent valde ex

[58] *post* fortitudo *del. A* eius
[59] secunda ... ea *mg. A*
[60] *post* impetuose *del. A* suum impetuose
[61] si fuerit *mg. A*
[62] *post* que *del. A* ipsum

magno potu; et habentes magnum appetitum comedendi letantur valde cum multum comedunt, necnon etiam ex esu dulcium et conditorum. Cognosco etiam nonnullos gaudentes plurimum ex multo et frequenti coitu. Decet etiam nos querere finem vere bonitatis et temperantie, sicut etiam predicti querunt et intendunt, quod est melius secundum opinionem ipsorum, licet apud virtuosos et sapientes intenta ab eis prava sint et illaudabilia. Et convenit quidem ut dimittat librum istum nec aspiciat ipsum ille qui noluerit[63] preeligere bonum vice mali, et quietam veram vice laboris, et temperiem corporis in electionibus suis. Et hoc quia non intendimus in eo ostendere qualiter[64] sit quis attentus ad lucrandum et adipiscendum particulariter mores laudabiles, sed intendimus solum manifestare in eo methodum qua diligens laudabiles mores, et circa hoc solicitus et attentus possit eos consequi et adipisci. Est tamen notum quod decet nos esse contentos in cibis et vestibus nostris et aliis rebus cum eo solum quod ex eis necessarium est nobis. Nec congregemus et emamus ea que omni ratione carent, sicut plures faciunt congregantes margaritas et gemmas seu lapides omnes, **quibus ornantur mulieres, intendentes et opinantes magnificare[65] inde earum decorem et pulcritudinem. Et similiter est de pannis deauratis in quorum compositione laborant longo tempore absque ratione.**

Explicit liber Galieni de cognitione propriorum defectuum et viciorum[66] translatus ab arabico in latinum a magistro Armengaudo[67] Blazii apud Montempessulanum xix^{no}[68] kl. septembris anno domini m°.cc°.xcix^{no};[69] transtuli autem ipsum in reverentiam magistri Guillermi de Mazeto, [70] reverendi et venerabilis magistri (231) in medicina periti. Ad reverentiam etiam aliorum laborare intendo in translatione plurium aliorum perutilium in medicina ab Averroy[71] et a Rabi Moyse Avensdare et pluribus aliis paulo ante hec tempora compilatorum—dum fuerint propriorum viciorum cognitiones, alias non; nam merito perditum reputatur quod ingrato datum est, aut dedisse penitet quod periisse videtur.

---

[63] *ante* noluerit *del. A* vo
[64] *post* qualiter *del. A* q
[65] *post* magnificare *del. A* e
[66] et viciorum *om. D*
[67] ermengando *D*
[68] xix *D*
[69] m°.cc [*vacat*] *D*
[70] mezeto *D*
[71] *ante* averroy *del. A* evi

THE MEANINGS OF NATURAL DIVERSITY: MARCO POLO ON THE "DIVISION" OF THE WORLD

KATHARINE PARK

> Lords Emperors, and kings, Dukes, and Marquesses, Counts, Knights, and Burgesses, and all people who wish to know the different [*deverses*] generations of men and the diversities [*deversités*] of the different [*deverses*] regions of the world, take then this book and have it read, & here you will find all the greatest marvels and the great diversities [*diversités*] of the Great Armenie and of Persie, and of the Tartars and Indie and of the many other provinces, just as our book will tell you clearly in order, as Master Marc Pol, wise and noble citizen of Venese, relates because he saw them with his own eyes.[1]

With this ringing sentence begins the protean text known variously as the *Livre des merveilles dou monde*, *De mirabilibus mundi*, *Milione*, or, perhaps most authentically (according to what appears to be the oldest.known version), the *Divisament dou monde*. Ostensibly written at Marco Polo's direction by one Rusticiaus (or Rustichello) of Pisa, with whom he was temporarily imprisoned in 1298, this work, as its proemium makes explicit, took as an overarching theme the diversity of the terrestrial world: its peoples, its flora, its fauna, and its mineral resources. Even the book's title emphasizes the earth's heterogeneity: in the thirteenth and early fourteenth centuries, *divisament* (more commonly, *devisament*) did not mean simply "division" or "narration," but also connoted distinction and difference—as principles of order, sources of beauty, and objects of desire.[2]

As these layered meanings indicate, diversity was a complicated, subtle, and highly charged idea in high and late medieval culture, with strong af-

---

[1] Marco Polo, *The Description of the World*, chap. 1, ed. and trans. A. C. Moule and Paul Pelliot (London: George Routledge, 1938), p. 73. Here and throughout, I have used the Franco-Italian version, commonly known as *F* (MS Paris, Bibl. Nat.: fr. 1116), which is generally agreed to be the earliest surviving text; I have consulted this in the edition of L. F. Benedetto, as revised by Ronchi: Marco Polo, *Milione, Le divisament dou monde: Il milione nelle redazioni toscana e franco-italiana*, ed. Gabriella Ronchi (Milan: Arnoldo Mondadori, 1982), pp. 303–662. The English translations are those of Moule and Pelliot, minus the italicized interpolations from later versions. On the complicated manuscript tradition of this work, which was translated with considerable freedom and numerous interpolations into various languages, see Polo, *Milione*, pp. 665–74. On French as a literary language in thirteenth- and fourteenth-century Italy, see John Critchley, *Marco Polo's Book* (Aldershot: Variorum, 1992), pp. 19–22.

[2] See Frédéric Godefroy, *Dictionnaire de l'ancienne langue française et de toutes ses dialectes du IXe au XVe siècle*, 10 vols. (Paris, 1883–1902; reprint, New York: Kraus Reprint, 1961); *Grande Larousse de la langue française*, 6 vols. (Paris: Librairie Larousse, 1971–78); and Paul Robert, *Le Grand Robert de la langue française*, 2nd ed., ed. Alain Rey, 9 vols. (Paris: Le Robert, 1985), s.v. *devise, deviser, devisement*.

fective and aesthetic as well as intellectual components. This paper explores the meanings of natural diversity—what it consisted of, whence it arose, and why it was important—in one text. But that text was itself an emblem of diversity in its medieval sense: the ostensible product of a collaboration between a Pisan writer of romances and the scion of a merchant family, who had lived and traveled in Asia for twenty-six years, it provided a place in which the courtly culture that was reflected (however indirectly) in romance came together with the mercantile culture of Venetian traders and the clerical and encyclopedic culture that provided the work's overarching structure and form. The idea of natural diversity had different meanings and associations in each of these contexts, and the interaction of these various meanings and associations is one of the principal sources of the book's coherence and power.

One of the most resonant expressions of the idea of natural diversity appears in the passage in which "Polo" (for so, for convenience, I will call the collaborative literary persona of Marco and Rustichello) described the Indian kingdom of Coilum. After enumerating Coilum's exotic vegetable wealth—brazil wood, pepper, indigo—Polo passed to its climate and its characteristic animals:

> There are many strange [*deverses*] beasts different [*devisees*] from all the others in the world. For I tell you that there are black lions without any other colour or mark. There are parrots of several kinds, for there are some all white as snow and they have the feet and the beak red, and again there are some red and white, which are the most beautiful thing in the world to see. There are some again very small which are likewise very beautiful. There are also peacocks much more beautiful and larger and of another sort [*devisee*] than are ours. And what shall I tell you about it? They have all things different [*devisee*] from ours, and they are more beautiful and better. For they have no fruits like ours, nor any beasts nor any birds; and this comes to pass through the great heat which is there.[3]

This passage incorporates a number of the elements that gave Polo's idea of diversity its complexity and force. In the first place, it portrays natural diversity as a highly positive and attractive quality, associating it specifically with beauty. This attractiveness functions on two levels. On the one hand, the invocation of difference, here as in the book's proemium, obviously aimed to seduce a reader hungry for novelty and exoticism. Indeed, it formed part of the conventional language of romance that was Rustichello's usual stock in trade. Not only Rustichello's own works, but those of many of his contemporaries used this language with numbing regularity, as in the introductory lines to the *Cléomadès* of Rustichello's contemporary, Adenet le Roi: "I have begun yet another book, one that is very wonderful [*mervoilleux*] and varied [*divers*].... The tale is of great worth and most

[3] Polo, *Description*, chap. 180, p. 415.

pleasing to hear. It is so varied and wonderful that I believe that no one ever heard one so unusual [*diverse*] as this."[4]

Like Polo's prologue, which described the *Divisament* as full of the "greatest marvels [*mervoilles*]," Adenet's also associated diversity with wonder. Indeed, these twinned concepts formed the core of a semantic cluster that lay at the rhetorical heart of twelfth- and thirteenth-century romance, where it received concrete form in rhapsodic descriptions of Eastern luxuries, thrilling quest and adventure, dazzling exotic settings, and objects with astounding powers.[5] The logic behind this association was made explicit in contemporary topographical writing: because wonder was the usual human response to the novel and unfamiliar, it was necessarily a function of common experience. As James of Vitry put it in his *Historia orientalis*,

> We know that all the works of God are wonderful, although those who are accustomed to look on them often are not moved by wonder. For perhaps the cyclops, who all have one eye, wonder as much at those who have two eyes as we wonder at them, or at others with three eyes. Just as we consider the pygmies dwarfs, so they would judge us giants, if they saw one of us among them.... We consider the black Ethiopians ugly, but among them, the blackest is judged the most beautiful. We do not wonder at many things in our lands that the peoples of the East, if they heard of them, either would not believe or would consider to be marvels."[6]

(The last included the absence of snakes in Ireland, the properties of quicksilver, and the midnight sun of Thule.) In other words, wonder was the natural emotional response to the natural diversity of the world.

In the context of Western experience, it was, of course, the East that was the emblem of difference and the home of wonders, and writers of romance filled their works with stories of the amazing things to be found there. The East of romance stood for novelty, opulence, luxury, and excitement—everything lacking in one's pedestrian home. "Crocodiles, parrots, ruby mines, artificial hills of lapis lazuli, golden birds that sing, trained leopards, postal systems, polygamy," as Mary Campbell has put it; "what makes it so poignantly unlikely, and wondrous, is that while it is not characteristic of

---

[4] Adenet le Roi, *Li roumans de Cléomadès*, ed. André van Hasselt, 2 vols. (Brussels: Victor Devaux, 1865), vol. 1, p. 1.

[5] See, in general, Katharine Park and Lorraine Daston, *Wonders and the Order of Nature, 1150–1750* (New York: Zone, forthcoming), chaps. 1–2; Edmond Faral, "Le merveilleux et ses sources dans les descriptions des romans français du XIIe siècle," in his *Recherches sur les sources latines des contes et romans courtois du Moyen Age* (Paris: Honoré Champier, 1913), pp. 307–88; Douglas Kelly, *The Art of Medieval French Romance* (Madison: University of Wisconsin Press, 1992), chap. 5.

[6] James of Vitry, *Historia orientalis*, chap. 92, in *Libri duo, quorum prior Orientalis, sive Hierosolymitanae, alter Occidentalis historiae* (Douai, 1597; rept. Westmead, England: Gregg International, 1971), pp. 215–16, commenting on the injunction to tolerance in Romans 14.

Marco's Europe, it is characteristic of European dreams."[7] But not of dreams alone. By the late thirteenth century, merchants like Marco Polo and his brothers were beginning to travel regularly between Europe and Asia, and the exotic gems, fragrant plantstuffs, and animal luxuries of romance were beginning to appear in the markets of Genoa and Venice.[8] As Polo emphasized, Coilum's diversity was available for consumption as well as wondering contemplation: "And again I make you know that the merchants come to this kingdom with their ships from Mangi and from Arabie and from the Levant and make there very great merchandise which they bring in from their country and which they carry away afterwards with the ships, of the merchandise of this kingdom."[9]

As this passage makes clear, diversity did not mark only the relationship between Europe and Asia, between "us" and "them": the provinces of the East were equally unlike one another, alimenting a thriving inter-Asian trade. The East of the *Divisament* is for the most part a patchwork of small provinces, each with its own wildly different animals, vegetables, and minerals, and with a topography that varies dramatically over even a few leagues. Taican, for example, has a great grainmarket and stands north of mountains made entirely of salt, "the best in the world." After three days travel to the east, through beautiful and richly cultivated country, full of fruits, grain, and vines, the traveler finds himself in a land of drunken hunters, who wear only skins and turbans and who apparently practice no agriculture at all. Three days beyond that, he reaches Scassem, notable for its cave-dwelling peasants and, especially, for its dangerous porcupines, which can project their quills like darts. Immediately afterward, he encounters completely deserted territory, with grass for horses, but no human food or drink.[10] Indeed, Asia exhibits considerably more variety than Europe. Whereas Italy has only one kind of crane, according to Polo, the province of Ciagannor has five: one large and black; one white, black, and red, with golden eyes on its feathers; one much smaller, in the same colors, but without the eyes; one grey, black, and red; and one "like ours."[11] In Polo's *Divisament*, the variety of the East stands as an emblem for the endless—and endlessly beautiful and desirable—diversity of the world.

[7] Mary C. Campbell, *The Witness and the Other World: Exotic European Travel Writing, 400–1600* (Ithaca, N.Y.: Cornell University Press, 1988), p. 111.

[8] On the explosion of Eastern trade in this period, see Critchley, *Marco Polo's Book* (n. 1), pp. 52–57; Luciano Petech, "Les marchands italiens dans l'empire mongol," *Journal asiatique* 250 (1962): 549–74; and, in general, Janet L. Abu-Lughod, *Before European Hegemony: The World System A.D. 1250–1350* (New York: Oxford University Press, 1989), chaps. 4–5.

[9] Polo, *Description*, chap. 180, p. 415.

[10] Ibid., chap. 46, pp. 135–36.

[11] Ibid., chap. 74, p. 185.

Polo's vision of the world as essentially and naturally diverse contrasts strongly with the two principal narratives produced by earlier European travelers to the Mongol court, those of the friars John of Pian di Carpini and William of Rubruck, who were sent in 1245–46 and 1253, respectively, as missionaries and ambassadors by the pope and the king of France. John's *Historia Mongolorum* and William's *Itinerarium* describe an earth structured not by a principle of variety, but by one of alterity or opposition—in every way contrary to Polo's view. William expressed this sense of opposition clearly when he described his party's first encounter with the Mongols: "When we arrived among these barbarians, it seemed to me, as I have already said, as if I were stepping into another world [*alter mundus*]."[12] William did not see the earth as characterized by constant and, for the most part, pleasing variation; for him, the difference between Europe and Asia, East and West, was binary, marked not only by their crucial difference in religion—he did not represent even the Muslims, the traditional bogeymen for Latin Christian writers, as a third entity, but incorporated them for the most part into the forces of Christian monotheism[13]—but by their physical features as well. John of Pian di Carpini expressed an even more extreme version of this view, emphasizing the utter inferiority of the Mongols' territory as well as of their religion. He described the people there as "quite different from all other men," being both "fewer in number and weaker in body than the Christian peoples,"[14] and their land and climate as "more wretched than I can possibly say."[15] Thus William and John divided the world into two opposing blocs, conceived in strictly hierarchical terms: healthy vs. degenerate, civilized vs. barbarian, temperate vs. frigid, pleasant vs. wretched, good vs. evil, true vs. false.

How can we explain the difference between this mid-thirteenth-century portrayal of the world and the celebration of natural diversity in Polo's book? In part, it represents a change of context: where William and John were missionaries and ambassadors to a militarily and (in their eyes) religiously hostile power, so that their world was organized around polarity and

[12] William of Rubruck, *Journey*, translated by an anonymous "nun of Stanbrook Abbey" in *The Mongol Mission: Narratives and Letters of the Franciscan Missionaries in Mongolia and China in the Thirteenth and Fourteenth Centuries*, ed. Christopher Dawson (London: Sheed and Ward, 1955), chap. 9, p. 103; see also chap. 1, p. 93 (the Latin text is edited in A. van den Wyngaert, ed., *Sinica franciscana*, vol. 1: *Itinera et relationes fratrum minorum saeculorum XIII et XIV* [Florence: Collegium S. Bonaventurae, 1929]).

[13] See, e.g., ibid. chap. 26, p. 141.

[14] John of Pian di Carpini, *History of the Mongols*, chaps. 2 and 8, translated in Dawson, *Mongol Mission* (n. 12), pp. 6 and 44 (a critical edition of the Latin text appears in John of Pian di Carpini, *Storia dei Mongoli*, ed. Enrico Menestò, trans. Maria Christiana Lungarotti [Spoleto: Centro Italiano di Studi sull'Alto Medioevo, 1989], pp. 227–333). For a similar view, see Simon de Saint-Quentin, *Histoire des Tartares*, ed. Jean Richard (Paris: Librairie Orientaliste Paul Geuthner, 1965), 3.71, pp. 31-32.

[15] Ibid., chap. 1, p. 6.

opposition, Polo was a merchant, for whom the earth offered a panoply of desirable natural commodities, the ground and the stuff of trade. In part, however, the contrast between them also testifies to an intellectual shift: the gradual diffusion, in both Latin and the vernacular, of a view of the world rooted in the texts of Greek and Roman antiquity, which emphasized its beauty, variety, and diversity. This view appears clearly in the clerical encyclopedic literature of the mid- and later thirteenth century: the *De proprietatibus rerum* of Bartholomaeus Anglicus, for example; the *De natura rerum* of Thomas of Cantimpré; and, especially, the *Speculum naturale* of Vincent of Beauvais.[16] Inspired as much by Pliny as by Augustine, these works presented the natural world as a spectacle of wonders, which reached their peak in the marvelous East. Pliny himself had written enthusiastically about these wonders. In one passage, for example, he described in admiring terms the exotic peoples of farthest India, like the Psilli, whose bodies secreted a poison deadly to snakes, or the Sciopodes, each with a single enormous foot. "These and similar varieties of the human race have been made by the ingenuity of Nature as toys for herself and marvels for us," he wrote. "And indeed who could possibly recount the various things she does every day and almost every hour?"[17]

Vincent of Beauvais applied this Plinian vision to the wonders of divine creation. Explaining why he had chosen the mirror as his governing metaphor, he noted: "mirror refers to almost everything worthy of contemplation [*speculationis*], that is of wonder or imitation."[18] Glorifying God's providence and omnipotence, in two chapters entitled characteristically "The Wonderful and Multiple Species of Things" and "The Various and Wonderful Quality of Things," Vincent described the extraordinary range of the physical world using predominantly Eastern examples: the hugeness of the elephant, whale, griffin, and dragon; the uniqueness of the fire-dwelling salamander and of the crocodile, unable, alone among all animals, to move its lower jaw; the extraordinary variety and beauty of the colors of flowers and precious gems.[19]

---

[16] On this literature see, in general, Maria Teresa Beonio-Brocchieri Fumagalli, *Le enciclopedie dell'Occidente medioevale* (Turin: Loescher, 1981), esp. pp. 35–39; Pierre Michaud-Quantin, "Les petites encyclopédies du XIIIe siècle," in *Encyclopédies et civilisations = Cahiers d'histoire mondiale* 9/3 (1966): 483–515; and W. J. Aerts, E. R. Smits, and J. B. Voorbij, eds., *Vincent of Beauvais and Alexander the Great: Studies on the Speculum Maius and Its Translation into Medieval Vernaculars* (Groningen: Egbert Forsten, 1986). Also see Park and Daston, *Wonders* (n. 5), chap. 1.

[17] Pliny, *Natural History*, 10 vols., Loeb Classical Library (Cambridge, Mass: Harvard University Press, 1935–63), 7.2.32, vol. 2 (trans. H. Rackham), p. 527. On the medieval *fortuna* of Pliny, see Arno Borst, *Das Buch der Naturgeschichte: Plinius und seine Leser im Zeitalter des Pergaments* (Heidelberg: Winter, 1994), chaps. 6–7.

[18] Vincent of Beauvais, *Speculum naturale*, 3 vols. (n.p., n.d.; copy on deposit in Harvard University's Houghton Library), prologue, vol. 1, fol. [1r].

[19] Ibid., 30.10–11, vol. 3, fol. [120v].

This view of the world as structured above all by diversity appears even more clearly in some of the smaller vernacular encyclopedias of the thirteenth century, such as Gossouin's hugely influential *Image du monde,* compiled in the 1260s. Like Vincent of Beauvais, Gossouin emphasized the variety of the physical world as a product of the omnipotence and charity of its creator.[20] The description of the regions of the earth that made up the second part of his book also hinged, like Polo's, on the ideas of division and diversity; Gossouin called this part "How the Earth Is Divided [*devisée*]," and his titles for its constituent chapters reiterate this overarching theme: "On the Diversities [*diversitez*] of India" (chap. 2); "On the Different [*diverses*] Islands of the Sea" (chap. 5); "Of the Diversities That Are in Europe and Africa" (chap. 6); "On the Diversities of Certain Common Things" (chap. 7). But Gossouin went beyond Vincent, relating this variety not only to divine omnipotence, but also to the realm of secondary causes: "All the diversities of peoples," he wrote, "which differ in their appearance [*faiture*] and their character [*courage*], and all [difference] that happens by nature in herbs, in plants, in animals, happens through the celestial power [*vertu*] that God gave to the stars, when he first formed the world.... So God, in whom all powers abound, wished to form the world."[21] Here Gossouin referred elliptically to ideas elaborated by contemporary scholastic philosophers, most notably Albertus Magnus and Roger Bacon, who related natural diversity to the difference between physical places, mediated by the influence of the stars.

Albertus formulated this theory most systematically in *De natura loci,* composed shortly after 1248. In one passage, he summarized the causal order described in this work:

> If anyone wishes to know particularly all the natures and properties of particular places in the water and air and earth, he will know that there is no point in [these elements] that does not have its special property from the power [*virtute*] of the stars shining down through the agency of the dwelling place [*habitationem*] of mixed beings. For to every dwelling place of animals and plants corresponds a different horizon circle, and to every horizon circle corresponds a different relationship between the heavens and the place. For this reason, the nature and properties and behaviors and species of even those things that appear to be generated in the same sensible place differ, so that different properties and behaviors are conferred even on animals and humans generated from twin seeds.[22]

[20] Gossouin, *L'image du monde de Maître Gossouin: Rédaction en prose*, ed. O. H. Prior (Lausanne and Paris: Payot, 1913), 1.1–2, pp. 61–62.

[21] Ibid., 3.6, pp. 173–74.

[22] Albertus Magnus, *De natura loci*, ed. Paul Hossfeld, in Albertus Magnus, *Opera omnia*, gen. ed. Bernhard Geyer, 37 vols. to date (Cologne: Aschendorff, 1951–), 1. 5, vol. 5, pt. 2, p. 8. I have not adopted the English translation by Sister Jean Paul Tilmann, which is based on the older Borgnet edition and is frequently in error: Jean Paul Tilmann, *An Appraisal of*

Here, Albertus signaled the extraordinarily fine-grained character of natural variation, which meant that even two plants of the same species grown from two adjacent seeds would not be identical. In general, however, he described earthly places as differentiated by their latitudes and longitudes, which thus served as a convenient reference system for cataloguing and organizing the variety of environments that accounted for the variety of terrestrial things. Of these two coordinates, by far the more important was latitude, since the most obvious effect of the influence of the planets (which included the sun) was to divide the earth into parallel bands, or "climates," distinguished by their relative heat, according to their distance from the equator and the tropics (and, hence, according to the elevation of the pole star over the horizon and the length of their days and nights).

Following Ptolemy, Albertus identified the inhabited world as what we would call the northern half of the Eastern Hemisphere, which included the three known continents of Africa, Asia, and Europe, and he divided it into seven climates, ranging from the sweltering tropics to the frigid north.[23] Each had its characteristic flora, fauna, minerals, and peoples. For example, he described the seventh climate as characterized by cold, damp, and fog, producing large, fleshy animals and people, mostly white in color. According to Albertus, the latter are blessed with excellent digestions, since the external cold concentrates their heat within their bodies; mentally, however, they are stupid and dull-witted, which accounts for the meager contributions of Dacians and Slavs to liberal studies and the arts.[24] In this last attribute, the inhabitants of the seventh climate differ markedly from those of the first three climates (corresponding to India and Ethiopia), who are dark-skinned, verging on black, with small, light, dry bodies, like the peppercorns that also thrive in those lands; although timid, they possess extraordinarily subtle and acute minds.[25]

An important corollary of the dependence of natural things on the place that produced them is their tendency to degenerate when removed from their proper "habitation." As Albertus put it, "the place and the thing in it are connatural [*connaturalia*], and therefore ... the thing contained is cor-

the Geographical Works of Albertus Magnus and His Contributions to Geographical Thought, Michigan Geographical Publication no. 4 (Ann Arbor: University of Michigan Department of Geography, 1971), pp. 25–145. See also Roger Bacon, *The Opus maius of Roger Bacon*, pt. 4, trans. Robert Belle Burke, 2 vols. (Philadelphia: University of Pennsylvania Press, 1928), vol. 1, pp. 159, 320–29.

[23] For more information on the nature and history of these divisions, see Ernst Honigmann, *Die sieben Klimata and die Poleis episemoi: Eine Untersuchung zur Geschichte der Geographie und Astrologie in Altertum und Mittelalter* (Heidelberg: Carl Winter, 1929); and Marian J. Tooley, "Bodin and the Medieval Theory of Climate," *Speculum* 28 (1953): esp. pp. 68–77.

[24] Albertus, *De natura loci* (n. 22), 1.8, p. 15; 2.3, p. 27.

[25] Ibid., 2.3, p. 26.

rupted by its container, when this is contrary to it."[26] The effects of removing things too far from their native places are therefore well known: animals lose their ability to reproduce, "so that elephants rarely generate outside the first and second climates and never beyond the fourth"; humans get sick; and plants, although perfectly healthy in one climate, when transplanted to another one wither and die.[27] Furthermore, if living beings manage to survive such a transmigration and reproduce, their stock inevitably alters in succeeding generations to conform to their new place; for example, if black people from the first and second climates move to the fourth or fifth climate, their descendants are "little by little altered toward whiteness."[28] Dramatic shifts of location can even produce transmutations of species, at least in certain plants; thus "wheat planted in earth above fifty degrees of latitude rarely flourishes, and if it comes up, within two or three generations it degenerates into a lesser grain, which is called rye and which flourishes very well in the seventh climate."[29] Fixated on the spiritual order, William of Rubruck and John of Pian di Carpini might dream of converting the world to Christianity, but no such homogeneity was even theoretically possible in Albertus's natural world.

Although Albertus's model would suggest at least some uniformity within climates, he emphasized that the world's diversity was in fact much greater, as I have already noted, depending on local topography—proximity to the sea, situation relative to mountains—and indeed on infinitely small variations in place. Roger Bacon described a similar causal order in his *Opus maius* (1267), where he took this vision to its ultimate conclusion: "since the objects of this world located in different places, however near they may be, receive the apexes of different pyramids [of rays] coming from the whole heavens opposite to them, an infinite diversity happens on this account."[30]

Neither Marco Polo nor Rustichello of Pisa was in any position to read Albertus Magnus or Roger Bacon, but they clearly shared these philosophers' view of the naturalness of diversity, most likely mediated by a vernacular encyclopedia like Gossouin's *Image du monde*. Indeed, the remarkable similarities to the language and content of Gossouin's account of the "division" of the world and the "diversities" of Asia strongly suggest that Rustichello was familiar with this particular source. Polo's *Divisament* not only underscored the variety of the natural world, but it attributed this va-

---

[26] Ibid., 1.2, p. 3 (drawing on the definition of place in Aristotle, *Physics* 4.1).
[27] Ibid., p. 4.
[28] Ibid., 2.3, p. 27.
[29] Ibid., 1.2, p. 24.
[30] Bacon, *Opus maius* (n. 22), 4.4.5, vol. 1, p. 159. Unlike Albertus Magnus, Bacon emphasized the dependence of philosophers on travelers—he specifically mentioned William of Rubruck—for information about this contingent order; see ibid., pp. 311–12, 324.

riety to the kinds of climatic and topographical factors described by Albertus (although Polo did not explore their celestial causes). If the unfamiliarity of Coilum's plants and animals came from its "extreme heat," as Polo indicated in the passage quoted toward the beginning of this paper, it was the cold of Badascian that accounted for its mineral wealth (rubies, lapis lazuli, silver), its magnificent horses, its falcons, and the fact that oil was extracted there not from olives but from sesame and nuts.[31] Scier, in contrast, was exceedingly dry, which meant that it produced sheep with horns instead of ears, who subsisted on a diet of fish.[32] Reobar was hot again, which gave it plants and animals entirely different from those in (relatively cool) Europe:

> The fruits of it are dates and apples of Paradise and pistacios and other fruits which are not in our cold place. And in this plain is a kind of bird which one calls francolin, which are different from the other francolins of the other lands, for they are black and white mixed together and they have the feet and the beak red. The animals are also different. And I will tell you of the oxen first. The oxen are large and are all white as snow. The hair they have is short and smooth, and this comes about from the hot place.[33]

Like Albertus, too, Polo knew that many things could not thrive and generate when removed from their "habitation." Thus brazil plants were easily sown and transplanted within the borders of the kingdom of Lambri, but Polo himself had tried unsuccessfully to introduce them into his native land: "Moreover I tell you quite truly that we brought away some of that seed to Venese and sowed it on the ground. Yet I tell you that it did not grow there at all; and this came about through the cold."[34]

As I have already noted, this natural diversity had special meaning for Polo, as a merchant. Unlike William of Rubruck and John of Pian di Carpini, who deemphasized variety in the service of a binary view that corresponded to military and religious opposition, Polo saw diversity as the motor of commerce and trade. If all regions of the globe produced the same animals, vegetables, and minerals, there would be no reason for any kind of exchange—indeed, for any travel at all. Although rich in pearls, for example, the kingdom of Maabar could not breed horses, so that "all the treasure of the revenue which they have each year, or the greater part, is consumed in buying horses" from Curmos, Aden, and other places where horses thrive.[35] In contrast, the dry land of Scier produces large amounts of white incense, dates, and horses, which they export to India, but "they have no

[31] Polo, *Description*, chap. 47, pp. 137–39. For an interesting passage (interpolated in a later version) that ascribes the character of peoples to the effects of their soil, see ibid. chap. 35, pp. 119–20.

[32] Ibid., chap. 194, p. 443.

[33] Ibid., chap. 35, p. 120.

[34] Ibid., chap. 169, p. 376.

[35] Ibid., chap. 174, p. 386.

corn but rice alone, and they have little of that; but I tell you that one carries corn there from abroad and they make great profit from it."[36] Polo's description of the island of Mogadexo makes a similar point:

> Moreover I tell you quite truly that more elephants are bred in this island than in other provinces.... Again you may know that there are trees of red sandal in this island as large as are the trees of our country, and these trees would be sufficiently valuable in another land. And they have woods of them as we have of other wild trees. They have ambergris enough because there are baleens in great abundance in that sea, and there are also capdoilles in plenty.... They have much merchandise and many ships come there with many wares, and these are cloth of gold and of silk, of several kinds and of many other things which we will not tell you here, and they sell all these and change them for the wares of the island.... For I tell you that the merchants make great profit there and great gain.[37]

Natural variety generates trade, in other words, and trade generates profit; thus Asia's extraordinary wealth, compared to the relative poverty of Europe, can be traced directly to its diversity. The association between commerce, money, and diversity was conventional, for it also appears in Gossouin's *Image du monde*; after explaining the role of the heavens in creating natural diversity, Gossouin related this to the invention of currency, which he described as "established for the peoples who didn't have everything necessary in one place. Some had grain, others wine, others cloth, and others animals. Who had wheat couldn't have wine without changing one into the other." Thus money was invented by philosophers "to save the state of the world."[38]

Gossouin's remark points to another context in which the idea of natural diversity acquired special meaning: the dramatic rise of trade and the transition to a money economy, which began to transform European society as early as the eleventh century, and which acquired increasing momentum in the twelfth and thirteenth. This dramatic shift required a revolution in values concerning commerce, wealth, and material acquisition. Christian tradition was notoriously harsh on these issues: "Corrigant se christiani; non negotientur," as Augustine had put it in his *Enarrationes in psalmos*.[39] Thus the growth of commerce evoked concern in ecclesiastical circles and anxiety among its practitioners; this period saw the promotion of avarice over

---

[36] Ibid., chap. 194, p. 442.

[37] Ibid., chap. 191, p. 429.

[38] Gossouin, *L'image du monde* (n. 20) 3.12, p. 187.

[39] Augustine, *Enarrationes in psalmos*, 70.1.17: ed. Eligius Dekkers and Johannes Fraipont, Corpus Christianorum, series latina, vols. 38–40, (Turnholt: Brepols, 1956), 39: 954. See, in general, Maurice Bloch and Jonathan Parry, "Introduction: Money and the Morality of Exchange," in *Money and the Morality of Exchange*, ed. Jonathan Parry and Maurice Bloch (Cambridge: Cambridge University Press, 1989), pp. 1–32.

pride as the most heinous sin.[40] This general Christian suspicion of commerce was reinforced by newly translated Aristotelian texts on the topic, for Aristotle had privileged the subsistence economy of the "household" and castigated all other forms of acquisition as contrary to nature. As he argued in the *Politics,*

> It is the business of nature to furnish subsistence for each being brought into the world; and this is shown by the fact that the offspring of animals always gets nourishment from the residuum of the matter that gives it its birth. The natural form, therefore, of the art of acquisition is always, and in all cases, acquisition from fruits and animals. That art, as we have said, has two forms: one which is connected with retail trade, and another which is connected with the management of the household. Of these two forms, the latter is necessary and laudable; the former is a method of exchange which is justly censured, because the gain in which it results is not naturally made [from plants and animals], but is made at the expense of other men.[41]

As Aristotle had already explained, it was long-distance trade, an "unnatural" form of acquisition, that had given rise to the invention of money and to the equally unnatural desire to accumulate, not the limited wealth of things necessary to life, but the unlimited wealth of currency.[42] Thirteenth-century Latin philosophers and theologians accepted and recapitulated this interpretation of the exchange of commodities for money in the service of profit as unnecessary and unnatural.[43]

The general castigation of trade by both Christian and Aristotelian authority posed an obvious problem for merchants, who were traditionally classed with soldiers, innkeepers, prostitutes, and others whose occupations necessarily entailed sin. Beginning in the twelfth and, especially, the thirteenth centuries, however, writers who represented contemporary merchants and their interests began to explore ways of legitimizing commerce and trade.[44] On the one hand, they justified the merchant's profit by stressing the personal danger and financial risks he incurred in the practice of his occupation. On the other hand, they emphasized his social utility and his contribution to the common good, using the loophole already provided by Aristotle when he exempted from blame the form of exchange in which

---

[40] Lester K. Little, *Religious Poverty and the Profit Economy in Medieval Europe* (Ithaca: Cornell University Press, 1978), pp. 35–40; Alexander Murray, *Reason and Society in the Middle Ages* (Oxford: Clarendon Press, 1978), pp. 78–80.

[41] Aristotle, *Politics*, trans. Ernest Barker (Oxford: Clarendon Press, 1946), 1.10 (1258a35–b2), p. 28.

[42] Ibid., 1.9 (1256b40–1258a19), pp. 22–27. See, in general, Scott Meikle, "Aristotle on Money," *Phronesis* 39 (1994): 26–44.

[43] See, e.g., Thomas Aquinas, *Summa theologiae* 2/2.77.4.

[44] See Jacques Le Goff, "Licit and Illicit Trades in the Medieval West," in his *Time, Work, and Culture*, trans. Arthur Goldhammer (Chicago: University of Chicago Press, 1980), esp. pp. 60–66; Little, *Religious Poverty* (n. 40), chap. 11; John W. Baldwin, *Masters, Princes, and Merchants: The Social Views of Peter the Chanter and His Circle*, 2 vols. (Princeton: Princeton University Press, 1970), vol. 1, pp. 261–69.

men bartered things they had in abundance for things they lacked—for example, wine for grain: "When used in this way," he wrote, "the art of exchange is not contrary to nature, nor in any way a form of the art of acquisition."[45]

Thirteenth-century writers expanded this loophole into a breach large enough to accommodate not only barter, but also trade for profit. I have already cited Gossouin's account of the invention of money for the benefit of "peoples who didn't have everything necessary in one place." But Gossouin was merely repeating what had become a commonplace in mid-thirteenth-century ethics and theology. Thomas of Chobham had sketched this position in his *Summa Confessorum* (comp. ca. 1215), arguing for the necessity of trade: "Otherwise, there would be great poverty in many regions, since merchants bring what abounds in one place to another where these same things are lacking," he wrote; "they can therefore justly receive the price of their labor."[46] Albertus Magnus repeated this idea in one of the principal arguments of his question concerning the licitness of trade: "Commerce is useful to the whole earth, taking abundance from one land and bringing back what is lacking."[47] He concluded that trade was legitimate if no specific sins were thereby committed—in other words, if the trader was not a Christian cleric, if he did not work on holidays, if he avoided deception and fraud, and if he maintained a just price. Such arguments gave great weight to natural diversity: the fact that divine decree and the forces of natural causation had ensured that plant, animal, and mineral species were unevenly distributed over the face of the earth. Diversity became, in effect, the reality that rendered commerce both necessary and natural. Far from violating the natural order, as Aristotle would have it, trade became a way of enhancing and perfecting that order.

Placed in this economic and ethical context, Polo's *Divisament* acquires special meaning. A hymn to the world's diversity, it is also a sermon naturalizing and justifying trade. From this point of view, the merchants of Mogadexo, Maabar, and the other great centers of Eastern trade described by Polo were performing vital services for their peoples and for the world. Polo himself emphasized that trade of this sort was both necessary and difficult. He noted that the merchants who plied the seas between Zaiton or Quinsai and the islands of Cin reaped "great profit and great gain," but he emphasized that they earned their money:

> I tell you that [these islands] are so distant that with great difficulty is it possible to go there.... Moreover I tell you that they work hard for a year to go.

---

[45] Aristotle, *Politics* (n. 41), 1.9 (1257a28–30), pp. 23–24.

[46] Thomas of Chobham, *Summa confessorum*, ed. F. Broomfield (Louvain: Nauwelaerts, 1968), 6.4.10a, p. 301.

[47] Albertus Magnus, *Commentarii in IV Sententiarum* 16.46, in his *Opera omnia*, ed. S. C. Auguste Borgnet, vol. 29 (Paris: Ludovicus Vives, 1894), p. 636.

> For they go in the winter, and come back in the summer. For the winds do not blow there but of two kinds, the one which takes them and the other which brings them back, and it blows the one in summer and the other in winter. And you may know that this country is distant from Indie a very great deal of way.[48]

Furthermore, he indicated that commerce could be conducted without sin, when he described the Braamans of Lar, one of the most admirable peoples he encountered, as "some of the best merchants in the world and of the most trustworthy, for they would not tell any lie for anything in the world."[49]

It is a long way from the wine and grain in Aristotle to the pearls and precious stones that Polo described as the Braamans' stock-in-trade. Nonetheless, the connection—at least in Polo's mind—was clear. In the *Divisament*, the theme of natural diversity—a staple of the literature of topography and romance, a commonplace of Augustinian encyclopedias of the wonders of creation, and a basic principle of thirteenth-century philosophical writing on the causal order—became a trope in the justification of commerce and trade. The work lay at the intersection of these diverse traditions and the different social and intellectual contexts that they represented, and it drew on and elaborated each in turn. But it was the last that lay closest to Polo's own interests, and his own vision found its best expression in the "Dit des marchants," composed a decade or so after the *Divisament* by Gilles le Muisit, abbot of St. Martin of Tournai:

> For itself no country can provide;
> For that merchants travel far and wide;
> Their work and toil feeds the nation
> So refrain from baseless fulmination.
> Merchants cross the seas and back
> To bring each nation what it lacks.
> No good merchant would reap blame
> But love, honor, and a just good name.
> They contribute to the love of nations;
> Thus their wealth is cause for jubilations.[50]

---

[48] Polo, *Description*, chap. 161, p. 365.
[49] Ibid., chap. 177, p. 401.
[50] Cited in Le Goff, "Licit and Illicit Trades" (n. 44), p. 66.

THOMAS BRADWARDINE'S *DE CONTINUO* AND THE STRUCTURE OF FOURTEENTH-CENTURY LEARNING

EDITH D. SYLLA

The late medieval Christian Aristotelian worldview, when pressed to its logical extremes, was incompatible with Euclidean geometry. In the standard medieval Aristotelian classification of the sciences, natural philosophy and geometry were separate disciplines, each with its own principles and conclusions. Geometry assumed as a principle, without proof, that continua must be infinitely divisible into ever smaller continua, while natural philosophy proved it.[1] Geometry included in its subject matter indivisibles such as points, lines, and planes—meanwhile asserting that higher-order continua are not composed of such indivisibles. Natural philosophy asserted that quantities—including indivisibles such as points, lines, and surfaces—really exist in substances as their accidental forms, although they are considered mathematically in abstraction from such substances.[2]

But when geometric indivisibles such as points and lines, or purported physical indivisibles such as instants or exact degrees of quality, were supposed to exist in reality (*esse in re*), paradoxes ensued. In view of these paradoxes, some idiosyncratic scholars opted to compose continua of indivisibles at the price of inconsistency with Euclidean geometry and/or Aristotelian physics. Mainstream scholars, including William of Ockham, John Buridan, and Thomas Bradwardine, often concluded that geometric indivisibles do not in fact exist, thus resolving the apparent paradoxes by denying the entities that led to them.[3] In either case, whether scholars adopted

---

[1] For Thomas Bradwardine's quotation of Averroes's statement of this view, see John Murdoch, "Thomas Bradwardine: Mathematics and Continuity in the Fourteenth Century," in *Mathematics and Its Applications to Science and Natural Philosophy in the Middle Ages*, ed. Edward Grant and John E. Murdoch (Cambridge, London, New York: Cambridge University Press, 1987), p. 136 n. 62. I discuss this passage below.

[2] For a discussion of John Dumbleton, whose ideas about the relation of mathematics to physics mirror part of the standard view, see Edith Sylla, "The Oxford Calculators and Mathematical Physics: John Dumbleton's *Summa Logicae et Philosophiae Naturalis*, Parts II and III," in *Physics, Cosmology and Astronomy, 1300–1700: Tension and Accommodation*, ed. Sabetai Unguru, Boston Studies in the Philosophy of Science, vol. 126 (Dordrecht, Boston, London: Kluwer, 1991), pp. 129–61.

[3] For reference to a number of sources on this point, see John Murdoch, "Philosophy and the Enterprise of Science in the Later Middle Ages," in *The Interaction between Science and Philosophy*, ed. Yehuda Elkana (Atlantic Highlands, N.J.: Humanities Press, 1974), p. 61. It is notable that Bradwardine and the nominalists, although often portrayed as belonging to opposing metaphysical camps, agreed on denying the existence of indivisibles. It should be noted that philosophers who denied the existence of geometric indivisibles nevertheless held the prime movers of the celestial spheres (or angels more generally) and human souls to be indivisible substances.

some sort of atomism or denied the existence of geometric indivisibles altogether, the previously standard Aristotelian understanding of the relation between mathematics and physics broke down: geometry could no longer be understood to deal with quantities existing in, but considered in abstraction from, physical bodies.

In this paper I want to use Thomas Bradwardine's *De continuo* as a source of evidence about the structure of fourteenth-century learning, particularly as it involved the relations between natural philosophy, mathematics, logic, and theology. Thanks to the wide-ranging research of John Murdoch, we have extensive knowledge about Bradwardine's text and about the earlier texts of such "atomist" authors as Henry of Harclay and Walter Chatton to which Bradwardine was reacting.[4] In addition, in a number of papers Murdoch has presented a broad picture of what he calls "the unitary character of late medieval learning," arguing for the identity of fourteenth-century science and philosophy and for a close connection of natural philosophy and theology through a group of shared tools, tactics, or conceptual schemes ranging from supposition theory in logic, through the mathematics of ratios, to the habit of arguing *de potentia Dei absoluta* or *secundum imaginationem*, and so forth.[5] What I want to do here is to put these two parts of Murdoch's scholarly contribution more explicitly together—that is,

[4] As evidence for my paper I have used both John Murdoch's published work and his unpublished draft editions of a wide range of texts, especially Thomas Bradwardine's *De continuo* and Henry of Harclay's question *Utrum mundus poterit durare in eternum a parte post.* In the 1960s Murdoch prepared critical editions and translations of Harclay, Chatton, Odo, Bonetus, and John the Canon, as well as of Bradwardine's *De continuo,* which were to form the basis of a several-volume study on the problem of the continuum and related issues in the Middle Ages; see John Murdoch and Edward Synan, "Two Questions on the Continuum: Walter Chatton (?), O.F.M. and Adam Wodeham, O.F.M.," *Franciscan Studies* 26 (1966): 215 n. 10 (hereafter Murdoch and Synan, "Two Questions"). At that time he gave me copies of his editions and translations. It is perhaps appropriate to my topic, if awkward in other ways, that my study is based in part on thirty-year-old manuscript copies of these editions and translations, most of which have yet to appear in print. I have not checked the editions, but I have sometimes made my own translations, because it was simpler to do so and because Murdoch's translations had not been published or were not available for citation. All credit for the texts and the substance of the translations should go to him, while any blame for erroneous deviations from the true sense should fall on me. The text of Thomas Bradwardine's *De continuo* in John Murdoch, "Geometry and the Continuum in the Fourteenth Century: A Philosophical Analysis of Thomas Bradwardine's *Tractatus de continuo*" (Ph.D. diss., University of Wisconsin, 1957), available in microfilm, is similar but not identical to the edition I am using.

[5] John Murdoch, "*Mathesis in philosophiam scholasticam introducta*: The Rise and Development of the Application of Mathematics in Fourteenth-Century Philosophy and Theology," in *Arts liberaux et Philosophie au Moyen Age* (Actes du Quatrième Congres International de Philosophie Médiévale) (Montreal and Paris, 1969), pp. 215–54; idem, "Philosophy and the Enterprise of Science," (n. 3), pp. 51–113; idem, "From Social into Intellectual Factors: An Aspect of the Unitary Character of Late Medieval Learning," in *The Cultural Context of Medieval Learning,* ed. J. E. Murdoch and E. D. Sylla (Dordrecht: Reidel, 1975), pp. 271–339.

to show how Bradwardine's *De continuo* exemplifies the unity of late medieval learning.

In a number of papers Edward Grant has examined late medieval Aristotelianism and has argued that it lasted as long as it did because, in a sense, it was too flexible—it bent, but it did not break as a good Kuhnian paradigm ought to do when confronted over decades and even centuries with intractable anomalies.[6] John Murdoch, for his part, has suggested that because late medieval science *was* philosophy, like philosophy it permitted the coexistence of competing views—e.g., Scotism and Ockhamism—rather than rejecting one view in favor of the monopoly of another.[7]

In the self-conscious understanding of scholastic Aristotelians, each scientific discipline has its own principles, and it is a mistake to transfer arguments from one discipline to another without special justification (unless, for instance, one science is subalternate to another).[8] But if the disciplines investigated by the arts faculties of medieval universities were divided into the seven liberal arts, natural philosophy, moral philosophy, and metaphysics, individual masters of arts were not so divided by discipline. And if members of the arts faculty at the University of Paris were not supposed to debate purely theological matters, there was less to prevent theologians from making use of arts subjects in theology.[9] Thus an important difference between modern science on Thomas Kuhn's model and medieval scholasticism is that in the Middle Ages there was nothing that matched the modern link between a scientific paradigm and a distinct community of individuals whose job it is to do science according to that paradigm—unless, that is, one supposes that Christian Aristotelianism in general was the paradigm and that university faculties as a whole, or the faculty of theology in particular, formed the relevant group of practitioners. Were "theologian–natural philosophers," to use Edward Grant's term, the maintainers of a Kuhnian paradigm before the Scientific Revolution?[10]

---

[6] See, e.g., Edward Grant, "Aristotelianism and the Longevity of the Medieval World View," *History of Science* 16 (1978): 265–302.

[7] Murdoch, "Philosophy and the Enterprise of Science" (n. 3), p. 90.

[8] See Steven Livesey, "Metabasis: The Interrelationship of the Sciences in Antiquity and the Middle Ages" (Ph.D. diss., University of California, Los Angeles, 1982).

[9] At Paris, there were occcasional complaints that theologians were making too much use of arts subjects, but restrictions were never made at Oxford. See Edward Grant, *The Foundations of Modern Science in the Middle Ages: Their Religious, Institutional, and Intellectual Contexts* (Cambridge: Cambridge University Press, 1996), p. 152.

[10] See ibid., p. 174. Amos Funkenstein writes of a secular theology, which might be supposed to have developed out of the work of these medieval theologian–natural philosophers, in *Theology and the Scientific Imagination from the Middle Ages to the Seventeenth Century* (Princeton: Princeton University Press, 1986), p. 3. If Funkenstein's secular theologians were more secular than theologians, the theologian–natural philosophers of the thirteen and fourteenth centuries were more theologians than secular.

In previous papers, I have examined the *Physics* commentaries of John Buridan, Nicole Oresme, Albert of Saxony, Marsilius of Inghen, and others to investigate how the writing of commentaries on the works of Aristotle served as a form of scientific practice in fourteenth-century Paris.[11] A clear difference between writing question commentaries on Aristotle and the writing of modern scientific papers is that medieval authors rarely claimed originality themselves and almost never gave credit to their contemporaries for innovative work. Even when the work of contemporaries was used, the authors were almost never cited by name. In this way the progress or development that occurred in fourteenth-century *Physics* commentaries was masked rather than emphasized—even when it in fact occurred, as, for instance, when Thomas Bradwardine's function relating forces, resistances, and velocities was rapidly and thoroughly assimilated.

In turning to Bradwardine's *De continuo*, then, I extend my investigation into fourteenth-century natural philosophy as a form of scientific practice by looking at a different genre of writing. In *De continuo*, a different dynamic occurred from that of an Aristotelian commentary, because the work was free-standing and written in axiomatic form. The axiomatic format allowed Bradwardine greater independence than a commentary format would have done. It likewise allowed a greater eclecticism, permitting him to draw upon many disciplines in a single work. If the medieval custom of avoiding what modern scientific writing accomplishes with footnotes tended to conceal Bradwardine's sources in *De continuo* as it obscured sources in Aristotelian commentaries, it also concealed the problem context in which various views had first arisen. The fourteenth-century problem of the relation of indivisibles and continua arose, I contend, not because of questions of mathematics or even of natural philosophy, but rather because of problems of theology. This is hidden by the form *De continuo* took, but we can discover it by looking at that work's sources.

In *The Guide of the Perplexed*, Maimonides opposed his own order of argument in proving the existence and unity of God to one that he traced back to such Greek Christians as John Philoponus. Philoponus, he said, and the Muslims and Jews who followed the same line of argument, did not conform their opinions to reality or to accepted philosophical opinion. Rather, they adopted premises with the goal of reaching conclusions that agreed with, or at least did not contradict, their religious beliefs. In doing so, they adopted philosophical or scientific opinions, such as belief in the existence of atoms and the void, that had already been philosophically re-

[11] Edith Sylla, "Aristotelian Commentaries and Scientific Change: The Parisian Nominalists on the Cause of the Natural Motion of Inanimate Bodies," *Vivarium* 31 (1993): 37–83; eadem, "The Transmission of the New Physics of the Fourteenth Century from England to the Continent," in *La nouvelle physique du XIVe siècle*, ed. Stefano Caroti and Pierre Souffrin (Florence: Olschki, 1997), pp. 65–110.

futed, and they made it appear necessary that all believers adopt such unconventional natural philosophical views.[12] Because philosophy had first arisen among the Greeks and Syrians, Maimonides argued, and the opinions of the philosophers were widely accepted in those communities—opinions that the Christians saw as opposed to Christianity—religious scholars thought it necessary to develop argumentation to defend their religious views.[13] Once they had found premises leading to conclusions in agreement with their religious views, they assumed that their premises were true. Later thinkers, not knowing the motivations of their predecessors, mistakenly believed that their inquiries had been disinterested.[14]

Interestingly, what Maimonides says about his own time and intellectual context applies also in its way to Bradwardine's *De continuo*: the most important of the "atomist" views that Bradwardine attempted to combat arose in the context of theological arguments, but the character of *De continuo* masks this origin. By choosing to write an independent treatise on the continuum, Bradwardine disconnected his work not only from any individual authoritative text, such as those of Aristotle or Euclid, but also from any one discipline, and in particular from theology.[15] Although the contemporary atomists whose work Bradwardine was primarily combatting wrote on the continuum in the context of commentaries on Peter Lombard's *Sentences*—and therefore, presumptively, in the context of theology—someone reading *De continuo* could easily fail to realize this.[16]

---

[12] Moses Maimonides, *Guide of the Perplexed*, trans. Shlomo Pines (Chicago: University of Chicago Press, 1963), bk. I, ch. 71, pp. 177–78.

[13] Ibid., p. 177.

[14] Ibid., pp. 178–79: "Men of later periods who study these books know nothing about all this and consequently find in these ancient books a vigorous argumentation and a powerful endeavor to establish the truth of a certain thing or to refute a certain thing and think thereupon that its establishment or its refutation is in no way required for these foundations of the Law that are required." In his introduction, Shlomo Pines (pp. lxxxiv–lxxxv) suggests that Maimonides's picture of the motivations of the Mutakallimun and the relation of their work to that of John Philoponus may have derived from a lost work of al-Farabi, *On Changeable Beings*, on the eternity of the world.

[15] John Murdoch, "Thomas Bradwardine: Mathematics and Continuity" (n. 1), p. 109: "there is still a decided disadvantage arising from his use of an axiomatic presentation. It is simply that such a format makes it quite difficult to see exactly how Bradwardine fit into the controversy over the continuum that was veritably raging in the earlier years of the fourteenth century." It might be noted that Murdoch has found only one minor citation of *De continuo* in the succeeding period (ibid., p. 104).

[16] It seems hardly likely to be accidental that the parallels between medieval Latin attempts to disprove the eternity of the world and Islamic attempts to do the same go far beyond their mutual use of John Philoponus's arguments involving paradoxes of infinites. In Maimonides's listing of the common premises of the Mutakallimun, the first premise states that "the world as a whole—I mean every body in it—is composed of very small particles that, because of their subtlety, are not subject to division. The individual particle does not possess quantity in any respect. However, when several are aggregated, their aggregate possesses quantity and has thus become a body.... these particles are alike in such a way that they are adjacent to one another" (Maimonides, *Guide of the Perplexed*, 1: 73, p. 195). "By

John Murdoch has characterized the major points of Bradwardine's *De continuo* as mathematical.[17] I argue below that the work was more fundamentally physical than geometrical. But because of its combatting of arguments coming from commentaries on the *Sentences*, *De continuo* was deeply, if invisibly, theological as well. It thus serves as a prime example of the weaker unity of late medieval philosophy, science, and theology that Murdoch has described so well—granted that the dynamic of the use of God's absolute power in late medieval theology meant that what God can do is often the same as what is logically consistent. In reference to problems of infinity and continuity, in particular, Murdoch has clearly explained the situation:

> The most influential theological element was the newly precise concept of God's omnipotence. Writers began to base their analyses of the infinite and the continuous upon what could obtain in respect of the absolute power of God. But since God's absolute power extended to everything that did not include a contradiction, to invoke this power in examining infinity or continuity was to transfer one's analysis from the realm of the physically possible (within the confines of Aristotelian natural philosophy) to the broader realm of the logically possible.
>
> Although this appeal to divine omnipotence arose within a theological context, it appears as well in works that are clearly devoted to natural philosophy. Yet whether such an appeal occurred within the one context or the other, it had the effect of extending the discussion at hand so that in principle it would take into account all logical possibilities. Invoking the absolute power of God thus supplied one with a warrant to reason *secundum imaginationem*, a manner of reasoning which, when applied to the infinite and the continuous (or to any other area of late medieval philosophy), permitted the

virtue of the first premise all geometrical demonstrations become invalid" (p. 198). "The tenth premise. It consists in the affirmation of admissibility that they mention. This is the main proposition of the science of kalam. Listen to its meaning. They are of the opinion that everything that may be imagined is an admissible notion for the intellect.... They also say with regard to all things that are existent and perceptible that supposing anything among them should be bigger than it is or smaller or different from what it is in shape or place ... all such differences would be admissible from the point of view of the intellect.... And they say this without paying attention to the correspondence or lack of correspondence of that which exists to their assumptions" (p. 206). This way of thinking could have helped furnish the background to the fourteenth-century habit of invoking God's absolute power and of thinking *secundum imaginationem*. One would not be claiming that the world has no stable natural order, but only saying that, in the terms of a commonly repeated phrase, one should not deny to God anything that does not involve a logical contradiction or that one does not see to involve a logical contradiction. Cf. Gabriel Biel, *Epitome et collectorium ex Occamo circa quatuor sententiarum Libros* (Tübingen, 1501; reprint Frankfurt/Main: Minerva, 1965), bk II, dist. I, q. 3: "Ratio pro ea principalis hec est. Nihil non includens contradictionem negandum est a potentia divina." See, however, William J. Courtenay, "The Critique on Natural Causality in the Mutakallimun and Nominalism," *Harvard Theological Review* 66 (1973): 77–94, and the sources cited there.

[17] Murdoch, "Thomas Bradwardine: Mathematics and Continuity" (n. 1), p. 103; but see also pp. 110–14.

> exercise of a number of new methods and tools of analysis in resolving (in some cases first creating and then resolving) problems.[18]

The recognition that the late medieval problem of the continuum has its roots most fundamentally in theology *cum* logic, while these roots are sometimes masked in the way that Maimonides describes, may help explain a very peculiar property of late medieval discussions of infinity and continuity—namely, that they repeat arguments whose origins extend back via Arabic authors to John Philoponus and other Greek authors, in seeming dissociation from these origins. For infinity, among the most frequently repeated arguments are paradoxes of the infinite that are said to arise if past time is infinite. John Murdoch, Richard Dales, and other recent historians typically cite Bonaventure's commentary on the *Sentences* as the proximate source from which later authors adopted these arguments, but the work of Richard Sorabji and Herbert Davidson has shown without doubt that these are arguments from Philoponus as repeated by a multitude of Islamic authors and by Maimonides as well.[19] Even if it has yet to be shown exactly how a Latin author who could not read either Greek or Arabic could have known of these earlier works, there is no doubt that these arguments were transmitted to Bonaventure from earlier sources and that *their repetition and transmission occurred precisely because of their use in discussing the possible eternity of the world.* According to both Maimonides and Averroes, al-Farabi dealt with problems of infinity and eternity in a work, now

[18] John Murdoch, "Infinity and Continuity," in *The Cambridge History of Later Medieval Philosophy*, ed. Norman Kretzmann, Anthony Kenny, and Jan Pinborg (Cambridge: Cambridge University Press, 1982), pp. 566–67.

[19] Ibid., p. 569, nn. 13, 14, lists Plutarch, Philoponus, Alexander of Aphrodisias, Proclus, and Lucretius among ancient sources of paradoxes of unequal infinites, saying that "only Lucretius and Proclus's *Elements of Theology* were available to the Latin West, but neither was cited relative to the paradox in question." Among medieval Jewish and Islamic sources, Murdoch lists al-Sharastani, Averroes, and Maimonides, saying: "Maimonides was translated into Latin, but was not cited in this context. The translation of Averroes's *Tahafut* occurred too late (1328) to enter effectively into the debate in the Latin West." For the role of Bonaventure, see ibid., pp. 569–70. See also Edward Grant, *Planets, Stars, and Orbs: The Medieval Cosmos, 1200–1687* (Cambridge: Cambridge University Press, 1994), p. 67 n. 16; Richard Sorabji, "Infinity and Creation," in *Philoponus and the Rejection of Aristotelian Science*, ed. Richard Sorabji (Ithaca: Cornell University Press, 1987), p. 167; Herbert Davidson, *Proofs for Eternity, Creation and the Existence of God in Medieval Islamic and Jewish Philosophy* (New York and Oxford: Oxford University Press, 1987), chap. 4, "John Philoponus's Proofs of Creation and Their Entry into Medieval Arabic Philosophy." Al-Farabi and Avicenna argued that the past eternity of the world is not disproved by the fact that if past time were infinite there would now be an actual infinite number of souls in heaven, because these souls would not be ordered. Al-Ghazzali argued against al-Farabi and Avicenna that the infinite souls would be ordered by their temporal origins (or by cause and effect), and hence that such an infinite and, along with it, a past infinite time, was impossible (Davidson, pp. 129–30).

lost, titled *On Changeable Beings*.[20] Other Islamic authors who may have transmitted the arguments are Avicenna and al-Ghazzali.

An analogous situation existed with regard to a set of standard arguments about indivisibles and geometry. Murdoch has argued that these arguments, going back at least to Avicenna and available in the Latin translation of al-Ghazzali's *Metaphysics*, were injected into later medieval discussions through their use in Duns Scotus's commentary on the *Sentences* with regard to the motion of angels.[21] It is most probable that these arguments for and against the composition of continua by indivisibles were invented and transmitted, not because of an independent interest in so-to-speak non-Archimedean geometries, but, as Maimonides suggested, because of their relevance to theological issues.[22] *The Latin authors who repeated these arguments were not, by and large, importing mathematics into theology, but perpetuating an ongoing intra-theological quasi-mathematical tradition.*

In the next section, I examine the structure of late medieval learning as exemplified in *De continuo* itself. I then turn to the sources of the views that Bradwardine attempts to combat and to what these sources reveal about the structure of scholastic practice. If we may suppose that Bradwardine was stimulated to write *De continuo* because a number of his contemporaries such as Henry of Harclay and Walter Chatton had espoused nonstandard views, why did Harclay and Chatton adopt such "atomistic" views? And, finally, what does the nexus between Harclay, Chatton, and Bradwardine tell us about the structure of fourteenth-century learning? I conclude with a brief discussion of Bradwardine's treatment of the eternity of the world in his *De causa Dei*, suggesting how this corresponds to the structure of fourteenth-century learning previously described.

THOMAS BRADWARDINE'S *DE CONTINUO*

The most obvious feature of the structure of *De continuo* is that it is written in an axiomatic format, with definitions, suppositions, and conclusions ostensibly based upon them.[23] Since Euclid's *Elements of Geometry* was by far the most famous work written in this format, the natural supposition is

[20] See Herbert Davidson, *Proofs for Eternity*, p. 128.

[21] Murdoch, "Infinity and Continuity" (n. 18), pp. 576, 579.

[22] In Islam there was a well-established connection between atomism and certain theological orientations such as that of the Asharites, and it has previously been suggested that Nicholas of Autrecourt's ideas about atomism may have come from such a source.

[23] My discussion of *De continuo* is based on John Murdoch's typescript edition of the 1960s, which is a revised version of the text appearing in his dissertation ("Geometry and the Continuum" [n. 4]). The definitions, suppositions, and conclusions are published in Murdoch, "Thomas Bradwardine: Mathematics and Continuity" (n. 1), pp. 119–30. For a discussion of the axiomatic format, see ibid., pp. 108–10.

that Bradwardine intended his work as mathematics. This supposition is called in question, however, by the fact that he similarly began his *De proportionibus velocitatum in motibus* in axiomatic format—and yet the later chapters of this earlier work are clearly physics rather than mathematics. Similarly, Bradwardine's *De causa Dei* is written deductively, but it is obviously theology. Even if the first sections of *De continuo* are axiomatic, moreover, the next several sections of the work are more dialectical than axiomatic: a typical section begins with a counterfactual hypothesis and then goes on to a series of conclusions beginning "If this is so ..."—i.e., if the counter-factual hypothesis is true, then such-and-such conclusion follows, where the conclusion is false or even impossible, and the intent is a reduction to absurdity of the hypothesis.

Thus *De continuo* begins positively, with definitions (Part I), suppositions (Part II), and thirty-four preliminary conclusions (Part III), but this positive beginning is followed by Parts IV–VI in which false hypotheses about the relations of indivisibles to continua are reduced to absurdity. *De continuo* then closes with positive parts asserting that all hypotheses assuming that a continuum is composed of indivisibles are false (Part VII) and going on to conclude that indivisibles do not in fact exist, let alone compose continua (Part VIII).[24] We can best understand what is fundamental to *De continuo* by looking at the positive parts, together with methodological remarks that Bradwardine makes along the way. In Part I, Bradwardine's definitions begin with the terms "continuum," "permanent continuum," "successive continuum," "body," "surface," "line," "point," "instant," "motion," "to-have-been-moved" (*motum esse*), "matter of motion," "degree of motion," and so forth. Thus already in his third definition, of a successive continuum (e.g., motion or time), he deals with a term that belongs to natural philosophy as well as, or instead of, mathematics. By implication, the term "continuum" itself is one that belongs both to mathematics and to natural philosophy, not to mention other disciplines. Likewise in the suppositions of Part II, by the third supposition if not before, Bradwardine has gone beyond mathematics into other disciplines. The suppositions, which he claims are self-evident, begin:

[24] Part IV, "On immediately joined indivisibles," ends with conclusion 56, "In no continuum are atoms immediately joined." Part V, on finite atoms, begins: "Assertio ponens continuum ex indivisibilibus finitis componi est omnibus scientiis inimica, omnes inpugnans, et ideo concorditer ab omnibus inpugnatur, cum qua mathematica primo congrediatur et vincat." Part VI, "De indivisibilibus infinitis," begins: "Rursum aliud principium sumentes dicamus: Si omne continuum ex indivisibilibus infinitis componitur," and works by reduction to absurdity. Part VII, bringing together the disproofs of the previous parts, ends with conclusion 141, "Nullum continuum ex athomis integrari. Unde sequitur et elicitur: omne continuum ex infinitis continuis similis speciei cum illo componi." Part VIII, "Utrum indivisibilia sint distincta a continuo" (conclusions 142–51), concludes that indivisibles are not distinct from continua.

1. Everything that is larger can be divided into what is equal and the difference by which it exceeds.
2. If a finite is added to a finite, the whole will be finite.

Like these first two, the final six suppositions are substantive, dealing with the subjects of mathematics or physics. But the third and fourth suppositions are more nearly methodological and extend the inquiry from subject matter to subject matter:

3. Where there is no cause of diversity or dissimilarity, the judgment is assumed to be similar.
4. Every science is assumed to be true where [i.e., insofar as] a continuum is not supposed to be composed of indivisibles.

The third supposition is crucial to Bradwardine's project, because he uses it to transfer the conclusions he reaches in natural philosophy from nature to mathematics. Thus his thirtieth conclusion states:

30. If one continuum were to have immediate atoms, whether finite or infinite, any continuum would have such.

The only justification of this conclusion is the statement, "This is clear enough by the third supposition." Conclusion 120, then, is that if every continuum is composed of infinite indivisibles, the atoms of any continuum will be immediately joined. The proof begins:

> Let two liquids come together to form a continuum. Then two indivisibles of prime matter, which were previously the termini of those liquid bodies, will not be corrupted nor will others be generated between them. Therefore, they remain immediate and also the points of quantity situated in them. Therefore by [conclusion] 30 this is so of every continuum.

In his commentary on Aristotle's *Physics*, Averroes had argued that physical continua and mathematical continua have different properties.[25] If one could suppose that at an interface between water and air there is a surface of water immediate to a surface of air, the same does not hold for the end points of two line segments brought into continuity with one another. If, that is, two points as indivisibles can only touch whole to whole—there being no parts of a point that could touch without the whole point touching—then the two end points of line segments brought into continuity must coincide and, in fact, be only one point. Thus, for mathematical entities such lines, in Averroes's opinion, contiguity and continuity are identical, while in physics they need not be.

In *De continuo*, by contrast, Bradwardine exhibits no reservations about analogies between mathematics and physics, even though he recognizes

---

[25] See Edith Sylla, "Infinite Indivisibles and Continuity in Fourteenth-Century Theories of Alteration," in *Infinity and Continuity in Ancient and Medieval Thought*, ed. Norman Kretzmann (Ithaca: Cornell University Press, 1982), pp. 233–34.

that there are potential problems concerning the relations between disciplines. In explanation of Supposition 4, he states:

> It says this because sometimes the statements of other sciences are used as if they are manifest, because it would take far too long to spell everything out. Where, however, other sciences treat of the composition of continua from indivisibles, they are not supposed to be true, in order to avoid a *petitio principii.*

Then at the end of Part VII, Bradwardine says:

> But a falsehood could occur concerning the preceding: Averroes in his comment on the *Physics* says that the natural scientist demonstrates the continuum to be divisible *in infinitum*, but the geometer does not prove this but supposes it as demonstrated in natural science. Therefore the geometrical demonstrations previously made can be challenged, saying that geometry everywhere supposes that a continuum is not composed from indivisibles and that it cannot be demonstrated.
>
> But this is not valid, because it makes a false supposition. Within geometrical demonstrations it is not posited that a continuum is not composed of indivisibles, nor is it dialectically required everywhere, since it is not required in Euclid, Book V.[26] And similarly, neither does a geometer suppose in some demonstration that a continuum is not composed of infinite mediate indivisibles, because, given the opposite, any demonstration proceeds nonetheless, as is inductively clear to anyone who knows how to demonstrate geometrical conclusions. It is true that Euclid in his geometry does suppose that a continuum is not composed from finite and immediate atoms, although he does not explicitly state this among his suppositions. If the opponent were to say the opposite and posit that a line was composed of two points, Euclid could not demonstrate his first conclusion, because an equilateral triangle could not be located on this line, because it would have no angle.... Similarly if the opponent said that a continuum was composed of immediate atoms, Euclid could not prove his fourth and eighth conclusions, for both are proved by superposition. Similarly in the proof of conclusion 23 of Book III.
>
> These conclusions are not proved by any previous conclusions, but are shown from immediate principles, and the rest of the conclusions are demonstrated through these. Almost the whole of Euclid's geometry depends on these three conclusions, and on this [Euclid's geometry] all other geometry is founded. It follows that geometry assumes that a continuum is not composed of finite and immediate atoms. That which is introduced, by omission, axiomatically, should not be attributed to an insufficiency of the arguer, for it is sufficient for geometry to demonstrate ostensively and by reduction to impossibility from its proper principles that none of its continua is composed of finite immediate indivisibles: ostensively, as is clear in conclusion 66 of the preceding and by conclusion 58 and the first part of its corollary; it can be proved *per impossibile* as is clear by conclusion 71 and many following.[27]

---

[26] The text here is problematic. I assume Bradwardine means that Book V of the *Elements* does not require the assumption that a continuum is not composed of infinite mediate indivisibles. See Murdoch, "Thomas Bradwardine: Mathematics and Continuity," (n. 1), pp. 117–18 and nn. 62, 64.

[27] Latin text in ibid., nn. 62, 65, 67.

Thus Bradwardine says that Euclid, and geometry in general, do not everywhere assume that continua are not composed of infinitely many mediate indivisibles, only that they are not composed of finite immediate indivisibles. Geometry can prove ostensively and by reduction to absurdity from its proper principles that its continua are not composed of finite immediate indivisibles.[28]

If Bradwardine believes that Euclidean geometry is consistent with the supposition that continua are composed of infinitely many mediate indivisibles, how can he claim to prove—such is the apparent upshot of Part VI—that the supposition that a continuum is composed of infinite mediate indivisibles implies that it is composed of finite immediate indivisibles? What is the status of the demonstrations of Part VI?

Bradwardine begins Part VI by arguing against the simple hypothesis that a continuum is composed of infinite indivisibles, no matter whether those indivisibles are supposed to be immediate to each other, as Henry of Harclay claimed, or mediate, as Robert Grosseteste stated. In agreement with Grosseteste and Harclay, Bradwardine argues in conclusion 115 that if every continuum is composed of infinite indivisibles, then for continua of the same type there will be a proportionality between the magnitudes of the continua, on the one hand, and the numbers of indivisibles composing them, on the other. As I describe below, this was the purpose of Robert Grosseteste's introduction of the hypothesis of indivisibles into his notes on the *Physics* in the first place—i.e., to use them as an absolute measure, for God, of the magnitudes containing them. Thus in conclusion 115 Bradwardine states something with which he knows Grosseteste and Harclay were in agreement. He himself rejects the consequent, however (i.e., he rejects the view that there will be a proportionality between the infinite indivisibles on the one hand and the magnitudes of the continua on the other), and so he rejects the antecedent, that a continuum is composed of infinite indivisibles. He says that conclusion 115 is clear from conclusion 57 and its

[28] Checking Bradwardine's references, we find that conclusion 66 ("Every straight line has infinitely many lines as parts") is shown by reference to the corollary to conclusion 20 ("Every straight line can be divided into many straight lines"), which is shown by considering the perpendicular bisector of a chord and constructing arcs of larger circles over it. Murdoch, "Thomas Bradwardine: Mathematics and Continuity" (n. 1), pp. 112–13, discusses the unexpected nature of the proof of the corollary to conclusion 20, and suggests that Bradwardine may have been interested in the indeterminacy of the circles that may be drawn and therefore of the divisions of the line in question. Conclusion 20 is the last of the preliminary conclusions that are labeled as geometrical. Conclusion 58 and its corollary, which are classified as arithmetical, state that "if so [i.e., if a continuum were composed of finitely many atoms] the atoms in a continuum are immediately joined. Whence it is manifest that all continua have infinitely many atoms and are not composed of atoms." Since the consequent of conclusion 58 is impossible (it has already been shown that immediate atoms are impossible), therefore the hypothesis of finitely many atoms is impossible (this is the corollary).

proof, although they apply only to finite magnitudes and finite numbers of indivisibles.[29] Then he goes on:

> And there is no other way of saving the ratio of continua to each other than by the ratio of the atoms first composing them. And Lincoln [i.e., Grosseteste] and Henry [of Harclay] concede this.

Thus Bradwardine's proof of conclusion 115 is dialectical—he assumes a point on which his opponents Grosseteste and Harclay agree, although he himself does not.

Conclusion 116 then begins an argument for the opposite conclusion. If continua are composed of infinite indivisibles, Bradwardine says, then condensation and rarefaction of prime matter by the loss or gain of indivisibles will be impossible, so that ultimately all continua will have equal numbers of indivisibles. When two liquids are separated or continued with each other, no indivisibles will be generated or corrupted (conclusion 118). In justification of these conclusions, he refers to conclusions 96 through 99, and on the basis of them he reaches conclusion 120, that if continuous natural substance is composed of infinitely many indivisible substances, then the atoms of any continuum will be immediately joined. As I have already indicated, Bradwardine's generalization from natural continuous substances to mathematical continua in conclusion 120 is based on conclusion 30, which in turn is based on supposition 3, a version of the argument from sufficient reason—ostensibly, there is no reason why mathematical continua should be different from physical ones; therefore, if indivisibles are immediate in physical continua, so are they in mathematical.[30]

Conclusions 96 though 99, on which Bradwardine bases his proof, are, as might be expected, all physical, stating that if substances were composed of infinitely many atomic substances, then the condensation or rarefaction of prime matter or of substance would not occur through there being fewer or more atoms than previously, and so forth. If there were to be fewer atoms, he argues, then some atom would have to be corrupted or would have to move away. Suppose a substance consisted of two atoms and that when it was condensed, these two atoms should be reduced to one. Since the two atoms are alike, why should one of them be corrupted rather than the other? If one supposes for reasons of symmetry that both are corrupted and that a new third atom is generated in their place, for which of the two original atoms should this new atom be generated? Again, by supposition 3 there is no more reason why one should be replaced than the other. In any case,

---

[29] Conclusion 57 states: "If a continuum were composed of finitely many atoms, then as the number of atoms of one continuum is to the number of atoms of the other, so is that continuum to the other." From which it would follow that if one continuum were incommensurable to another, then one number would be incommensurable to another, which is absurd.

[30] In these arguments, Bradwardine is assuming the standard Aristotelian view, i.e., that mathematical indivisibles are accidents of physical substances.

why would two atoms be corrupted if one of them was to be replaced with an identical one? Nature does not act in vain, Bradwardine says, nor does it accomplish by more what could be accomplished by less. The physics of such corruption is implausible and, moreover, prime matter is supposed to be incorruptible. Similar arguments hold for rarefaction. Thus, he concludes, indivisibles are not corrupted in condensation, nor generated in rarefaction.

It follows, says Bradwardine, that every substance contains an equal number of indivisibles, which means that every substance is of an equal quantity, thus making conclusion 115, and the whole purpose for which Grosseteste introduced these indivisibles in the first place, otiose. Harclay, Bradwardine reports, says that indivisibles are not corrupted, but that they come to occupy the same position. But this cannot be, because then the indivisibles of different bodies could occupy the same place and one body could enter another, with their dimensions overlapping (something that Aristotelians uniformly rejected). Therefore the hypothesis of infinite indivisibles leads to contradictory conclusions, so that the hypothesis itself must be false.

Where does all this leave Bradwardine's argument that geometry does not require the assumption that continua are not composed of infinite mediate indivisibles if he himself has shown, albeit using physical arguments, that the assumption of infinite mediate indivisibles implies finite immediate indivisibles? Unless a geometer rejects Bradwardine's third supposition and what follows from it, the geometer will be compelled to reject not only finite immediate indivisibles, but also infinite mediate indivisibles, since the latter—with the aid of supposition 3, according to Bradwardine—implies the former. But then it follows that Bradwardine in *De continuo* as a whole is not arguing as a geometer, but rather as a natural philosopher. It is natural philosophy that proves that a continuum is infinitely divisible into divisibles—geometry assumes this, but does not prove it. In *De continuo*, natural philosophy ostensibly proves that the hypothesis of infinitely many mediate indivisibles implies finitely many immediate indivisibles. If geometry were to accept this proof as valid, it would have to reject infinitely many mediate indivisibles as well as finitely many immediate ones.

Having proved using physics that continua are not composed of indivisibles, Bradwardine concludes *De continuo* by arguing that indivisibles in fact do not exist. Something can be continuous and finite without any indivisible continuing or limiting it, he argues (conclusion 145). He ends: "Surface, line, and point are altogether nonexistent, from which it is manifest that a continuum is not continued or terminated by such but by itself" (conclusion 151).

What, then, follows about the structure of fourteenth-century learning, from the explicit contents and structure of Bradwardine's *De continuo*?

Bradwardine is here revealed as a typical fourteenth-century scholastic, operating in many disciplines while attempting to keep them distinct. Bradwardine, indeed, accepted the Aristotelian distinction of the disciplines. He was keenly aware of the possibility of fallacious arguments involving *petitio principii*. He knew that in trying to prove that continua are not composed of indivisibles he should not make use of any discipline, or take over as true without proof any proposition, that assumes either that continua are composed of indivisibles (this would defeat his purpose) or that they are not (this would be a *petitio principii*). Yet he assumed the truth of certain principles of Aristotelian physics. He did not assume the truth of the assertion that continua are not composed of indivisibles—this, indeed, would be to assume what is to be proved—but he did assume the sort of principles on the basis of which Aristotle too had proved that continua are not composed of indivisibles. So he assumed, for instance, that given any velocity, a velocity half as fast is possible—by which he could design a case in which a mobile would stop in the middle of a supposed indivisible, and so forth.

Because we think of the continuum as an entity proper to mathematics, we think that the scholastics embarrass themselves when they mix physical arguments into discussions of the continuum, but for them it was a legitimate strategy if they were operating as natural philosophers rather than as geometers. Thus *De continuo* is, fundamentally, a natural philosophical work, although it has sections that belong to other disciplines and are labeled as such. There is more connection between its parts than may at first appear. At the same time, it is notable that Bradwardine has silently excluded the arguments of Grosseteste, Harclay, and Chatton that mention God. Since references to God play a central role in the ideas of Bradwardine's atomist opponents, it appears that it must have been a conscious decision on his part not to mention God or anything theological. More on this below.

Bradwardine's conclusion that surface, line, and point are altogether nonexistent obviously could not be adopted within geometry; indivisibles belong to the subject matter of geometry, even if higher-order continua are not composed of them. A geometer could reject Bradwardine's conclusion and adopt something analogous to the so-called doctrine of double truth, claiming that what is true in natural philosophy can be false in geometry and vice versa. Alternatively, geometry could be seen as a self-consistent system the premises of which are conceptual, fictional, or imaginative. In either case the upshot of Bradwardine's ostensible defense of the traditional Greek and scholastic, Aristotelian and Euclidean, doctrine of the continuum, was to disengage Euclidean geometry from physical reality. It followed that geometry is not about things existing *in re*, but rather about fictions, or things existing only in the mind, perhaps more particularly in the imagination.[31] The conclusion that most supposed geometrical entities are entirely nonexistent may be connected to the fact that, *in fourteenth-century natural philosophy, logic and not mathematics was and remained the most important conceptual tool.*[32] Perhaps it is consistent with what goes on in *De continuo* that an Aristotelian sort of approach to mathematics would lose ground in later scholastic natural philosophy, to be replaced, not by a different, more Platonic kind of mathematization, but rather by further use of the methods of logic.

BRADWARDINE'S SOURCES AND 14TH-CENTURY LEARNING

Now I want to look at the work of the "atomists" whose views Bradwardine combatted in *De continuo*, and in this way to push farther the examination of what can be learned about the structure of fourteenth-century scholasticism starting from *De continuo* as a text. In the preceding section I have suggested that *De continuo* itself shows that, in fourteenth-century scholastic Aristotelianism, geometry and natural philosophy were considered as separate disciplines but were nevertheless brought into juxtaposition, with natural philosophy playing a more fundamental role. The upshot was to undermine the standard Aristotelian understanding of the relation of mathematics to physics, and to deny that some central items in the subject matter of geometry—surfaces, lines, and points—actually exist in the real world. Now I want to argue that theology played a crucial role in generating the paradoxes that led, on the one hand, to "atomism" and, on the other, to the rejection of the reality of mathematical indivisibles.

There were two main routes, I believe, by which nonstandard views of infinity and continuity entered fourteenth-century thought, and both are theological. The route of longest standing, which involved infinity more than continuity, had to do with arguments about the eternity of the world.

[31] How this might fit with Bradwardine's high regard for mathematics is a puzzle. Perhaps his view of mathematics should be characterized as more Platonic. See Murdoch, "Thomas Bradwardine: Mathematics and the Continuum" (n. 1), p. 134 n. 31, quoting Bradwardine on mathematics, "Ipsa enim suis ceteris sororibus acutius contemplatur, inflexibilius telum iacit et se protegit clyppeo tut[iu]s. Nullus enim physico certamine se speret gravisurum triumpho nisi mathematice utatur consilio et auxilio confortetur. Ipsa est enim revelatrix omnis veritatis sincere et novit omne secretum absconditum et omnium litterarum subtilium clavem gerit. Quicunque igitur ipsa neglecta physicari presumpserit, sapientie ianuam se nunquam ingressurum agnoscat."

[32] See John Murdoch, "*Scientia mediantibus vocibus*: Metalinguistic Analysis in Late Medieval Natural Philosophy," in *Sprache und Erkenntnis im Mittelalter* (Acts of the Sixth International Congress of Medieval Philosophy, Bonn, 29 August–3 September 1977), in *Miscellanea Mediaevalia* 13 (1981): 73–106; idem, "The Analytic Character of Late Medieval Learning: Natural Philosophy without Nature," in *Approaches to Nature in the Middle Ages*, ed. Lawrence Roberts (Binghamton, N.Y.: Center for Medieval and Early Renaissance Studies, 1982), pp. 171–213.

Briefly put, there were a number of arguments that used paradoxes of the infinite to prove that an eternal world, if it implies that past time is actually infinite, is impossible. These arguments can be traced back as early as Philoponus, as indicated by Maimonides in *The Guide of the Perplexed*, and they were widely discussed by Islamic and Jewish authors. In the end, the arguments failed if and when thinkers accepted their conclusions as true, arguing, for instance, that there can be two infinites of which one is twelve times another, or of which one adds something beyond the other.

Robert Grosseteste

The second route by which nonstandard views of infinity and continuity entered into fourteenth-century discourse was apparently of more recent origin and immediately less threatening. The proximate source of these arguments was a suggestion made by Robert Grosseteste: in the margins of his copy of Aristotle's *Physics*, he wrote that if there were no external standard, God could measure magnitudes by means of the infinitely many indivisibles contained within them.[33] Although Grosseteste himself never put

[33] Richard Dales, ed., *Roberti Grosseteste Episcopi Lincolniensis Commentarius in VIII Libros Physicorum Aristotelis* (Boulder: University of Colorado Press, 1963), pp. 90–93. Grosseteste's idea had such repercussions that it merits quotation here: "Sed cum linea numeretur, mirum est quo numeretur. Ponamus enim solam unicam lineam esse, et intelligamus eam abstractam ab omni materia. Hec iam metiri potest non per aliam lineam nec aliud longum, quia non sit alia dimensio longitudinis ab ipsa, nec per seipsam metiri potest.... Unde igitur scietur et metietur mensura linee unice? Unde aut quomodo mensuravit natura in intencione sua quantam voluit, aut oportuit facere primam lineam quam fecit, aut quomodo mensuravit? Non enim dici potest quod mensuravit eam corpore naturali subiecto. Similiter intelligamus unicum motum tantum, ut motum diurnum primi celi, et circumscribatur penitus omnis alius motus. Quero quomodo scietur an iste motus sit velox vel tardus, et quomodo scietur an una revolucio sit facta in brevi tempore aut longo? Unde metietur tempus huius motus unici? Hec difficilia sunt valde ad quem vix pervenit humanus intellectus.

Credo tamen, sicut alibi diximus, quod unus numerus infinitus ad alium numerum infinitum se potest habere in omni proporcione numerali et non numerali. Aliquis enim infinitus est numerus duplus ad alium numerum infinitum et triplus et sic secundum ceteras species proporcionis, et eciam aliquis numerus infinitus se habet ad alium sicut diameter ad costam; et hoc alibi probatum est. Et iterum audaciter dico quod omnis numerus infinitus ipsi deo cuius sapiencie non est [finis non est] numerus infinitus plus quam binarius. [Q]ui est mihi infinitus est illi finitus: numerus infinitus collatus ex omnibus paribus, et similiter omnis numerus infinitus collatus ex omnibus imparibus, et similiter omnes numeri infiniti qui infinicies dividi possunt. Sicut enim que vere in se finita sunt, nobis sunt infinita, sic que vere in se infinita sunt, illi sunt finita [my proposed emendation in brackets].

Ipse autem omnia creavit numero, pondere, et mensura. Iste autem mensurator primus et certissimus, iste numeris infinitis sibi finitis mensurat alias lineas quas creavit; numero alico infinito certo sibi et finito mensurat lineam cubitalem; et numero infinito duplo lineam bicubitalem; et numero infinito subduplo lineam semicubitabilem; et unus est numerus infinitus punctorum omnium linearum unius cubiti quo numero certissime et finitissime mensurat omnes lineas unius cubiti.

Quomodo ergo mensuratur et numeratur primo linea prima mensurata? Puto quod numero infinito punctorum illius linee finite ... qui numerus punctorum non est in aliqua alia linea maiori vel minori, sed in maiori omni est maior numerus infinitus punctorum et in minori minor."

his marginalia on the *Physics* into a connected commentary, others soon did, and the result had a surprisingly widespread influence. And although Grosseteste's proposal about infinites appeared in connection with Aristotle's *Physics*, it involved God in a central role. Grosseteste's concern was to find a foundation for absolute measurement when no external standard existed, which led him to argue that God can "measure" magnitudes by the number of indivisibles they contain even though each contains infinitely many. God can see all the points in a one-foot line and in a two-foot line, Grosseteste argued, and, although we cannot understand it, God knows that there are twice as many points in a two-foot line as in a one-foot line.

Essential to Grosseteste's proposal was the assertion that God now knows or sees all the points of a line. For an Aristotelian, the points of a line, corresponding to the places at which it may be cut or divided, are potentially infinite, in the sense that no matter how many times one has cut a line, it can be cut still further without ever coming to anything indivisible. In the Aristotelian understanding, at any given time the cuts in a line are finite. One can never cut a line an actually infinite number of times. Paradoxes of actual infinites are avoided.

In Grosseteste's conception, on the other hand, the points that God sees or knows in a line are not potentially, but actually, present, so that they are not potentially, but actually, infinite in number. Since God is not in time, he does not consider a process in which the line may be cut some finite number of times and then more times and so forth; he eternally knows all at once all the points at which a line could be cut, i.e., actually infinitely many points.[34] According to Scripture, from God nothing is hid; to God everything is open.[35] According to St. Augustine, God knows infinitely many numbers.[36] Grosseteste thought along these lines most likely not because he

---

[34] Exactly what entitled medieval thinkers to conclude that God now sees all the infinitely many points of a line is not absolutely clear, but it may have involved the temporal construal of modality, as described by Jaako Hintikka. See, e.g., Jaako Hintikka, "Aristotle on the Realization of Possibilities in Time," in his *Time and Necessity: Studies in Aristotle's Theory of Modality* (Oxford: Clarendon Press, 1973), pp. 93-113.

[35] Wodeham quotes Hebrews 4:13 in support of this view: "omnia nuda et apperta sunt oculis eius secundum apostolum" (Murdoch and Synan, "Two Questions" [n. 4], p. 274).

[36] Saint Augustine, *The City of God*, Book XII, Chapter 18: "As for their other assertion, that God's knowledge cannot comprehend things infinite, it only remains for them to affirm, in order that they may sound the depths of their impiety, that God does not know all numbers. For it is very certain that they are infinite; since, no matter at what number you suppose an end to be made, this number can be, not only increased by the addition of one more, but however great it be, and however vast be the multitude of which it is the rational and scientific expression, it can still be not only doubled, but even multiplied. Moreover, each number is so defined by its own properties, that no two numbers are equal. They are therefore both unequal and different from one another; and while each number is finite, collectively they are infinite. Does God, therefore, not know numbers on account of this infinity; and does His knowledge extend only to a certain height in numbers, while of the rest He is ignorant? Who is so demented as to say so? Yet they can hardly pretend to put numbers out of the question, or maintain that they have nothing to do with the knowledge of God.... Far be it, then, from

wanted to introduce theology into physics, but because he wanted to preserve measure when only one object or one motion existed. Shlomo Pines has described a lost work of Thabit ibn Qurra that covered a range of topics closely related to Grosseteste's concerns in this instance, and it is tempting to believe that Grosseteste could have been influenced by Thabit's work, although he cites no names.[37] Once Grosseteste introduced the consideration of God's knowledge of the indivisibles within a continuum into physics, from whatever source, it had repercussions beyond anything he may have expected.

Henry of Harclay

The most important opponent of Bradwardine's *De continuo* was Henry of Harclay, in whose work these two traditions concerning unequal infinites converged.[38] In particular, in Harclay's question on the possible future

---

us to doubt that all number is known to Him 'whose understanding,' according to the Psalmist, 'is infinite.' The infinity of number, though of infinite numbers there is no number, is yet not incomprehensible to Him of whose understanding there is no number. And thus, if everything which is comprehended is defined or made finite by the comprehension of him who knows it, then all infinity is in some ineffable way made finite to God, for it is comprehensible by his knowledge" (modified [by comparison with Latin] from the translation of Marcus Dods, in *Great Books of the Western World*, vol. 18, *Augustine* [Chicago: Encyclopaedia Britannica, 1952], p. 354).

[37] Al-Ghazzali's *Incoherence of the Philosophers*, which begins with the question of the eternity of the world in the past and goes on, like Harclay, to discuss its possible duration into an infinite future, also includes the sort of consideration that motivated Grosseteste. Is there some absolute standard by which the size of things in the universe is measured? What difference would it make if the things in the cosmos were bigger or smaller than they actually are? Thus the problem context that Grosseteste was addressing in his marginalia to the *Physics* resonated with al-Ghazzali's *Incoherence of the Philosophers* in more than one regard. Rather than inventing his idea of ratios of infinites known to God, Grosseteste may have been influenced by earlier discussions such as al-Ghazzali's. For Thabit ibn Qurra, who earlier advocated ratios between infinites, see Shlomo Pines, "Thabit B. Qurra's Conception of Number and Theory of the Mathematical Infinite," in *Actes du XIe Congrès International d'Histoire des Sciences* (Warsaw, 1968), 3: 164, reprinted in Pines, *Studies in Arabic Versions of Greek Texts and in Mediaeval Science* (Jerusalem: Magnes Press; Leiden: Brill, 1986), Part 2, pp. 423–29. Thabit in the lost work that Pines studies was also interested in absolute measure. Thabit says, e.g. (*Actes*, pp. 161–62; reprint, pp. 424–25): "Similarly [the thing which is] the double or the half [of the other would be in fact] the double or the half [*correxi ex* fact] of the other even if there were no one at all who would know this." As Pines comments, Thabit is here taking an anti-Aristotelian position on the relation of numbers to things. With regard to infinites, Thabit held, for instance, that all the even numbers equal all the odd numbers and that there are twice as many numbers altogether as odd numbers. Theologically Thabit held that God knows particular things, and that he knows an actual infinity of things (p. 165). Thus, although Thabit's work is now lost, his position is so similar to that of Grosseteste that it is plausible to suppose that there was some kind of link.

[38] For Harclay see John Murdoch, "Henry of Harclay and the Infinite" (hereafter Murdoch, "Henry of Harclay"), in *Studi sul XIV secolo in memoria di Anneliese Maier*, ed. A. Maieru and A. Paravicini-Bagliani (Rome: Edizioni di Storia e Letteratura, 1982), pp. 219-61, which contains detailed outlines of both of Harclay's questions on the (past and future) eternity of the world. See also Richard Dales, "Henricus de Harclay *Questio 'Utrum mundus*

eternity of the world, the long-standing tradition of arguments about infinites in connection with an eternal world came together with Grosseteste's proposal concerning infinite indivisibles within magnitudes. John Murdoch first identified Harclay's question whether the world could endure infinitely into the future by recognizing that it held arguments criticized by Harclay's opponents, especially William of Alnwick.[39] Harclay seems to have been the first to make his contemporaries take indivisibilism seriously, arguing that continua are composed of infinitely many immediate indivisibles.

But first let me discuss Harclay's question about past eternity. In this question, on the possible eternity of the world in the past, Harclay argued, like Thomas Aquinas and others, that the world could have existed from eternity in the past. To the standard paradoxes of the infinite frequently opposed to the possible eternity of the world, Harclay responded, moreover, by asserting—with Aquinas, he said—that there can be proportionalities between infinites on the one hand and finites on the other: just as one finite can be double another, so one infinite can be double another.[40] As in a single year there is one rotation of the sun through the stars and twelve rotations of the moon, so in infinite past time there could be infinitely many rotations of the sun and moon and yet twelve times as many rotations of the moon as of the sun.

If, as Harclay claimed, there was no reason why infinites could not have ratios to each other, so too, he claimed, it was possible (though harder to construe) that one infinite could differ from another by a finite amount and hence not equal the other, though both were infinite. Consider all the numbers higher than 100 and all the numbers higher than 2, he said. If one insists that the two infinite sets of numbers are equal, then subtract the two infinite sets from all the numbers. By Euclid's axiom that equals subtracted

potuit fuisse ab eterno,'" Archives d'histoire doctrinale et littéraire du Moyen Age 50 (1983): 223–55 (hereafter Dales, "Henricus de Harclay").

[39] See Murdoch, "Henry of Harclay," p. 220.

[40] Dales, "Henricus de Harclay," pp. 247–48: "Praeterea, iste doctor frater Thomas, sic respondens, concedit alibi quod finiti ad infinitum potest esse proporcionalitas, etsi non proporcio. Est autem proporcionalitas secundum eundem Euclidem, 5 libro, similitudo proporcionum. Unde in libro suo quem vocat *De veritate* in questionibus de sciencia dei, questio prima: 'Utrum Deus cognoscit alia a se,' respondendo ad 4m argumentum, concedit quod proporcionalia sunt finita et infinita. Igitur qualem proporcionem habet unum finitum ad alium finitum, talem habere potest unum infinitum ad aliud infinitum. Sicut igitur 8 sunt in dupla proporcione ad 4, sic infinita octonarii sunt in dupla proporcione ad infinitos quaternarios. Igitur unum infinitum est maius alio. Preterea, ipse concedit quod unum infinitum est equale alteri. Possum autem de infinito aliquid subtrahere certum est, vel intelligere aliquid esse subtractum. Aliter consisteret in indivisibili. Fiat igitur illa subtraccio non in uno termino nec in alio, quia terminum non habet, sed ubicumque volueris, certum est quod residuum post subtraccionem adhuc est infinitum, nam infinitum non fit finitum per subtraccionem finiti. Quero igitur tunc utrum sit equale alteri infinito cui prius fuit equale, aut minus illo. Si minus habetur propositum. Si equale, hoc non est intelligibile. Probacio: quia tunc pars addita non faceret maius, nec subtracta minus." Of course, Harclay could also derive these notions from Robert Grosseteste.

from equals give equals as a result, it will follow that the numbers up to 100 equal the numbers up to 2, which is obviously false. Therefore the infinitely many numbers over 100 must be fewer than the infinitely many numbers over 2.[41]

Not only can one infinite past time differ from another at their finite terminations in the present, Harclay said, they can even differ in the direction in which they are infinite—that is, toward the remoter past. Consider, he says, all the past generations of fathers and sons. If each son generated is preceded by his father and the sun, and if this is true for an actual infinity of fathers and sons, then even though the total time in which humans have existed is infinite, the sun's existence will precede that infinite time. This inference and conclusion perplexed Harclay. Some of his opponents considered it to be fallacious or sophistical. Nevertheless, as far as he could understand, it was a valid argument. If a universal affirmative proposition is true in every instance, he said, and if there is no reason why it cannot be true collectively as well as divisively, and if, moreover, the cases are ordered—if, for example, when the sun precedes the father, it must also precede the son—then, Harclay says, the inference from finite to infinite should be valid.[42]

---

[41] Ibid., pp. 245–46: "et sic respondetur ad 7 quod illa inconveniencia procedunt ex falsa ymaginacione, quando scilicet infertur quod pars erit equalis toti.... Unde concedit [following variant reading] quod omne infinitum est equale infinito, et ponitur exemplum huius, nam hoc modo probarem quod numerus non crescit in infinitum, quia pars est equalis toti. Arguitur enim sic: omnes numeri excedentes centenarium excedunt eciam binarium, et non econtrario. Multi enim numeri excedunt binarium et non centenarium. Igitur sunt plures numeri excedentes binarium quam centenarium, et infiniti utrobique. Igitur sunt equales, et sic potest probari quod X [*sic*, should be C] sunt equales duobus, quia si ex equalibus equalia demas, et cetera. Contra hoc ostendo quod argumentum per responsionem istam confirmatur, nec vitatur inconveniens quin C sint equalia X [*sic*, should be 2] quando aliqua condicio opposita numero vel quantitati non aufert ab eo quin sit numerus vel quantitas.... sed quantitati et numero per propriam racionem in eo, scilicet quod quanta sint, convenit hec condicio vel hoc principium, scilicet quod totum est maius sua parte. Similiter, illud, 'si ab equalibus equalia,' et cetera, cum igitur processus et addicio in infinitum, non tollat a numero quin sit numerus nec a quantitate quin sit quantitas, sicut dicetur infra. Igitur non tollet huiusmodi principia. Ad propositum ergo, cum addicio ista fuit ab eterno tempore infinito, quod uni revolucioni solis addite sunt 12 revoluciones lune, sequitur quod nunc sit maior numerus revolucionum lune quam solis, et similiter nunc maior est numerus in potencia numerorum excedencium denarium [*sic*, binarium] quam centenarium, quia 'si ab equalibus' et cetera. Que racio tenet necessario, etiamsi fiat processus in infinitum, quod nulli dubium est nisi volenti protervire." There is a confusion in the example as to whether one is comparing the numbers over 100 to the numbers over 10 or 2, but the point is clear.

[42] Ibid., p. 248: "Praeterea, probo quod si mundus fuisset ab eterno, quod aliquis precessisset infinitum duracionem, ita quod excessus sit in utroque latere. Et videtur michi quod potest probari demonstrative, licet aliis videatur racio sophystica. Ego autem nullam responsionem videre possum, nec potero in eternum, ut credo. Racio est ista: suppono quod mundus fuit ab eterno secundum condicionem quam modo habet, quod fuisset generacio hominis ab homine per accionem solis sicut modo est, ita quod sicut modo pater meus et sol precesserunt me duracione, ita pater illius et sol precesserunt eum duracione, et sic ab eterno.... Dico tunc quod si sol precessit patrem meum et me, sicut precessit infinitos genitos et generantes,

Thus the conclusions that Harclay reached about infinite past time were troubling, yet he did not believe that the arguments that led to these conclusions were fallacious, and therefore he accepted them for the time being. They were not conclusions about the actual physical world in any case, but about what is possible for God, or logically possible.[43] Straight lines and time and motion, Harclay concluded, can be actually infinite, but not bodies or curved lines, because if a line is curved it will enclose a figure, which cannot be actually infinite.[44]

In asking whether the world could endure eternally into the future, then, Harclay again considered problems of the inequality of infinites, but he also introduced questions of the composition of continua from indivisibles—the connection being that the differences between infinites that occur for the infinity of time also occur for the composition of continua from indivisi-

---

eodem enim modo se habuit respectu infinitorum sicut respectu duorum, igitur precessit duracione infinitos generatos et generantes. Cum igitur infinitis hominibus correspondent necessario duracio infinita per positum, oportet quod sol precesserit duracionem infinitam et per consequens infinitum potest excedi a parte ante." In a long discussion of possible objections to this inference, Harclay says, for instance: "Probo quod illa multitudo sit aliqua una signata multitudo ... et non est finita; igitur infinita," and "si aliquod predic[a]tum conveniat supposito singillatim, et illud predic[a]tum natum sit convenire pluribus coniunctim, necessario si signatur convenire eis singillatim, eo ipso convenit eis coniunctim, etsi omnibus singillatim, convenit omnibus coniunctim" (p. 249). "Preterea, quelibet singularis est vera, igitur universalis est vera. Antecedens patet inducendo per infinita" (p. 250). "Concedo quod ... si mundus fuisset ab eterno, quod duracio solis prolixior esset duracione totali infinita omnium hominum, et quod unum infinitum unius racionis potest esse maius alio infinito eiusdem racionis, sicut una linea alia si utraque esset infinita, et una multitudo alia multitudine, licet utraque sit infinita. Et videtur michi non esse dubium quin esset quantitas infinita, si esse posset, nichilominus in illa esset totum et pars, et per consequens excedens et excessum, propter raciones superius tactas.... Dico quod quantitas 4 pedum non est divisibilis in tot partes infinitas eiusdem quantitatis in quo quantitas 8 pedum" (p. 251). [My emendations in brackets.]

[43] Ibid., p. 253: "Nunc autem dico quod infinitas in potencia cuiusmodi est in continuo respectu divisionis in infinitum.... Unde non est dare maximam speram inquantum spera est, nisi per accidens, inquantum scilicet est res naturalis, que non permittit magnitudinem quancumque indifferentem. De quo tamen dubito utrum deus posset facere, sed non repugnat inquantum mathematicum et abstractum est a re naturali, tamen non potest esse infinitum in actu, sicut nec continuum potest esse actu divisum in infinitum, quia hoc includit contradictionem in omni figura, et in omni corpore inquantum corpus, sicut demonstrative probari potest." P. 244: "Aliter potest responderi, quod non repugnat multitudini in quantum multitudo est infinitas [?esse infinita?] in actu, sicut dicetur inf[ra]. Unde credo quod modo sint infiniti numeri apud deum noti, et per consequens sunt numeri; nam, ut credo, numerus non dependet ex continuo, nec ex re numerata, sed numerus quo numeramus est quedam mensura apud mentem existens, per quam numerans devenerit in cognicionem rei numerate, sicut patet in questio de numero, et tales sunt modo infiniti in mente divina secundum Augustinum 12 *De civitate*, capitulo 18." I have made the suggested emendation in brackets after comparing a copy of MS Assisi 172.

[44] Dales, "Henricus de Harclay," pp. 252–54; Murdoch, "Henry of Harclay," pp. 232–33. Although infinite time would result from the rotation of the *primum mobile*, which is a sphere, motion and time are not circular but linear, since they are not the same motion or time when the sphere returns to the same position; since they are successive entities, only one indivisible of motion or time actually exists all at once.

bles. The question begins with principal arguments with which Harclay agrees, and there is no core response to the question or determination, but only arguments *ad oppositum* and replies.[45] Given this structure, it is perhaps not surprising that the question contains more on other issues than on the possible future eternity of the world—something that was, after all, to be accepted according to faith. One might well wonder what Harclay's major interest in this question was.[46]

In an article on Henry of Harclay for a volume in memory of Anneliese Maier, who had written about Harclay as an actual infinitist, John Murdoch emphasized Harclay's attention to the inequality of infinites, remarking:

> six out of the seven arguments cited by Harclay as putative refutations of his view that continua are composed of indivisibles are seen as also being directed against, even primarily against, his yet more dearly held belief in the possible inequality of infinites.[47]

The point, I think, is that either the inequality of infinites succeeds in all cases or it fails in all cases. Harclay's crucial methodological tool in this question, as Murdoch also points out, is reference to what God could do—or, more especially, to God's knowledge:

> In order to establish the inconclusiveness of Aristotle's strictures against the infinite and the greater persuasiveness of his own arguments in favor of the infinite, in appropriate fourteenth-century fashion Harclay repeatedly invokes God's absolute power. Aristotle's arguments may indeed hold *in rebus naturalibus* or in quantities that are possible for *us* to comprehend, but not when we consider what infinites can exist or be created by divine power.... And the same appeal to divine power and knowledge provides a way to consider all of the indivisibles in a continuum or to consider merely one indivisible apart from all of the rest. It is not simply, however, that such things are possible under the assumption of divine power because they involve no contradiction; it is because God has a perfectly determinate knowledge of the infinite itself.[48]

The dynamic of Harclay's thought, I believe, is that he starts with the question of the possible eternity of the world, which leads him to consider what

[45] For comments on the idiosyncratic structure of Harclay's questions, see Franz Pelster, "Heinrich von Harclay, Kanzler von Oxford, und seine Quästionen," in *Miscellanea Francesco Ehrle*, vol. 1 (Rome: Biblioteca Apostolica Vaticana, 1924), pp. 333–36.

[46] Michael McVaugh has suggested to me that Harclay could have been interested in the question for the practical reason of combatting millenialist predictions of the end of the world or of the coming of the Antichrist in 1300. The first of Harclay's questions in MS Worcester F. 3 is "Utrum astrologi vel quicunque calculatores possint probare 2^m^ adventum Christi." See our introduction to this volume.

[47] Murdoch, "Henry of Harclay," p. 249 n. 81.

[48] Ibid., pp. 230–32. Pelster, "Heinrich von Harclay" (n. 45), p. 325, indicates that Harclay argued for God's knowledge of individuals in a question, "Utrum ad distinctam cognicionem quam habet Deus ab eterno de rebus creabilibus, requirantur in eo distincte rationes cognoscendi, per quas cognoscat ipsas res creabiles." Pelster (p. 325) also reports that Alnwick writes, "Deus intelligit alia a se per alia et quod intelligit res per res et non per essenciam suam," in the margin of which is written "Opinio Cancellarii Oxoniensis"—i.e., Harclay.

is possible for God, understanding that everything is possible for God except what is known to involve a logical contradiction.[49] He concludes that there is no logical contradiction in the conclusion that infinites can have ratios to each other, and hence that an eternal world is possible. Here Harclay noticed that he could recruit support for the idea that there can be ratios between infinites by referring to Robert Grosseteste's ideas about God's knowledge of ratios between infinite indivisibles. Influenced by Grosseteste, he assumed that God now sees or intuits all the points of a line. This led him ineluctably not only to the conclusion that there are more points in a longer line than in a shorter, as Grosseteste had asserted, but also to the conclusion that these points are immediate.

In both of Harclay's questions on the possible eternity of the world there are extensive discussions of whether there is something fallacious about arguing from individual instances to an infinite multitude of such instances, for example from individual points in a line to all the infinitely many points that God sees or knows in the line other than the end point. In each case, Harclay argued that the reasoning is valid, thus arriving at the view, which Bradwardine later attempted to disprove, that continua are composed of infinitely many immediate indivisibles. The key argument leading to this view was as follows:

> It is certain that God now intuits every point that can be designated in a continuum. Take, then, the first inchoative point of a line. God sees that point and any point in this line different from it. Then, [from the first point of the line] up to the more immediate of the points God sees, there is either a line or not. If not, then God sees this point to be immediate to the first one. If a line does intercede, then, since points can be assigned in the line [which falls between the first inchoative point and the other point], these mean points were not seen by God. Proof of this inference: by hypothesis a line falls between the first point and any point other than this point that God sees. Therefore, according to you, God did not see the point just found [but it was asssumed that He saw all the points of the line, so there is a contradiction].[50]

---

[49] Dales, "Henricus de Harclay," p. 242: "Potentie autem divine attribuendum est totum quod scitur non includere contradiccionem vel quod non scitur includere contradiccionem."

[50] Latin of John Murdoch, quoted in part in Murdoch, "Henry of Harclay," pp. 228–29 n. 24: "Respondeo ad argumenta que probant quod continuum non componitur ex indivisibilibus sustinendo opinionem Lincolniensis. Et videtur michi breviter quod, licet ista argumenta videantur concludere humano intellectui quod indivisibile non potest esse continuum et inmediatum indivisibili, tamen raciones in contrarium magis movent meum intellectum et movere debent intellectum cuiuslibet, ut michi videtur. Licet enim meus intellectus non comprehendit quomodo continuum componitur ex indivisibilibus inmediate se habentibus, tamen intellectus divinus hoc necessario comprehendit. Cuius est una racio hec: Certum est quod Deus modo intuetur omne punctum quod possit signari in continuo. Accipio igitur primum punctum in linea incoativum linee; Deus videt illum punctum et quodlibet aliud punctum ab isto in hac linea. Usque ad illud punctum inmediatiorem quem Deus videt intercipit alia linea aut non. Si non, Deus videt hunc punctum esse alteri inmediatum. Si sic, igitur cum in linea possint signari puncta, illa puncta media non erant visa a Deo. Probacio huius consequencie:

Harclay found this argument difficult, but more persuasive than Aristotle's arguments against immediate indivisibles:

> These arguments ... persuade me of the truth of the matter more than the arguments for the opposing point of view, howsoever little they may agree with Aristotle's way of doing things and in spite of the fact that the [contrary] arguments of the Stagirite and others are extremely difficult [to refute], undoubtedly because our intellect is dull when it comes to understanding the infinite or the infinite division of a continuum.[51]

As Harclay himself admitted, although the arguments for immediate indivisibles in a continuum were more persuasive to him than the opposite, he did not understand them all perfectly well. The crucial step was the assumption that God now sees/intuits/knows all the points that could be designated in the continuum (*Deus modo intuetur omne punctum quod possit signari in continuo*). Considering all the remaining points in the line other than the first, are the closer points immediate to the first point or not? It seems to follow that they must be immediate, because if not, then there will be a line between the first point and the other points of the line and then it will be possible to designate points in this line. But then God would not have seen these newly designated points, which contradicts the hypothesis that God now sees all points that might be designated.

Harclay reinforced this line of thinking by a second argument that the indivisibles must be immediate:

> Furthermore, God can corrupt or annihilate the end point of a line without corrupting the line. First proof of this assertion: For suppose that some continuous line is later broken at its mid-point and then the two extremities of the line are joined [again]. Here we have two actually separated points which can later be joined to one another by God's power, and become two points in potency (for they are "in potency" when they are not separated). It is possible that numerically the same point, which now exists potentially in the line and which first existed actually and was the beginning of the right part of the line, is now the end of the left part [of the line]. The proof of this is as follows: A line can be broken at all [i.e., any one] of its points in such a way that any point can indifferently be the beginning of one or the other lines [resulting from the break], though it cannot really exist in both at the same time (for it is impossible that the same point really exists in two non-continuous, broken and separated lines [at the same time]). Consequently, the point that previously began one line and was its intrinsic terminus, is now the intrinsic terminus of another line. Therefore, since one line can be corrupted while the other remains an integral whole, it follows that the point that was the terminus of the former line is now corrupted while the latter line remains an integral whole. Thus, it is manifestly evident that God can corrupt a point without corrupting any part of a line. Assume, then, that God corrupts only the first point

Nam per positum linea cadit inter hunc punctum priumum et quodlibet aliud ab hoc puncto quod Deus videt; et ideo, per te, modo inventum punctum medium Deus non videbat."

[51] Latin text of Murdoch, "Henry of Harclay," p. 230 n. 27.

> of a line; the line remains and has an actually existing point at its terminus. Moreover, it is certain that there was no mean line between the latter point and the corrupted point, since that line would then have been corrupted, which is contrary to that assumed. Therefore, since God could corrupt the other point without corrupting the line, it remains that point is necessarily immediate to point.[52]

Here Harclay imagined that God could divide a line making a certain point the first point of the right side, then rejoin it, and then divide it again making the same point the last point of the left side. If, then, God annihilated the line on the left, there would remain the line on the right missing only its first point. But this remaining line would have, Harclay asserted, a new first point—which point must previously have been immediate to the point corrupted along with the line on the left. As in his first argument, he again concluded that between the first removed point of a line and all the other points there is no line or interval, so point must be immediate to point. Since the argument holds for any point at which a line may be cut (that is, for all its points), it follows that point is immediate to point everywhere.

Thus Harclay was persuaded to accept the conclusion that indivisibles are immediate to each other by, in effect, considering an infinite multitude of ordered indivisibles from which a single maximal or minimal member has been removed, namely that corresponding to the end point of a line. Whereas we might argue that the remaining line is open-ended, that it no longer has a terminal point, Harclay is persuaded by various arguments that it necessarily still has an end point. God will clearly see all the points of the line other than the end point, so he will necessarily see the extreme point, the one closest to the original end point. Since that point cannot have been distant from the end point, it must have been immediate to it. Lest the argument should be made that even God cannot remove only the end point of

[52] Ibid., p. 229 n. 25: "Propterea, Deus potest corrumpere vel adnichilare punctum in extremitate linee, linea non corrupta. Probacio istius primo: Nam ponamus aliquam lineam continuam, postea fractam in puncto medio; tunc due extremitates linee coniungantur. Et sunt duo puncta in actu separate, et postea possunt coniungi virtute Dei et sunt duo puncta in potencia (in potencia, inquam, quando non sunt separata). Et possibile est quod idem punctum numero qui modo est in potencia in linea, qui primo fuit in actu et fuit principium partis dextre linee, sit modo terminus partis sinistre. Cuius probacio est: Quia linea potest frangi in omni puncto ita quod quilibet punctus in linea potest esse indifferenter vel principium unius linee [vel] alterius, et simul in ambobus non potest esse realiter (inpossibile est enim quod idem punctus realiter sit in duabus lineis non continuis sed fractis et separatis). Igitur, ille punctus qui incoavit et fuit terminus intrinsecus huius linee prius, modo est terminus intrinsecus alterius linee. Cum igitur una linea possit corrumpi, alia tota integra remanente, sequitur quod punctus ille, qui fuit terminus alterius linee, modo est corruptus, et linea est tota integra. Igitur, manifeste patet quod Deus potest corrumpere punctum non corrumpendo aliquam partem linee. Ponamus igitur Deus corrumpere punctum primum linee tantum; manet linea et habet punctum in actu in suo termino. Et certum est quod inter punctum illud et punctum corruptum non fuit linea media, quia tunc fuisset corrupta illa linea, quod est contra positum. Igitur, sicut aliud punctum corrumpere posset sine linea, et relinquitur necessario punctus inmediatus puncto."

a line, Harclay has shown in his second argument how it may be done by dividing a line once with the dividing point going to one side and then again with the same point going to the other side.

Again, we may say with the hindsight of modern set theory that once the last point of a line is removed, the set of remaining points has no last member in that direction, but for Harclay this did not follow. God, who is assumed to have infinitely powerful intuition or sight, intuits individually all the points that exist. Seeing all the points in order, God sees the one closest to the end of the line. Why not, if these points all actually exist simultaneously and if everything that exists is open (*nuda et apperta*) to God? Even if human beings cannot see or order the points of a line, God can do it if it does not involve a logical contradiction—and no such logical contradiction was apparent to Harclay.[53]

Harclay's argument became the talk of the town in Oxford—*commune in villa*, in Walter Chatton's term.[54] It intrigued people the way Zeno's paradoxes intrigue people.[55] Adam Wodeham, who thought the argument fallacious, clearly was glad that he could avoid the tedium of an exact refutation, since he denied the existence of indivisibles in the first place.[56] Harclay, he acknowledged, found the argument demonstrative.[57] William of Alnwick, as he thought, refuted the argument, but the argument refused to die.[58]

---

[53] Here and elsewhere, a difference between medieval discussions and modern ones is that medieval authors do not clearly or consistently distinguish a set from the members of a set. This is discussed further below.

[54] Murdoch and Synan, "Two Questions" (n. 4), p. 252 n. to lines 77-79: "Argumentum est ad hoc commune in villa: Inter primum punctum huius linee et quodlibet aliud punctum eius aut est aliquod medium aut non. Si sic, illud mediabit inter primum punctum et se ipsum; si non, ergo primum punctum et aliquod aliud sunt immediata. Item clarius in ista forma: Inter primum punctum et quodlibet aliud punctum huius linee visum a Deo aut est medium aliquod punctum aut non. Si sic, et cetera ut prius. Respondetur quod hec particularis est falsa: 'Aliquod punctum est medium inter primum et quodlibet aliud, et cetera'; quia quelibet singularis eius est falsa. Set hec responsio non sufficit, quia accipio totam multitudinem punctorum preter primum, quia potest demonstrare quodlibet tocius multitudinis simul et signare [in] hoc punctum vel inter hoc punctum et primum etiam aliquod medium. Completa ergo tota multitudine per talem demonstrationem, ipse inveniet aliquod punctum inter quod et primum non est medium."

[55] John Leland, in his *Commentarii de scriptoribus Britannicis* (Oxford, 1709), says of Harclay, "fuitque ex ea philosophorum secta, qui elencis freti subtilius argumenta implicant" (as quoted in Pelster, "Heinrich von Harclay" [n. 45], p. 321).

[56] Murdoch and Synan, "Two Questions," p. 283 n.: "Istud argumentum est tediosum quoad quedam, et aliqui pro argumento negant puncta, aliqui, que ponunt continuum componi ex punctis, concedunt. Et si tenerem puncta vel talia indivisibilia, libenter studerem exquisite circa solucionem bonam; sed quia indivisibilia non pono, ideo minus curo."

[57] Ibid., p. 282: "ad quod respondentes dicut aliqui quod non sunt aliqua puncta, alii vero, propter argumentum, concedunt quod continuum componitur ex indivisibilibus, asserentes hoc argumentum demonstrative hoc probare, sicut hercele."

[58] See John Murdoch, "Naissance et développement de l'atomisme au bas moyen-âge latin," in *La Science de la nature: théories et pratiques*, Cahiers d'études médiévales, 2 (Montreal: Bellarmin, 1974), p. 26 n. 14.

What was Harclay's fallacy in the eyes of his opponents? He assumed that between the first point and all other points there is either some middle point or not—but, the opponents said, this did not exhaust the possibilities. While no one point is between the first point and all other points, nevertheless, between each of the remaining points and the end point there is some middle point or other (not always the same one). To this Walter Chatton responded: it is possible to demonstrate the whole multitude of points of the line except the end point, and it is possible to choose points in the multitude and to note whether between that point and the end point there is a middle point or not. When this has been completed, one will find some point between which and the end point there is no mean point:

> I take [*accipio*] the whole multitude of points except the first, since it is possible to point out [*demonstrare*] any one of the whole multitude at the same time and to designate [*signare*] this point or between this point and the first point also some medium. Having surveyed [*completa*] the whole multitude by such demonstration [*demonstrationem*], the same one [*ipse*] will find some point between which and the first there is no medium.[59]

If the opponents argued that it is not possible to infer from "any other point except the first" to some particular point (i.e., in the terminology of supposition theory, if they argued that when "any" stands for the other points in a "merely confused" way, then it is not possible to descend to individual points), Chatton responds that it is possible to particularize the distribution, so that the supposition becomes determinate, and then conversion of the proposition will be possible and the proof follows. Somewhat like Harclay, Chatton seems to be thinking that if it is true for all points seen by God that between the first point and the given point seen by God there is something between, then it is true that there is something between the first point and all points seen by God. Harclay and others similarly try to vary the argument, for instance by talking in the plural rather than the singular, in order to show that there is no fallacy or that the argument works. Even Bradwardine, rather than countering Harclay's and Chatton's arguments as such, seems to have been forced ultimately to destroy their position by denying the existence of indivisibles altogether.

---

[59] Text in Murdoch and Synan, "Two Questions," p. 252 n. to lines 77-79, cited above in n. 54. The text seems somewhat garbled—it would make more sense if Chatton said that for any point chosen either there will be another point between that point and the end point or not. Elsewhere Chatton expressed a similar argument by saying that if between the first point of the line and any other point seen by God there is a mean point, then this mean point will be between itself and the end point "because the word 'any' is distributed for any point other than the first given in this line" (ibid., p. 251: "inter primum punctum huius linee et quodlibet aliud punctum visum a deo in hac linea est aliquis punctus medius vel non; si sic, illud mediabit inter primum punctum et seipsum, quia deus videt utrumque, quia ly *quodlibet* distribuit pro quolibet puncto alio a primo data in hac linea; si non, igitur primum punctum et aliquod aliud visum a deo sunt inmediate, et propositum").

One Thomas of Felthorp resolved the problem, as Adam Wodeham reports, by claiming that the multitude of all the points other than the end point is immediate to the end point, but that (all) the (individual) points themselves are not immediate to the end point.[60] Wodeham himself argued that the following reasoning is fallacious: these points are not immediate to point *A*, but these points are this multitude, therefore this multitude is not immediate to *A*. The reasoning is false, Wodeham said, because the antecedent is true, but the conclusion false.[61] Thus one might distinguish the points from the multitude (read "set" in modern terms) of points. The problem for nominalists and Ockhamists, however, was that for them a whole is generally nothing but the sum of its parts—or, in other words, that there is no difference between a multitude of points and the points that make up the multitude. This made it difficult to construe Thomas of Felthorp's argument. Moreover, if God knows individuals directly and not via general or abstract concepts, then God will know the individual points and not the multitude. Thus the opponents of Harclay's conclusion who tried to argue that his reasoning was fallacious were not, apparently, able to convince either Harclay or Chatton.

Of course, claiming that a continuum is composed of immediate indivisibles leads to problems in terms of the standard Aristotelian and Euclidian views, which everyone realized. Harclay therefore went on to discuss arguments against the composition of continua from indivisibles, often ones previously made by Aristotle, which attempted to prove that indivisibles cannot be immediate or next to each other. To the argument that indivisibles can touch only as wholes, since they have no parts, but that if this is the case, then two indivisibles are no larger than one, Harclay first replied that this may be the case for two or a finite number of indivisibles, but it need not be the case for an actually infinite number of indivisibles. Thinking that this might not be a totally satisfactory reply, he then argued that the indivisibles may touch, but not be in the same position, and in that way produce an increase in size. Even surfaces or three-dimensional extensions would not create a larger extension if they were superimposed upon one another, but if they are in separate places, a larger dimension will occur. Nevertheless, for points in different positions to create a perceptible or comprehensible dimension, there must be an infinite number of them.[62] Harclay's conception of the composition of a continuum from indivisibles

[60] Ibid., p. 282: "et dicit felthorpus quod multitudo omnium punctorum istorum est inmediata *a*, ista tamen puncta omnia non sunt inmediata *a*, quia tunc sequeretur quod quodlibet [?aliquod?] istorum esset inmediatum *a*."

[61] Ibid., pp. 282–83: "sed hec responsio falsa est apud me; quod potest probari expositorie sic; ista puncta non inmediata sunt *a* puncto; sed ista puncta sunt ista multitudo; igitur, ista multitudo non est immediata *a*, quia antecedens est verum et consequens falsum."

[62] See Murdoch, "Henry of Harclay," pp. 242–44.

was not particularly compelling, but that should not surprise us. This was a conclusion he held because his reasoning about God's knowledge of all the points of a line apparently forced him to it—not because he found the conclusion in itself particularly attractive. Harclay's goal was not to show that continua are composed of indivisibles, but rather to show that infinites can differ in size, with the implication that an infinite duration in time, whether in past or future, is not impossible, *de potentia Dei absoluta.*[63]

Without trying to unpack this further, let me summarize. Reasoning about God's knowledge of all the points of a line led Harclay to conclude that the points must be immediate. Because the conclusion contradicted Euclid and Aristotle, a number of Harclay's contemporaries tried to determine where the fallacy in his reasoning lay, but Harclay himself and some others like Chatton were not convinced. The difficulty seemed to involve reasoning between what are, in our terms, infinite sets (such as all the points of a line), on the one hand, and their members, on the other. If all the points of a line save the end point are supposed to exist, and if God sees all the individual points absolutely clearly, then, it was reasoned, God must see the point closest to the end point. If there is a line of points and one touches the end, then one must touch the last point.[64] If modern set theory rejects these conclusions, it does so using axioms unknown to the fourteenth century. Lacking such axioms, fourteenth-century thinkers could not definitively resolve the paradox. Failing a resolution within the parameters of the problem, some like Wodeham avoided it by denying that indivisibles exist. Walter Chatton, on the other hand, resolved the situation by admitting indivisibles, but denying infinity. Before I conclude, then, let me turn briefly to Chatton's finitism.

Walter Chatton

An exposition of Chatton's views appears in a question on the continuum, most likely by Chatton, published by John Murdoch and Edward Synan. The question asks whether a continuum is composed of indivisibles as integral parts, and it concludes that all the indivisibles of a continuum are not

[63] Cf. ibid., pp. 242, 247–48, and esp. 259.

[64] Compare Chatton, in Murdoch and Synan, "Two Questions," pp. 241–42, par. 38: "capio aliquod quantum quadratum vere tangens fines illius quanti, et video ex omni parte, igitur omnes res que sunt extra animam in hoc quanto sunt finite, et propositum. consequencia patet, quia contradiccio est quod vere tangantur fines istarum rerum contentarum intra fines et, tamen, quod res contente sint infinite. probacio: quia, si habeant fines, igitur finiuntur, et si sunt infinite, non finiuntur—sed quod finitur et non finitur contradiccio est generaliter." We might perhaps argue that no point of the line is so close to the end point that there is not a point closer to the end. The medieval authors might respond that it is impossible for us to find the extreme point of all the points, but it is not impossible for God.

only immediate to each other, but also finite in number.[65] A crucial distinction in Chatton's thought was between what is only in the mind or the imagination of mathematicians and what is outside in the real world;[66] he wanted to know, he said, what could be said about the real world or things *in re*. And he made extensive use of theological arguments. In a typical argument, he supposed that his cheek was pricked with the tip of the sharpest needle God could make and that God then separated off the smallest possible end point of the needle that did the pricking. Are there as many parts in the tip of the needle that God separates off as in the whole rest of the needle, or even in the whole cosmos, he asked? If yes, then the part will equal the whole with regard to parts, which is implausible. If not, then there will be only a fixed number of parts in the tip, as he proposed.[67]

In the conception of Walter Chatton, then, there are no real infinites. If God had taken some action in every instant of eternal past time—for instance, if in every instant God divided a millet seed—by now the division would be complete, but the indivisible parts that remained would have some small size and total some large but not infinite number. In line with such thinking, Chatton replaced talk of points with talk of, for instance, small bits of meat. But he deployed God's power in ways less persuasive than Harclay's. Few would have denied that God knows everything that can be known, so God must know all the points that really exist, as Harclay assumed. Chatton, on the other hand, argued that a continuum is composed of indivisibles because it is not a contradiction that God could remove only one part of a quantity, leaving all the rest. But then that one part is indivisible. Why? Because if it were divided, it would be more than one part, and it has been conceded that God can remove only one part.[68] Or, God could make only two or three or four things and form a continuum of them; therefore there would be a continuum with only finitely many parts.[69] To the modern reader, such arguments are not persuasive, because we assume that the word "part" has only a conventional sense. We may consider a

---

[65] The context in which Chatton discussed this question in his *Sentence* commentary is not clear from the separate question on the continuum published by Murdoch and Synan.

[66] See, e.g., Murdoch and Synan, "Two Questions," p. 235: "tu, qui dividis inter porciones et proporciones, divides inter partes quanti extra animam vel inter aliqua quorum aliqua sunt in quanto et aliqua solum in ymaginatione mathematicorum"; "quia ymaginata non sunt partes rei, et tantum est hic sermo de vere quanto et de vere partibus eius integralibus ipsum."

[67] Ibid., p. 249: "Capio punctus acus pungentis me; fiat ita accutus sicut deus potest; tunc, una res tangit me—sensus tactus hoc dicit michi; quero: quid inmediate tangit me? non oculus acus, nec medium, sed tantum pars extrema. [S]eparet deus quantum potest illud quod inmediate tangit me; hec res est ferrum vel calibs; quero: utrum sint tot partes in hac re quot in toto residuo acus vel non? si sic, igitur pars equalis toti in partibus; si non, propositum, scilicet, quod in hac re sunt ferra to<t> quod non plura."

[68] Ibid., p. 234, par. 2.

[69] Ibid., p. 241, par. 36: "deus potest facere tantum duas res, vel 3, vel 4, et continuare simul, igitur erit continuum, et non habebit partes nisi finitas tantum."

body divided into as many parts as we please. Objectively, there is no fixed number of parts unless, indeed, the world is composed of atoms. Chatton's argument seems to imply that God makes atoms by designating something as one part after which it becomes, in effect, indivisible, but in an Aristotelian world, a continuum has no objectively determined parts—even for God. Thus, as far as I can see, Chatton deployed the modal logics of God's absolute power in ways that would have been dubious even in the fourteenth century.

BRADWARDINE'S MOTIVATIONS

Thus, Henry of Harclay's argument from God's present knowledge of all the points of a line to the immediacy of points had a Zenoesque flavor—the argument seemed reasonable, but the conclusion contradicted accepted views. This caused a minor industry in attempts to deal with it, ranging from Chatton's finitism to Bradwardine's (and the nominalists') rejection of indivisibles altogether. Why did this sequence of events happen and what was its result? The events happened, I believe, because there were a number of arguments about infinity that continued to be a part of the discourse due to their use in discussions of the possibility of an eternal world. These arguments, which went back to Philoponus and must have entered Latin thought through Arabic sources, were accepted by some, for instance Bonaventure, and rejected by others, such as Harclay. The arguments had been invented by Philoponus for the purpose of proving that Aristotle could not have been correct in arguing for an eternal world, that is, for an infinite past time. When the arguments were accepted, it was, at least in part, because they disproved an eternal world. When they were rejected, it was usually by those who wanted to argue, not that past time is infinite, but rather that, although it is finite, it could have been infinite had God chosen to create the world from eternity, *de potentia Dei absoluta*. No medieval thinker developed a really satisfactory mathematics of infinites or explained how one infinite could be both equal, as infinite, to another infinite, and yet unequal—for instance, by being twelve times as large, or containing all of the other and more besides.

The medieval interest, in the end, was not in the mathematics of infinites itself, but rather in the theological problems whose resolution required that problems of infinites be addressed. Grosseteste used unequal infinites as a foundation for absolute measure, but Harclay took up Grosseteste's proposal about ratios of infinites because of its bearing on ratios of infinites in infinite past or future time. Chatton's motivation is harder to discern, partly because his atomistic system is more peculiar from a modern point of view, making it more difficult for us to take him sympathetically, but more importantly because so far we have examined here only Chatton's separate question on the continuum and not his related question as it appeared in his commentary on the *Sentences*.[70] Let me leave Chatton for a future occasion, then, and turn back to Bradwardine.

Why did Bradwardine invest his energy in refuting the atomists' views of the continuum in his *De continuo*, and why did he write a work that must have consciously restricted itself to arts faculty disciplines, leaving out the mention of God that was central to the atomist strategies of Grosseteste as well as of Harclay and Chatton? Perhaps he wrote *De continuo* as a book to be read by those in the arts faculty, but he could not have written it without being familiar with arguments stemming from theology. If Bradwardine had not become acquainted with the arguments in their theological context, then they must previously have caused such excitement as to spill out from the theological faculty to a wider arts audience.

We are lucky to know relatively much about the biographies of Grosseteste, Harclay, and Bradwardine as compared to other medieval authors. We know that Grosseteste wrote his marginalia to the *Physics* rather late in life, when he was already a theologian.[71] Harclay's question discussing the continuum and indivisibles was explicitly theological. We know, moreover, that Harclay argued elsewhere for God's knowledge of individuals—which may indeed have been relevant to his claim that God sees/intuits/knows all the points of a line. When Bradwardine wrote *De continuo*—after *De proportionibus velocitatum in motibus*, which it cites—he must already have been studying theology for some years (beginning perhaps in 1325).[72] Bradwardine's own lectures on the *Sentences* are supposed to have taken place in 1332–33 or the year after.[73] Thus even if Bradwardine was not entitled to *teach* theology when he wrote *De continuo*, he would already have heard theological lectures for several years. If we take Part V of *De*

---

[70] A significant part of the question from the *Sentences* is provided by Murdoch and Synan in footnotes as parallel passages to the question they publish, but this does not provide the context.

[71] See, e.g., Richard Dales and Servus Gieben, eds., *Robert Grosseteste: Hexaëmeron*, Auctores Britannici Medii Aevi, vol. 6 (London: Oxford University Press for the British Academy, 1982), p. xii. Dales and Gieben say that Grosseteste was compiling his notes on the *Physics* in the period 1228–35, but abandoned the project unfinished when he became Bishop of Lincoln. Chapter VIII.5 of the *Hexaëmeron*, which they date to shortly after 1230, addresses the "Causae erroris philosophorum in ponendo mundum aeternum." See also R. W. Southern, *Robert Grosseteste: The Growth of an English Mind in Medieval Europe* (Oxford: Clarendon Press, 1986); and James McEvoy, *The Philosophy of Robert Grosseteste* (Oxford: Clarendon Press,1982).

[72] All that is proved about the dating of *De continuo* is that it was after *De proportionibus*, which it cites, and which is known to have been written in 1328. It is commonly supposed that *De continuo* was written before 1335, when Bradwardine left Oxford, but there is no hard evidence to prove this.

[73] Jean-François Genest, *Prédétermination et liberté créé à Oxford au xive siècle: Buckingham contra Bradwardine* (Paris: Vrin, 1992), pp. 13–14.

continuo seriously, Bradwardine's motivation could have been that some theologians like Chatton were taking the composition of continua from finite numbers of atoms seriously, whereas, as Bradwardine could persuasively argue, finitist atomism would undermine the foundations of all the arts disciplines.

Bradwardine's *De causa Dei*, however, casts new light on the situation, for there we see that he sides with Bonaventure (and the tradition of thought extending back to Philoponus) in arguing that paradoxes of infinites prove that the world could not have had an infinite past—in other words, Bradwardine rejects the idea of ratios between infinites common to the atomist ideas of Grosseteste and Harclay, as well as the finitist composition of continua upheld by Chatton. The argument Bradwardine favors in *De causa Dei* is the argument of Avicenna concerning the actual infinity of souls that would now be in heaven if the world had existed for an infinite time.[74] Bradwardine rather astutely, from a modern point of view, uses one-to-one mappings as a way of measuring infinites. If past time were infinite, he says, then there would have been infinitely many souls and infinitely many bodies, and one could match each soul to its proper body in a one-to-one match—but one could also match all the souls to only every second or every tenth body, or, conversely, one could match a thousand souls to each body and not run out of souls before bodies.[75] But if past time is infinite there will have been infinitely many top

[74] Thomas Bradwardine, *De causa Dei contra Pelagium et de virtute causarum* (London: ex officina Nortoniana apud Ioannem Billium, 1618), p. 120A–B: "Nonne ergo modo sunt actualiter huiusmodi animae simpliciter infinitae? Quare & ille Algazel, sectator tuus [i.e., Aristotle], et filius in hac parte, infinitatem animarum concedit. Primo namque *Metaphy.* suae, definitione sexta, sic ait: Ens dividitur in finitum & infinitum; Infinitum vero dicitur 4 modis; quorum duo non sunt; duos vero esse argumentatio deprehendit. Dicitur enim quod motus coeli non habet finem, quia non habet principium, & hoc iam deprehendit argumentatio. Dicitur etiam quod animae humanae que a corporibus separantur sunt infinitae: Hoc autem necessario verum est, si removeatur finitas a tempore, & a motu coeli, quod est remotio incaeptionis. Et infra, Animas humanas quae sunt separabiles a corporibus per mortem, concedimus esse infinitas numero, quamvis habeant esse simul; quoniam non est inter eas ordinatio naturalis, qua remota desinant esse animae, eo quod nullae earum sunt causae aliis." Bradwardine attributes the argument to al-Ghazzali, but the work of al-Ghazzali cited (i.e., the Latin translation that went under the title *Metaphysics*) did not contain al-Ghazzali's own views but those of the philosophers he intended to refute in *The Incoherence of the Philosophers*—in this case, Avicenna (the Arabic title of the work is *Maqasid al Falasifa*). Cf. Terry Hanley, "St. Thomas's Use of al-Ghazzali's *Maqasid al-Falasifa*," in *Medieval Studies* 44 (1982): 243–70.

[75] Bradwardine, *De causa Dei*, pp. 121C–22B, "Amplius autem si occupatio continua fineret, & prolixitas non obstaret, videtur quod ista sententia posset per alias rationes quasi mathematicas quodammodo mathematice reprobari; quod ideo per aliqua brevia breviter faciendum, exquisitiorem diligentiam aliis, vel alteri tempori relinquendo. Si igitur infinitae sint animae, habuerunt, & habent, vel habebunt corpora propria infinita per proximam partem huius. Ponantur igitur illa corpora infinita, vel aliqua infinita ex eis secundum unam seriem consequenter, ad modum loquendi Philosophi, 6 Phys. 2. ita quod quodlibet corpus praecedens, habeat aliquod proximum ipsum sequens, nullo huiusmodi corpore intercepto. Et sit illa series ordinata secundum unam lineam rectam, veram vel imaginariam, incipientem hic in centro horizontis, & ad occidentem in infinitum extensam, vel secundum lineam girativam,

rulers, popes or emperors. Let's call them all popes for convenience's sake, Bradwardine says. Suppose that at the time of each pope there were a thousand or more people. Then take the souls of the infinitely many departed popes and match them with the bodies of the infinitely many departed commoners on a one-to-one basis. Since there are infinitely many popes' souls, there will be enough papal souls for every human body.[76] If were were enough papal souls for every human body, why would God have superfluously created more souls, when papal souls would have been sufficient?[77]

---

ab eodem centro per partes proportionales inaequales diametri, aut semidiametri horizontis, versus eius circumferentiam infinities circumductam; vel adhuc secundum partes proportionales inaequales alicuius lineae rectae, finitae, puta semidimetri horizontis ad occasum tendentis. Nec corpulentia horum corporum quenquam conturbet; Ponantur enim vere vel imaginarie in illa linea recta vera, vel imaginaria infinita, vel si in illa linea girativa, aut recta finita ipsa volueris situare; ponantur posteriora semper minora prioribus proportionaliter ut oportet, vel loco eorum ponantur modo praedicto superficies, lineae, sive puncta, quae tamen corpora nominentur. Ratio quoque ista de multitudine sola procedit; & ideo de magnitudine nullus curet. His itaque per Dei omnipotentiam vere, vel per imaginationem ficte ita dispositis, disponantur & animae, contra sua corpora correspondenter omnino; & sit A tota multitudo omnium animarum, & B tota corporum multitudo; A igitur multitudinis singulae unitates; B singulis unitatibus, & omnes omnibus, & econtra, aequaliter correspondent: quaelibet enim anima habuit unicum corpus suum, & quodlibet corpus unicam animam propriam, sicut ex proxima parte patet; & ista responsio confitetur, quod & ostendunt series illae totales, & ipsarum unitates sibi invicem mutuo aequaliter respondentes. Quod & ut patule demonstretur, his ita dispositis distribuantur animae per Dei omnipotentiam, vel per imaginationem hoc modo: Prima, primo corpori; secunda, secundo: & ita deinceps, qua distri/butione completa quaelibet anima unicum corpus habebit, & quodlibet corpus unicam animam. Haec igitur singillatim atque coniunctim mutuo sibi aequaliter correspondent. Sed quomodo stabit istud: Dispositis etenim A & B ut prius, detur prima anima primo corpori, secunda tertio, vel decimo, vel quantum volueris distantia a primo; & tertia anima corpori tantum distanti a secundo corpore animato, quantum illud a primo, & ita deinceps donec tota distributio huiusmodi compleatur. Quo facto vel singulae & omnes sunt corporibus talibus distributae vel sunt aliquae remanentes; si singulae & omnes sunt corporibus talibus distributae, tota A multitudo illi parti B divisim & coniunctim correspondet aequaliter & e contra. Si aliqua anima remanet, cum ab illa ad primam sint tantum finitae, & omnia talia corpora praeaccepta B multitudinis sunt totidem & finita; quare & tota B multitudo similiter est finita, quae posita fuerat infinita: Ad aliud vero latus intuitum convertendo, A & B dispositis sicut prius, dentur primo corpori mille animae, & secundo totidem, & deinceps quamdiu multitudo sufficit animarum.... Quomodo ergo A B multitudines & unitates earum singulae singulis singillatim, & omnes pariter omnibus aequaliter sibi invicem mutuo correspondent, sicut superius monstrabatur, nec opinio diffitetur."

[76] Ibid., p. 122B–C: "Adhuc autem ut ratio dilucidetur apertius, ponantur quod sicut mundus & species humana sunt aeterni anterius, ita aeternaliter habuerit unicum principem omnium, puta Imperatorem vel Papam, qui & omnes exempli causa Papae vocentur. Sicut igitur secundum hypothesin infiniti homines praecesserunt, sic & Papae similiter infinita; quare & nunc sunt solorum Paparum animae infinitae, quae si distribuantur corporibus modo praedicto, sufficiunt ad singula & omnia corpora animanda; imo & sufficiunt ut corporibus singulis mille animae, & plures quocunque volueris tribuantur, quod potest demonstrari clarissime sicut supra."

[77] Ibid., p. 123A: "Quod & potest demonstrari similiter de Papalibus animabus. Simili quoque modo potest ostendi, quod quantalibet multitudo corporum communium, & etiam Papalium correspondet singulis unitatibus animarum. Sed quis inconvenientias & repugnantias tantas feret: ut quid etiam, quaeso, tanta, & tam excessiva superfluitas animarum pro toto

With its typical verbosity, *De causa Dei* contains a plethora of such arguments, all concluding that if the world were eternal there would be excessively many souls—more than necessary, disorderly, and inconsistent with God's wisdom and propriety.[78] (There are so many of these arguments, all making similar points, that *De causa Dei* itself is excessive!) With thousands of papal souls and the like being bandied about, the effect is rather irreverent. Two points, however, are quite clear. First, on each page of this section Bradwardine repeats the set phrase "by the omnipotence of God or by imagination," or "really by the omnipotence of God or fictionally by imagination," as his instrument for setting up paradoxes of infinity.[79] And second, Bradwardine's purpose is to prove the Aristotelian philosophers wrong who argue that the world has always existed and will always exist. The infinite is not more perfect than the finite, Bradwardine says, but less perfect.[80]

---

corporibus seu hominibus animandis, quando plene sufficiunt pauciores, etiam animae Papales, & adhuc quantumlibet pauciores, sicut praemissa demonstrant?"

[78] Cf. ibid.: "ut quid ibi superfluunt animae infinitae, & infinities infinitae, ut patet perspicue ex praemissis? ut quid etiam tanta & tam excessiva superfluitas corporum, pro tot spiritibus hospitandis, pro tot hominibus componendis, quando plene sufficerent pauciora, etiam corpora Papalia, & adhuc quantumlibet pauciora? ut quid ergo ibi superfluunt corpora infinita, & infinities infinita, sicut ex prioribus clare patet? hoc Deum sapientissimum nusquam decet, sicut prima Suppositio & partes sequentes ostendunt, hoc naturae non convenit, hoc omnes Philosophi detestantur."

[79] Ibid., p. 121E: "per Dei omnipotentiam vere, vel per imaginationem ficte"; p. 126C: "imaginarie vel vere secundum hypothesin per omnipotentiam magnam Dei." Cf. also pp. 126D, 128D, 129D, 130C, 134C, etc.

[80] Ibid., p. 120B–D: "Dices fortassis, quod Animae infinitae ponuntur propter perfectionem universi maiorem: Alias enim non esset ita perfectum. Sed cur magis animae infinitae, quam Angeli infiniti, quam sphaerae, quam stellae, quam caetera huiusmodi entia infinita? Imo & multitudo finita videtur perfectior infinita: Omnis namque numerus perfectus, sphaericus, circularis, cubicus quadratus pariter sive impariter, par aut impar, digitus, articulus, compositum sive cuiuscunque alterius dispositionis laudibilis est finitus, sicut Arithmetica clare monstrat; Numerus autem infinitus nullius huiusmodi dispositionis laudabilis esse potest. Eodem quoque modo videtur esse de numeris, sicut de corporibus & figuris; sed corpus finitum est perfectius infinito, si esset: Habet enim figuram determinatam & certam; Infinitum, nequaquam. Imo & omne corpus regulare, & apud naturam & Geometras laudabile & famosum est finitum; Omne namque corpus huiusmodi est pyramis, cubus, octoedron, duodecaedron, Icosaedron, vel sphaera; Figura autem sphaerica ratione docente & Mathematicis, atque Philosophis naturalibus testantibus, est perfectissima figurarum: quare & coelum corpus perfectissimum habet eam; Infinitum vero nullam habet figuram. Nonne & ordinatum & determinatum est melius & pulchrius inordinato & confuso? quantum namque bonum est ordo? quantam pulchritudinem affert rebus? Annon tu ipse Aristoteles.... Boni autem maxime species & ordo, commensuratio & determinatum, quod maxime ostendunt Mathematicae scientiae.... Valde autem excedens numerus non potest participare ordine." P. 121B–C: "Infinitus autem numerus ... ex sua inordinata confusione & confusa inordinatione, indecens, indelectabilis, & turpis potius videretur.... Si igitur haec ita se habent, quomodo arbitrandum, quod Deus sapientissimus sapientissime cuncta disponens, sicut praemissa ostendunt, civitatem suam coelestem, propriam domum suam, suam familiam specialem, ex sanctis animabus seu hominibus constitutam secundum proximam partem huius, inordinatam faceret & confusam, indispositam imperfectam, incompositam, indecoram, in multitudine scilicet actualiter infinita? Imo rationabilius aestimandum, quod ipsam in aliquo numero

Bradwardine's opponents might defend their position by saying, for instance, that all souls of common people, if infinite, are not more than all papal souls, because they are both infinite.[81] This argument is not convincing to Bradwardine, because it is clear that every whole is greater than its part, and this is so because the whole contains the part and something more besides. Surely there are twice as many human souls altogether as there are masculine souls, even if there are infinitely many masculine souls, infinitely many feminine souls, and infinitely many souls altogether.[82] Or consider God's infinite knowledge, and consider the same knowledge minus the knowledge of geometry—will the two infinite knowledges be equal?[83] If they are, then some other being who had all the knowledge of God except geometry, and all the

---

sapientiae suae, noto, perfecto, electo, convenienti, decenti, pulchre & suaviter ordinato sapientissime disponebat."

[81] Ibid., p. 123C: "Et si falsigraphus forsitan de maiori voluerit cavillare, dicatur illud maius alio quod continet illud totum & aliquid amplius, seu aliud quantum ultra seu extra.... Et si quis dixerit ... neque ista responsio evacuat argumentum."

[82] Ibid., p. 132D–133A: "Adhuc [*or* ad hoc] autem & huiusmodi multi multipliciter fatagunt respondere, quidem namque non verecundantur negare, Omne totum esse maius sua parte, neque concedere totum esse aequale sue parti; ut si A sit tota multitudo infinita omnium animarum, B vero una earum, C autem tota residua multitudo, dicunt quod A non est maior C sed aequalis, quod & dicunt, sicut & habent necessario dicere consequenter, de quibuslibet duobus quantis infinitis ad invicem comparatis. Sed nonne Euclides .1. Elementorum suorum supponit istud principium tanquam per se notum cuilibet, Omne totum est maius sua parte, quod de parte quantitativa intelligit, quod & omnes Mathematici atque naturales Philosophi concorditer profitentur? cui & videtur cuiuslibet animus sponte notis terminis consentire; quod & videtur clarere ex significationibus terminorum. Nonne illud est maius alio, quod continet illud & amplius, seu aliud quantum ultra vel extra: cuius animus contradicit? cuiusve animus non sic dicit? quare & forte sicut dicit Alacen in perspectiva sua, illud famosum principium per talem cognoscitur syllogismum. Omne continens aliud & amplius est magis illo. Omne totum contenet partem suam & amplius; omne ergo totum est maius sua parte. Sed quia notis terminis praemissae & consecutio statim sunt notae, non percipitur forsitan talis syllogizatio aut discursus; & ideo reputatur conclusio principium per se notum. Nonne & multitudo quaecunque puta omnium animarum, additis semper unitatibus augmentatur, sicut & subtractis minuitur, praesertim si addatur alia multitudo infinita aequalis priori, aut etiam subtrahatur? Quis enim ignorat duplationem & dimidiationem duplus & dimidium ibi esse, sicut definitiones eorum notis terminis cuilibet per se note, cuilibet clare monstrant? cuius sensus dissentit? cuius conscientia non consentit? tota igitur multitu/do omnium animarum simpliciter est dupla ad totam multitudinem omnium animarum masculorum, & similiter ad totam multitudinem animarum omnium foeminarum, cum utraque harum sit infinita; quare & aequalis alteri per praemissa."

[83] Ibid., p. 134B: "Adhuc autem & secundum istam sententiam insensibilem insensatam, cum secundum praemissa scientia Dei sit infinita simpliciter; & et vere vel imaginarie detraheretur ab ea aliqua scientia finita vel infinita quam habet, puta aliquorum vel omnium theorematum geometriae, non esset minor quam prius, nec ipse minus sciens quam prius. Eodem modo de eius potentia infinita, si detraheretur ab ea aliqua potestas finita, vel etiam infinita, puta sciendi geometricalia, vel creandi, adhuc maneret aeque potens & omnipotens sicut prius."

power of God except the ability to create, and so forth, would equal God—how blasphemous a conclusion![84]

Noting that Grosseteste supports the view that infinites may have ratios to each other,[85] Bradwardine concludes this section by admitting that many other objections, even very probable ones, can be made, if it is assumed, *per impossibile*, that there is an actuallly infinite multitude, to which it is difficult to respond either truthfully or plausibly. He will, however, hasten on to other things, leaving such arguments to those who have more leisure, having already said enough to refute Aristotle and al-Ghazzali (really Avicenna) using arguments involving infinites.[86]

To conclude, the upshot of this fourteenth-century discussion was to weaken the connection of mathematics to the physical world, or in other words, to undermine the standard Aristotelian conception of the relations between mathematics and physics. The force driving this result was the deployment of arguments involving God's absolute power. Aristotle had argued that there are only potential and not actual infinites in the world, and that for mathematicians potential infinites are sufficient. When Grosseteste and Harclay, among others, argued that for God absolute infinites are possible, then, in effect, mathematicians had to choose whether mathematics (arithmetic and geometry) are limited to the physical world or not. Bradwardine, Ockham, Buridan, and others chose to stick with an Aristotelian conception of mathematics, but for those who, like Grosseteste and Harclay, had tasted the delights of a mathematics of infinites, another road lay ahead.

If the result of fourteenth-century discussions of indivisibles was a weakening if the Aristotelian bonds between mathematics and physics—so that, at least for some people, mathematics was no longer seen as concerned with quantities really existing in, but in abstraction from, things, but rather as concerned with entities existing in the mind or imagination—there was no intellectual community whose profession it was to develop pure mathe-

---

[84] Ibid., p. 134C: "erit perfectionis aequalis simpliciter, & virtutis cum Deo, quod facinus tam blasphemum prima suppositio, & partes sequentes non sinunt, sed evidenter redarguunt, & refellunt, sicut praemissa per viam omnipotentiae Dei perspicue demonstrabant."

[85] Ibid., p. 134C–D: "Robertus quoque Lincolniensis, Grosseteste, subtilis ingenii & profundi in quodam tractatu suo in luce incipiente (primam formam corporalem lucem esse arbitror) ostendit omnes proportiones inter finita repertas, inter infinita similiter reperiri; Tota namque multitudo omnium binariorum, ut docet, est dupla ad totam multitudinem omnium quaternariorum, & tota multitudo omnium septimanorum quadrupla ad totam multitudinem omnium mensium, & similiter in similibus universis: quod verum esse videtur, supposito secundum opinionem communem, quod actu sit multitudo talium infinitarum."

[86] Ibid., p. 134D: "Praeter haec autem ista falsigraphia, etsi impossibili isto dato, aliquas rationes praemissas posset utcunque definire; ad multas tamen nec vere nec apparenter sufficit respondere, sicut considerare volentibus facile est videre. Scio quod multae aliae responsiunculae, & multae obiectiunculae, imo & multae obiectiones multum probabiles sunt possibiles in haec parte, sed festinans ad alia, aliis abundantioribus otio, vel alii tempori has relinquo. Haec igitur per viam infiniti contra Aristotelem & Algazelem hac vice sufficiat protulisse."

matics in this direction. In the twentieth century, infinity and continuity belong to the subject matter of mathematics, and we are satisfied that modern set theory has at last solved the problems of infinity and continuity that troubled Harclay and Bradwardine. The fourteenth century did not create a mathematics of infinity and continuity at least in part because it was theologian–natural philosophers who were concerned with the problems that might have led to such mathematics. To repeat: the mathematics of infinity was not developed independently and then applied to theology. Rather, theologians whose goal it was to solve theological problems tried to clarify the mathematics of infinity and continuity while they were doing theology. If, because of their intrinsic interest, questions on infinity and on the continuum sometimes were excerpted from commentaries on the *Sentences* to lead an independent life, nevertheless there was no independent community of mathematicians whose task it was to create a new mathematics of infinites.

Thus the fact that Thomas Bradwardine and other fourteenth-century scholastics went so far and no further in their development of a mathematics of infinity and continuity was, in a general sense, a result of the unitary character of late-medieval learning. For a theologian, Bradwardine was exceedingly well trained and active in logic, mathematics, and natural philosophy—few theologians today are comparably advanced in such a wide range of disciplines—but the Bradwardine who wrote *De continuo* was a theologian and not a mathematician. A fuller development of the mathematics of infinity and continuity had to await the coming into existence of a community of individuals whose primary goal was the development of mathematics.

NICOLE ORESME, ARISTOTLE'S *ON THE HEAVENS*, AND THE COURT OF CHARLES V

EDWARD GRANT

Nicole Oresme was not only one of the most extraordinary scholastic minds of the Middle Ages, he also left an exceptional intellectual legacy that has thus far escaped attention. He left two strikingly different kinds of interpretations of Aristotle's *De caelo*: one in the form of a traditional Latin *questiones*, written sometime between 1346 and 1356;[1] the other a French commentary based on Oresme's own French translation of the Latin *De caelo*, both completed in 1377, and together forming *Le livre du ciel et du monde*,[2] his last known work, produced five years before his death in 1382. To add to the exceptional nature of Oresme's achievement, both the translation and the commentary were made at the explicit command of King Charles V of France. What Oresme did may well be unique in the history of medieval scholasticism. Not only did he leave us two different interpretations of *De caelo* written twenty to thirty years apart, but he did so in two radically different formats, the first in the mode of a questions treatise, the second in the form of a lengthy commentary. Further striking dissimilarities are manifested in the fact that the two treatises were written in two different languages, and for two different audiences.

Oresme's Latin treatise on *De caelo* was written while he was an arts master teaching at the University of Paris, and it was intended for a university audience of masters and students. The modern editor, Claudia Kren, regards it as a very early work because of "the awkwardness of its style and the unfelicitous and confused manner in which Oresme often expresses his ideas"; it is the work "of a young master of arts whose expression of the learning imparted by his masters or even of his own ideas was still inarticulate."[3] By contrast, Oresme's *Livre du ciel et du monde* is the last known treatise of an experienced theologian–natural philosopher, whose attitudes toward faith and natural knowledge shaped his final work into which he

[1] For an edition and English translation, see Claudia Kren, ed. and trans., "The Questiones super De celo of Nicole Oresme," (Ph.D diss., University of Wisconsin, 1965) (henceforth, Kren, "Oresme: De celo"). For the conjectural date of composition, see p. xi.

[2] For the French text and an English translation, see Albert D. Menut and Alexander J. Denomy, eds., and Albert D. Menut, trans., *Nicole Oresme: Le livre du ciel et du monde* (Madison: University of Wisconsin Press, 1968) (henceforth Menut, *Oresme: Le livre du ciel*).

[3] Kren, "Oresme: De celo," p. xii. For a work written early in Oresme's career, the treatise seems quite articulate and sophisticated. Some of the interesting ideas that Oresme fully developed in the later *Le livre du ciel* are briefly treated in the *Questiones*.

poured the cumulative knowledge and understanding of a lifetime. A comparison of these two treatments of Aristotle's *De caelo*, one written at the beginning of his career, the other at the end of it, should convey a good sense of the changes that influenced Oresme's ideas about natural philosophy at the end of his career. For even where he retained the same opinion or viewpoint in both treatises, he often found occasion to alter, or expand, his thoughts about that opinion.

The *Questiones super De celo* is a straightforward scholastic treatise in the questions genre, the standard format of medieval natural philosophy; hence, it requires no special discussion. By contrast, the *Livre du ciel et du monde* is unusual, if not unique, in the Middle Ages. Although at least one partial French translation of an Aristotelian scientific work had been made before this one,[4] Oresme's was the first complete translation into French, the first to be commissioned by a monarch, and perhaps the first nonmedical scientific treatise in any vernacular language. The context in which Oresme translated *Le livre du ciel* constitutes the vital background against which the relationship between the *Livre du ciel* and the *Questiones super De celo* must be viewed.

The relationship between King Charles V (1338–1380) and Nicole Oresme began when Charles was a youngster. By 1356, Oresme had already come into contact with the royal family when King John II requested his assistance in dealing with problems of national finance confronting the monarchy, a request that prompted Oresme to compose a Latin treatise on money (*Tractatus de mutationibus monetarum*), which was followed by a French version (*Traictié de la monnoie*). Presumably at around this same time Oresme established an advisoral relationship with the dauphin, the future Charles V, who was still a youth, perhaps around fifteen or sixteen years of age.[5] Charles became king in 1364 and reigned until his death in 1380. One of his first acts was to appoint Oresme to an important post, Dean of the Cathedral Church at Rouen. Over the years, Oresme served Charles faithfully, performing various missions for his royal master.

---

[4] Most of Aristotle's *Meteors* had been translated from Latin into French around 1270 by Nahieu le Vilain. Menut calls this the "earliest vernacular translation of an Aristotelian scientific treatise"; see Menut, *Oresme: Le livre du ciel*, p. 5 and n. 6. There were other, earlier, partial vernacular translations of Aristotle's genuine treatises, as well as of those falsely ascribed to him.

[5] The most extensive modern account of the meager details of Oresme's life is in Albert D. Menut, ed., *Maistre Nicole Oresme "Le livre de politiques d'Aristote," Transactions of the American Philosophical Society*, n. s., 60/6 (1970): 5–22. For briefer summaries, see Edward Grant, ed. and trans., *Nicole Oresme, "De proportionibus proportionum" and "Ad pauca respicientes"* (Madison: University of Wisconsin Press, 1966), pp. 3–10; Marshall Clagett, ed. and trans., *Nicole Oresme and the Medieval Geometry of Qualities and Motions, A Treatise on the Uniformity and Difformity of Intensities Known as "Tractatus de configurationibus qualitatum et motuum"* (Madison: University of Wisconsin Press, 1968), pp. 3–5; and idem, "Oresme, Nicole," in *Dictionary of Scientific Biography*, 10: 223.

Their most enduring relationship, however, was intellectual. It has been said that the reputation of Charles V was "founded both on his prudent conduct of public affairs and his love of learning."[6] Both of these characteristics were manifested in the magnificent library that Charles collected in the tower of the Louvre. It included more than one thousand books, most copied and bound by talented artisans who also included miniatures of all kinds. Many of these books were in French—indeed, quite a few had been commissioned by Charles himself—and they ranged over natural philosophy, medicine, astrology, and numerous other subjects, including works on public affairs (for a further description of the library, see Joan Cadden's article in this volume).

Of the works that Charles commissioned for translation into French, four are of special interest because they involve Aristotle and concern both science and public affairs. Sometime around 1370, Charles prevailed upon Nicole Oresme to translate four Aristotelian treatises from Latin into French. These were the *Nicomachean Ethics, Politics, Economics*, and *On the Heavens* (*De caelo*).[7] There is little doubt that the first three were intended to make Charles's councillors and courtiers better at the art of government, as is evident from Oresme's remarks in the proem of the *Ethics*, where he explains that

> because the books on moral subjects written by Aristotle were originally in Greek and have come down to us in Latin quite difficult to understand, the King desired, for the common good, to have them translated into French so that he and his counsellors and others may understand them better, particularly the *Ethics* and the *Politics*; the first teaches how to be a good man, and the other how to be a good ruler of men.[8]

Indeed, Charles was eager to apply Aristotle's ideas in the *Politics* to his own government. As Albert Menut describes the situation, Charles, at the apparent urging of Oresme,[9] "wished to try out the idea of sharing a small part of his responsibilities as a monarch with his Council, a body of some 200 citizens of some distinction chosen from the nobility and upper ranks of the bourgeoisie to assist the king in his undertaking and to represent the

---

[6] Claire Richter Sherman, "Some Visual Definitions in the Illustrations of Aristotle's *Nicomachean Ethics* and *Politics* in the French Translations of Nicole Oresme," *Art Bulletin* 59 (September 1977): 321.

[7] It is not clear whether Aristotle's *De caelo* was among the original list of treatises to be translated for Charles (see Menut, *Oresme*: *Le livre du ciel*, p. 7). According to Menut, *Oresme "Le Livre de Politiques"* (n. 5), pp. 19–20, Oresme began translating the *Ethics* in 1370, probably completing a first redaction in that year. He began and completed a first redaction of the *Politics* and *Economics* (the latter was usually appended to the former) in 1371; and he completed *On the Heavens* in 1377, perhaps beginning it in 1374.

[8] Translated by Albert D. Menut in *Nicole Oresme: Highlights from His French Commentary on Aristotle's "Politics"* (Lawrence, Kans.: Coronado Press, 1979), p. 4.

[9] Menut, *Oresme "Le Livre de Politiques"* (n. 5), p. 9, col. 2.

public sentiment and reaction to royal measures of government."[10] To achieve this goal, Charles urged Oresme to quicken the pace of his translation so that relevant Aristotelian passages might be available for his councillors to consider at a meeting scheduled for 21 February 1372.[11]

SALIENT FEATURES OF ORESME'S TRANSLATION

If Charles V wanted the *Ethics, Politics,* and *Economics* translated for reasons of state, what motivated him to have Oresme translate Aristotle's *De caelo* into French? There is no indication of the king's motive in the *Livre du ciel* itself. In the prologue, Oresme tells us only that he has translated the work "at the command of the all-powerful and most excellent Prince Charles, the fifth so named, by God's grace King of France, seeker and lover of all noble wisdom";[12] and at the conclusion he informs his readers that "with God's help, I have finished the book on *The Heavens and the World* at the command of the very excellent Prince Charles, the fifth of this name, by the grace of God, King of France, who, while I was doing this, has made me Bishop of Lisieux."[13] In fact, despite Menut's assertion that "there can be little doubt that it was Charles's interest in astrology that prompted the selection of this work"[14]—an implausible claim[15]—it may have been Oresme himself who suggested the translation to Charles. At the end of his great translation and commentary, Oresme proclaims that "no mortal man has ever seen a finer or better book of natural philosophy in Hebrew, in Greek or Arabic, in Latin or French than this one."[16] And as if this were somehow an inadequate testimonial, he reinforces this sentiment, in Latin, a few lines below, concluding the entire treatise with these glorious words about Aristotle's *De caelo*: "never in this world was there a book on natural philosophy more beautiful or more powerful."[17]

Since the first three French translations were intended for the edification of Charles's councillors, Oresme may also have thought it advisable to make available to them the most "beautiful" and "powerful" book in the world on natural philosophy, Aristotle's *De caelo*. If Charles had been convinced, perhaps by Oresme, that it was useful for his courtiers and

[10] Menut, *Oresme: Highlights* (n. 8), p. 5.
[11] Ibid.
[12] Menut, *Oresme: Le Livre du ciel*, p. 39.
[13] Ibid., p. 731.
[14] Ibid., p. 7.
[15] Other than an emphasis on the incomparable superiority of the celestial region over the terrestrial, an idea that was commonplace in medieval natural philosophy and hardly peculiar to astrology, the *De caelo* has nothing to do with astrology. Charles also knew that Oresme was an ardent foe of astrology.
[16] Menut, *Oresme: Le livre du ciel*, p. 731.
[17] Ibid.

councillors to learn about the operations of government from Aristotle's *Ethics*, *Politics*, and *Economics*, perhaps Oresme also convinced Charles that his courtiers and councillors should learn something useful about the operation of the physical cosmos from the *De caelo*. After all, the physical universe was the grand and constant context within which government and politics were carried on. Perhaps Charles thought good government was effected against the backdrop of cosmic regularity, and that his courtiers would be well served by having a reasonable degree of familiarity with it.

The Challenge to Authority

In her paper in this volume, Joan Cadden has emphasized that Charles V was interested in encouraging his courtiers and councillors to engage in limited debate on various issues, and that he regarded such debate as prestigious for his court. A passage at the end of Oresme's commentary and translation suggests that Oresme may have reflected that attitude in *Le livre du ciel et du monde*. Following the anouncement of his appointment as Bishop of Lisieux as reward for his translation and commentary of *De caelo*, he explains that he made the translation and commentary "for the purpose of animating, exciting, and moving the hearts of those young men who have subtle and noble talents and the desire for knowledge to prepare themselves to argue against and to correct me because of their love and affection for the truth."[18]

In this extraordinary and significant passage, Oresme encourages his audience, presumably the young men of the royal court, to disagree with him because of their "love and affection for the truth." Although he does not mention Charles in this context, nor does he encourage the young men to argue and debate with each other, only with him, he may, nevertheless, have intended to communicate Charles's desire to stimulate debate and discussion among the varied members of the court. Indeed, Oresme may have had something more daring in mind. By requesting the young courtiers to challenge him and disagree with him in the interests of truth, he was actually asking them to challenge authority: a daring proposal, but one that Charles might well have approved—at least in the realm of natural philosophy and perhaps to a lesser extent in the operations of government.

Oresme's exhortation to the young to challenge authority was not mere empty rhetoric inserted for dramatic effect at the end of a lengthy treatise. On the contrary, anyone who read *Le Livre du ciel et du monde* with a modicum of care would soon realize that Oresme himself was the role model for challenging authority—in this case, the authority of Aristotle, creator of the greatest work ever written on natural philosophy (a judgment conveyed to the reading audience by Oresme himself). For those young

[18] Ibid.

readers who might somehow have missed his role as challenger of authority, he made certain that his efforts would stand revealed even to the unattentive. Toward this end he compiled an "Analytical Table of Contents" ("Table des matieres"),[19] to call attention to "many things worthy of note" ("choses bien notables"). In this unalphabetized list of noteworthy topics organized by order of occurrence, there are 133 unnumbered entries. In 35 of these, Oresme cites Aristotle by name and explicitly disagrees with his arguments. Many of the entries are presented in the form "Opposing Aristotle" (*contre Aristote*), as for example: "Opposing Aristotle, who says that if God exists, it must follow that the motion of the heavens is perpetual—Chapter 8," which is followed by: "Opposing Aristotle, who says that, if the heavens move, the earth must necessarily be at rest in the middle of the heavens; the contrary is demonstrated by several arguments—Chapter 8."[20] If we add 4 entries specifically opposed to some claim made by Averroes, we obtain 39 entries that Oresme directed against the two greatest medieval authorities in natural philosophy. The percentage of critical sections would then seem to be 39 out of 133.

But this is misleading, because a number of the entries are purely theological, unrelated to themes that were discussed by Aristotle or Averroes. For example, entries such as "On the Holy Trinity"; "What existed before the world existed"; "How an angel can be at the same time in the heavens and in more than a hundred thousand distant places on earth without occupying more space than the equivalent of one foot"; "How bodies of the blessed will be situated in Paradise"; and "In what manner the body of Jesus Christ is present on high in Paradise."[21] If we eliminate the 14 entries in this category, we are left with 119 relevant entries. We may therefore claim that approximately one-third of Oresme's entries on themes in natural philosophy were in explicit opposition to Aristotle and/or Averroes. That is a high percentage indeed. I know of no other medieval scholastic commentator on Aristotle who explicitly enunciated disagreements with Aristotle in a comparable manner, and to so great an extent.

Theology and Natural Philosophy in "Le livre"

As Oresme grew older, the relations between theology and natural philosophy seem to have become ever more important, and they played a large role in his last work, written for the royal court and for an essentially non-university audience. Moreover, the *De caelo*, the work on which he expended considerable theological effort (as is evident from the theological

[19] Ibid., pp. 732–49.

[20] Ibid., p. 739. This is ch. 8 of bk. 2.

[21] In Oresme's Analytical Table of Contents (ibid.), see, respectively, pp. 733, 737, 739, and 749.

topics cited in the preceding paragraph), was not one on which medieval theologians chose to exercise their theological skills. In this regard, Oresme may have been unique in the Middle Ages. Why did he do it?

Around 1370, approximately seven years before completion of his translation of *Le livre du ciel*, Oresme revealed the attitude that would partially shape his French commentary on *De caelo*. In his lengthy *Quodlibeta*, he asserts that although faith assumes miracles, such as the Trinity and the Resurrection,

> it seems to me that many things equally miraculous are assumed in philosophy and [are even] less demonstrated because they are customary [or familiar], as, for example, what is prime matter, and [how] the production of a new form [occurs] of which at first nothing exists...; how fire heats [or burns] and what is the nature and quiddity of things.... Surely, if you consider the matter properly, [you will realize that] many such natural things are unknown, more [unknown] than many articles of the faith. And therefore, indeed, I know nothing except that I know that I know nothing.[22]

In addition to encouraging the young courtiers who would read his treatise to challenge his authority, and perhaps authority in general, Oresme also seems to have had a desire to convey an additional message: natural knowledge itself is often as unknowable, mysterious, and elusive as are articles of faith. Thus, as the paragraph from the *Quodlibeta* reveals, one might legitimately challenge the authority of natural philosophers because natural knowledge itself is far from certain and, therefore, ought not to be employed to question the faith. In at least two places in *Le livre*, Oresme emphasizes the uncertainty of knowledge. After completion of the last theme in the book, he declares that he has "set forth all this with great humility and fear in my heart, respecting always the majesty of the Catholic faith, in order to subdue the inquisitiveness or presumption of those who, by chance, would heap calumny upon it or impugn [*impugner*] it or press too far their curiosity to their own confusion."[23]

A little earlier in the fourth and final book, he confronts the question as to how human bodies will dwell with Jesus Christ in the region beyond all the heavens. He imagines three ways in which they might exist there. But "which of these three ways is true or not true no mortal man can know without a revelation from God," he concludes, and then quickly adds that "even the greatest philosophers who ever lived are ignorant of many things in nature and were mistaken and erred many times, as I have often pointed

[22] My translation is drawn from Edward Grant, *Planets, Stars, and Orbs. The Medieval Cosmos, 1200-1687* (Cambridge: Cambridge University Press, 1994), p. 646. The Latin text appears in Lynn Thorndike, *A History of Magic and Experimental Science*, 8 vols. (New York: Columbia University Press, 1923–58), 3: 469 n. 128.

[23] Menut, *Oresme: Le livre du ciel*, p. 731.

out in many chapters and passages of this book *On the Heavens*."[24] Oresme's message seems to be that although God, if He chooses, can give us a revelation to elucidate some point of theology or religion, mortal philosophers, "even the greatest" of them, are frequently unable to provide us with analogous "revelations" about the workings of the natural world.

There Is Only One Truth

A third message that attentive readers of Oresme's *Livre du ciel* would have received is that there is only one truth: what is true in natural philosophy must also be true in religion and theology, and vice versa. As will be seen below, Oresme was making this point when he insisted that Aristotle erred by arguing that whatever had a beginning must have an end, and that what has no end, cannot have had a beginning.[25] Oresme thought it unacceptable that in natural philosophy the world might be regarded as eternal, without beginning or end, but that in religion and theology it must be perceived as created, that is, as having a beginning in time. On such matters, faith and natural philosophy must be in agreement.

Let us now compare a few themes from the analytical table of contents that were considered to some extent in both treatises.

A Comparison of the Two Treatises

Apart from the distinctive formats of the two treatises, they are also distinguished by a number of major differences. Perhaps the most striking contrast, as we saw, is the intrusion of theology into *Le livre du ciel et du monde*—a subject that Oresme could not have discussed in the *Questiones*, which he presumably wrote while he was an arts master, when he would have been forbidden to introduce theology into his deliberations. Differences also occur in the subjects discussed. Themes that are included in the one treatise—even at great length—may appear only cursorily or not at all in the other. Furthermore, the *level* of treatment often differs substantially: what is presented in technical detail in one treatise may be greatly simplified in the other.

On the basis of the considerable differences between the intended audiences of the *Questiones super De celo* and the *Livre du ciel*, we might infer that Oresme would have sought to be less technical and "scholastic" for his

[24] Ibid., p. 725.

[25] For Anneliese Maier's support of this interpretation, see her *Metaphysische Hintergründe der spätscholastischen Naturphilosophie* (Rome: Edizioni di Storia e Letteratura, 1955), pp. 27–28. See also Peter Marshall, "Nicholas Oresme's Questiones super libros Aristotelis De anima" (Ph.D. diss., Cornell University, 1980), p. 526, where Marshall cites Maier approvingly.

courtly audience in *Le livre du ciel* than he was for his university audience in the *Questiones*. In truth, the situation is more complicated. Oresme did indeed reduce the complexity of some topics in *Le livre du ciel*, but he then chose to enlarge upon some sophisticated and technical themes in *Le livre* that received only cursory treatment in the *Questiones*. Why he chose one course or the other for a particular topic can only be surmised and is perhaps ultimately indeterminable, but the differences remain.

Avoidance of Complexity in "Le livre"

Let us examine a few examples in which Oresme chose to avoid complexities. In bk. 1 of his *Questiones,* he devoted three lengthy questions (20–22) to the Aristotelian problem of maximum and minimum powers that were required to lift or move weights, a subject that had both physical and mathematical aspects in the fourteenth century.[26] In *Le livre du ciel*, he reduced the earlier discussion to two simplified paragraphs, resting content to mention that

> certain active forces can be compared either to the effect of their accomplishment or to the velocity or speed of their motion or to the time it takes them to act or to other things in many ways according to varying circumstances. In this regard there can be many wonderful considerations which I have previously dealt with; I pass over these now since they are not pertinent to our principal subject matter.[27]

In bk. 1, qu. 2 of the *Questiones*, Oresme considered "whether there may be only three dimensions."[28] The discussion that follows is highly technical, with no mention of God or of the significance of the number three. Oresme's treatment of this problem in *Le livre du ciel* (bk. 1, ch. 1) bears little resemblance to his earlier discussion. In *Le livre*, he ignores the technical aspects of the concept of dimensionality and focuses instead on the importance of the trinitarian aspects of the number three, explaining that "the laws whose purpose is to magnify the Creator are related to this number three," so that "the trinity is a natural component of things and creatures," and that "nature teaches us to praise God according to this number."[29] Indeed, Oresme goes on to distinguish "three kinds of trinity,"[30] concluding the entire discussion on dimensionality by moving from trinity to Trinity, explaining that

> we must necessarily assume that God, who is the exemplar of creatures, possesses in Himself incorporeal trinity, for it is impossible that the divine sub-

[26] See Kren, "Oresme: De celo," bk. 1, qq. 20, 21, and 22, pp. 295–390.

[27] Menut, *Oresme: Le livre du ciel*, bk. 1, ch. 29, pp. 193–95. We have here a probable reference to the earlier *Questiones super De celo*.

[28] Kren, "Oresme: De celo," p. 16.

[29] Menut, *Oresme: Le livre du ciel*, bk. 1, ch. 1, p. 51.

[30] Ibid., p. 55.

> stance should be corporeal in and of itself. And this trinity of God, which nature demonstrates to us, was perceived formerly in natural light by many solemn philosophers, although it was obscurely seen, as one sees a thing from a distance. The holy patriarchs and the prophets knew it by inspiration of the Holy Spirit, but did not explain it to us clearly, but in figurative language and obscure words. Yet our Savior Himself and His disciples and the holy doctors of the church have clarified it plainly for us in sufficient manner to lead us to believe our salvation, and with this to understand how an incorporeal creature is the image of this divine trinity.[31]

Oresme may have thought that the technical, formal, mathematically oriented presentation about dimensionality in the *Questiones* was inappropriate for Charles's courtiers. Perhaps he regarded it as of little relevance, or thought it too abstract, or sensed that it was too difficult. He substituted for it a readily comprehensible, theologically oriented disquisition on the role of the trinity in nature and in the world, culminating in the Trinity itself. In this one example, Oresme made two significant changes from his earlier *Questiones*: he not only rendered the subject much simpler, but he also seized the opportunity to theologize.

More Complex Themes in "Le livre"

Among the complex themes that Oresme included in *Le livre*, themes that were often more elaborate and extensive there than in the *Questiones*, are the incommensurability of the celestial motions, the possibility of infinite bodies in a variety of manifestations, and the meanings of the terms "possible" and "impossible"[32]—to say nothing of the numerous theological topics he chose to include. From the fact that Oresme made a French translation and commentary of Aristotle's *De caelo* at the royal court by royal command, not much can be inferred about the reasons that prompted him to make some topics less complex, and others more complex, than in the much earlier *Questiones*, although there is no doubt that he did so.

The Analytical Table of Contents of *Le livre* reveals, however, that there Oresme chose to emphasize how things could occur otherwise than as described in Aristotle's philosophy, a feature that differentiates *Le livre* from the earlier *Questiones*. A major topic that illustrates this feature and that Oresme treated only briefly in his *Questiones* concerned the possibility of other worlds. Because it was inherently a fascinating topic, and because he might have sensed its appeal for the courtiers of Charles V, he greatly expanded and elaborated his treatment of this theme in *Le livre du ciel et du monde*, where he distinguished three types of plurality: (1) successive worlds; (2) concentric worlds, or "worlds within worlds"; and (3) separate,

---

[31] Ibid., p. 57.

[32] For the discussion on "possible" and "impossible," see bk. 1, ch. 30, pp. 211–19.

distinct, and simultaneous worlds in space.[33] In treating these three themes, Oresme emphasizes that "it is good to consider the truth of this matter without considering the authority of any human but only that of pure reason."[34] Ignoring the first of the three alternatives,[35] he focuses wholly on the other two. In the Analytical Table of Contents, he subsumes most of his counterarguments to Aristotle under the heading of "Replies to the arguments Aristotle offers to prove that there can be but one world."[36]

The "worlds within worlds" option is of great interest. Oresme recognizes the speculative nature of this possibility, but he wishes to "toy" with it "as a mental exercise."[37] In the course of a lengthy discussion, he probably impressed, and perhaps even dazzled, his audience with a detailed exposition of the concept of relativity of size. As Oresme explains, "'large' and 'small' are relative, and not absolute, terms used in comparisons. For each body, however small, is large with respect to the thousandth part of itself, and any body whatsoever, however large, would be small with respect to a larger body."[38] Also, if a concavity of air developed in the center of our earth, does this imply that the earth would, according to Aristotelian principles, no longer be in its natural place at the center of the world? Not at all. For its center of gravity is still the center of the world, even if its geometric center is not. Hence, "if a world were enclosed within a concavity inside our earth, nevertheless our earth would be in its natural place since the center of the world would be the middle or center of its weight."[39] At the end of his discussion of concentric worlds, Oresme insists that there are no proofs from reason or experience that can validate or invalidate claims for the existence of such worlds. But we should not assume their existence, because "we should not guess nor make a statement that something is thus and so for no reason or cause whatsoever against all appearances; nor should we support an opinion whose contrary is probable."[40] And yet, "it is good to have considered whether such opinion is impossible."[41] Indeed, contrary to

---

[33] To these three, Oresme added a fourth in the *Questiones*: worlds that contain each other eccentrically, rather than concentrically. He does little more than mention this option and declares that although "this is possible for God nor does it imply [a contradiction]; nevertheless it ought not to be posited in fact nor is it likely" (Kren, "Oresme: De celo," bk. 1, qu. 18, p. 278).

[34] Menut, *Oresme: Le livre du ciel*, bk. 1, ch. 24, p. 167.

[35] Oresme says that "this opinion is not touched upon here" and rests content to assert that "God could do it and could have done it in the past by His own omnipotence, or He could annihilate this world and create another thereafter" (ibid.).

[36] Ibid., p. 735.

[37] Ibid., p. 167.

[38] Ibid., p. 169.

[39] Ibid. Oresme makes no mention of this argument in his *Questiones*.

[40] Ibid., p. 171.

[41] Ibid.

Aristotle, Oresme makes it abundantly clear that such an opinion *is* possible.

In bk. 1, ch. 24 of *Le livre du ciel*, Oresme treats ideas about extracosmic existence that he had earlier discussed in three successive, but related, questions of the *Questiones super De celo*. In the first question of the latter, he asks whether "if there were another world, the earth of that world would be moved to the center of this one" (bk. 1, qu. 17).[42] He then specifically inquires about the possibility of a plurality of worlds (bk. 1, qu. 18), and then raises the even broader question, whether something—body or vacuum—might exist beyond our world (bk. 1, qu. 19). In his discussion of a possible plurality of worlds (bk. 1, qu. 18), Oresme presents six arguments against the possible natural existence of other separate worlds and then concludes with the customary refrain that "God, who is of infinite power can make another world or many."[43] Despite this concession to God's absolute power, he defends Aristotle's position in the *Questiones* and rejects the arguments for other worlds.

But in *Le livre du ciel*, Oresme begins the discussion about the existence of simultaneous, separate worlds with an assertion that Aristotle's "arguments are not clearly conclusive"[44] and then shows which of Aristotle's arguments against the possibility of other worlds are inconclusive. That is, he shows that the contraries of Aristotle's positions, which were presented as impossibilities, are possible. Among these we find Aristotle's claim, as Oresme expresses it, that "it would follow that the earth in the other world would tend to be moved to the center of our world and conversely, etc."[45]—a claim that he refutes by redefining "up" and "down" independently of natural places.[46] In this re-definition, a discussion that is absent from the *Questiones*, Oresme defines "up" and "down" in terms of light and heavy bodies. He speaks of the "natural law concerning heavy and light bodies, which is that all the heavy bodies so far as possible are located in the middle of the light bodies without setting up for them any other motionless or natural place."[47] In effect, he abandons Aristotle's absolute sense of the terms 'up,' 'down,' 'light' and 'heavy,' and argues instead that a body is judged to be 'heavy' and said to be 'down' when it is surrounded by light bodies, where the surrounding light bodies are said to be 'up.' Thus 'heavy' and 'light,' with the associated concepts of 'down' and 'up,' could be con-

---

[42] Kren, "Oresme: De celo," bk. 1, qu. 17, p. 244.

[43] Ibid., bk. 1, qu. 18, p. 276.

[44] Menut, *Oresme: Le livre du ciel*, bk. 1, ch. 24, p. 171.

[45] Ibid., p. 173.

[46] See Grant, *Planets, Stars, and Orbs* (n. 22), pp. 164–66. Aristotle's claim that an earthy body would move from the center of one world to that of another was also refuted by John Buridan, William of Ockham, and many others; see ibid., pp. 159–60.

[47] Menut, *Oresme: Le livre du ciel*, bk. 1, ch. 24, p. 173.

ceived independently of the natural places of bodies.[48] From this "placeless" definition of heavy and light bodies, Oresme concludes a few lines below that, "if God in His infinite power created a portion of earth and set it in the heavens where the stars are or beyond the heavens, this earth would have no tendency whatsoever to be moved toward the center of our world."[49]

Other of Oresme's counterarguments about other worlds, which emphasize God's omnipotence, are also absent from the *Questiones*. One such is Aristotle's insistence, as Oresme puts it, "that there cannot be more than one God and, therefore, it seems there can be only one world."[50] For Christian theologians, the response was obvious: "I reply," Oresme argues, "that God is infinite in His immensity, and, if several worlds existed, no one of them would be outside Him nor outside His power."[51] As for Aristotle's argument that "this world is composed of all the matter available for the constitution of the world, and outside this world there can be no body or matter whatsoever,"[52] Oresme again offers a customary rebuttal that "in truth, God could create *ex nihilo* new matter and make another world."[53]

Apart from other worlds, Aristotle had also insisted, as Oresme put it in *Le livre du ciel*, that "outside this world there is no place or plenum, no void, and no time."[54] Although, at first glance, it seems that Oresme responded similarly in both treatises, there is a considerable difference between the two replies. In the *Questiones*, he declares that although

> according to faith there is no space outside the heaven ... one can concede that outside the heaven there may be a vacuum [*et sic potest concedi quod extra celum sit vacuum*] because God can create a body or place there. Therefore, if it is asked what is that vacuum outside the heaven, one should reply that it is nothing but God Himself, Who is His own indivisible immensity and His own eternity as a whole and all at once.[55]

Here, Oresme is tentative. Because God has the power to create a body beyond the world, a vacuum may actually exist there, because Oresme defines

[48] I have here followed Edward Grant, "The Condemnation of 1277, God's Absolute Power, and Physical Thought in the Late Middle Ages," *Viator* 10 (1979): 222.

[49] Menut, *Oresme: Le livre du ciel*, bk. 1, ch. 24, p. 173.

[50] Ibid., p. 175.

[51] Ibid. This argument does not appear in the *Questiones*.

[52] Ibid.

[53] Ibid.

[54] Ibid. See Aristotle, *De caelo*, 1.9.279a.10–17. "Aristotle's argument may be summarized as follows: since body cannot exist beyond the world, neither place nor vacuum could exist there, because 'in all place there is a possibility of body' (no body, therefore no place) and because 'void is defined as that which, although at present not containing body, can contain it' (no body, therefore no vacuum)" (Grant, *Planets, Stars, and Orbs* [n. 22], p. 169 n. 1).

[55] Kren, "Oresme: De celo," bk. 1, qu. 19, p. 288 (for the Latin text, see p. 287).

a vacuum as "where a body can be and is not."[56] This vacuum, if it exists, must be identified with God's immensity. Although there is no space, beyond the world, there is possibly a void, and perhaps Oresme even meant that there is actually a void space to be identified with God's immensity.

In *Le livre du ciel*, all tentativeness has vanished. There is no doubt where Oresme stands. He insists that if two distinct and separate, spherical worlds existed, a void space would lie between them. To buttress this assertion, he boldly proclaims that "the human mind consents naturally, as it were, to the idea that beyond the heavens and outside the world, which is not infinite, there exists some space whatever it may be [*est aucune espace quelle que elle soit*], and we cannot easily conceive the contrary."[57] This space outside the heavens "is an empty incorporeal space quite different from any other plenum or corporeal space.... Now this space of which we are talking is infinite and indivisible, and is the immensity of God and God Himself, just as the duration of God called eternity is infinite, indivisible, and God Himself."[58] Although we cannot directly perceive this extracosmic space with our corporeal senses, "reason and truth ... inform us that it exists."[59] Oresme was thus convinced that God could make other worlds and that Aristotle could not prove the contrary.[60] Nevertheless, he was firmly convinced that "there has never been nor will there be more than one corporeal world."[61]

One of two themes in the *Livre du ciel* to which Oresme gave the greatest emphasis was the eternity of the world. He concluded the first book with it, dedicating the last seven chapters (24 to 30) to this perennially absorbing topic, which had been a major theme of the Condemnation of 1277. The basic conclusions he reached were much the same as those he had arrived at many years before in his *Questiones super De celo*, where he considered the following question: "Whether something created *de novo* could be perpetuated and also whether something that existed from eternity could be corrupted."[62] To both parts of this question, he responded affirmatively, thus concluding that Aristotle was wrong to deny them. In the course of his technical, scholastic treatment of this important question, Oresme offers the following two (of six) conclusions in support of each claim made in the enunciation of the question. In the third conclusion, he proclaims:

> If the motion of the sun and moon are incommensurable, as is likely, then the pyramid of the earth's shadow always crosses the sphere of the moon by some

[56] Ibid., p. 286.
[57] Menut, *Oresme: Le livre du ciel*, bk. 1, ch. 24, p. 177 (for the French text, see p. 176).
[58] Ibid.
[59] Ibid.
[60] Ibid., p. 179.
[61] Ibid.
[62] Kren, "Oresme: De celo," bk. 1, qu. 24, pp. 410–27.

> path through which it will never cross [again], nor would be able to cross naturally nor could be able [to cross] in the future, therefore continuously some light is categorically and wholly corrupted which has existed as an entity from eternity, and also by the same [reasoning], something is generated which will never be corrupted, and this could be mathematically demonstrated.[63]

And if this is not sufficient to convince his readers, he invokes the faith in the sixth conclusion:

> I say that according to faith, God from perpetuity could have been able to have created any individual or any species without a beginning and at some time He may have wished to annihilate [them] and similarly [this would be true of] whatever [species or individual] He wished to produce and to perpetuate.[64]

A glance at Oresme's list of questions in the *Questiones super De celo* reveals that he treated the eternity of the world in a single, average-sized question, devoting much less space to it than he did to the problem of the possible existence and behavior of infinite bodies, to which he devoted four questions (bk. 1, qq. 13–16). In the *Livre du ciel* he reversed this relationship, devoting approximately twice as much space to the eternity of the world as to the possibility of infinite bodies. (In his French treatise, Oresme chose to lavish the greatest attention in approximately equal amounts on two themes: the eternity of the world and the status of the earth—primarily its shape—and whether it rests or moves.)

At the core of his treatment of the eternity of the world in *Le livre* lay the example Oresme had used in the *Questiones*, to which he would add other illustrations. Once again, solely on grounds of natural philosophy and mathematics rather than faith, he sought to make it "clear that Aristotle's arguments are not conclusive."[65] To show that Aristotle erred when he argued that whatever had a beginning must have an end, and that what has no end, cannot have had a beginning, Oresme assumes the eternity of the world. At the root of his counterargument, he also assumes the incommensurability of the celestial motions—as he did in the *Questiones super De celo*—drawing specifically upon his earlier *Treatise on the Commensurability or Incommensurability of the Celestial Motions* (*Tractatus de commensurabilitate vel incommensurabilitate motuum celi*).[66] Oresme presents

[63] Ibid., p. 424.
[64] Ibid., pp. 424–26.
[65] Menut, *Oresme: Le livre du ciel*, bk. 1, ch. 29, p. 197.
[66] Although the assumption of celestial incommensurability in the *Questiones super De celo* may have been Oresme's first public mention of this concept, he worked out the details in two subsequent treatises, *Tractatus De proportionibus proportionum* and *Tractatus De commensurabilitate*. For editions of these two treatises, see *Oresme "De proportionibus proportionum"* (n. 5) and Edward Grant, ed. and trans., *Nicole Oresme and the Kinematics of*

not only the single example from the *Questiones super De celo*, but others as well. Indeed, he assumes that Saturn, Jupiter, and Mars have three motions each, and that the centers of these bodies are "fixed upon some point, line, or place," that is, they are in conjunction. [67] He concludes that "it is impossible for these three bodies ever again to be joined thus or to be everlastingly joined together in this or in any other place."[68] Taking these planets one at a time, and assuming that one of its three (or more) motions is incommensurable with the others, he argues that "this body is continually in a new disposition and its center in such a spot or motionless point as it never was in before nor will be again, describing a new line forever and perpetually."[69] The incommensurability of one of the sun's three motions also produces phenomena that are contrary to Aristotle's claims. Under these circumstances,

> it follows necessarily that in each movement the center of the sun is in a new position where it never was previously and that the apex of the earth's shadow shifts continually to some place or point where it had never been before and will never be again. Consequently, some light in the sky ceases to exist as such or ceases entirely, which light never had a beginning, while some light begins which will never cease.[70]

Oresme presents similar examples for the moon and for both sun and moon taken together in eclipses. Using a diagram of a series of wheels, he then shows that, for circular motion, "some motion has a beginning and lasts forever,"[71] and then presents a similar illustration for rectilinear motion involving the fall of a heavy body through a successively more resistant medium.[72]

That Oresme designed his French commentary on *De caelo* to counter the excessive influence of Aristotle by showing that the latter's arguments and conclusions were often inconclusive and even mistaken is nowhere better illustrated than at the end of the long discussion about the kinds of eternities just described, which concludes the first book of *Le livre du ciel*:

> Now, therefore, with God's help and to His glory, I have shown clearly in many ways by the light of human reason and by natural light how it is not impossible that any destructible thing can be perpetual and that anything which has had a beginning can endure forever. However, as clearly shown above, both Aristotle and Averroes would like to prove the opposite, which is contrary to truth and to our faith. I have, however, loyally presented their ar-

Circular Motion: "Tractatus De commensurabilitate vel incommensurabilitate motuum celi" (Madison: University of Wisconsin Press, 1971).

[67] Menut, *Oresme: Le livre du ciel*, bk. 1, ch. 29, pp. 197–99.

[68] Ibid., p. 197.

[69] Ibid., p. 199.

[70] Ibid.

[71] Ibid., p. 201.

[72] Ibid., pp. 203–5.

> guments and the evidence to show that their arguments are not sound and that some of them are purely sophistic; for, although Aristotle was an excellent philosopher, nevertheless it is clear from what Eustrathios says concerning the First Book of the *Nicomachean Ethics* that Aristotle was sometimes unduly harsh in his criticism of Plato, whose opinions[73] he hated unreasonably. As we have said many times, he was here arguing against Plato, whom St. Augustine prefers and recommends above all others, along with Plato's followers in philosophy, in the eighth and ninth books of *The City of God*, and he holds that their teachings are more congruous and more in harmony with Catholic faith than those of other philosophers. Thanks be to God. Here ends the first book of *The Heavens and the World*.[74]

Such criticisms of Aristotle were rather unusual during the late Middle Ages. Judging from the tone of this passage, Oresme appears to have had as one of his objectives a theological mission: to subvert, on behalf of the Catholic faith, Aristotle's arguments on the eternity of the world. His praise of Plato, who accepted the creation of the world, underscores Oresme's disagreement with Aristotle.

Although the content of Oresme's discussions in the *Questiones* and the *Livre du ciel* differed in many ways, the conclusion about the two oppositely directed eternities was essentially the same. On the other hand, in treating the possible daily axial rotation of the earth, Oresme's opinions changed dramatically from the earlier to the later treatise. As happened in a number of instances, many of the arguments appear in both works. In the *Questiones*, after presenting a number of powerful arguments in favor of a rotating earth,[75] Oresme opts for the "more common opinion" of Aristotle that the heaven and not the earth rotates daily. He offers counterarguments against four *persuasiones* in favor of the earth's rotation, as well as opposing other more substantial arguments in favor of axial rotation. His rejection of the fourth "persuasive" argument is perhaps the most telling blow against the earth's rotation. The argument was that:

> it is vain to do by many [operations] what can be done by fewer, so it is less [complicated] to posit that the earth which is so small be moved in diurnal motion than the whole heaven [which is] so large and [has] such innumerable stars, any one of which is larger than the earth. Further, one need not posit that the heaven is moved with so great a velocity nor with so many motions, but every motion will be from the west, so that the earth will make one complete circle in a day and the moon one in only a month, in such a way that [the moon] once in a month and the sun in a year circles the center, and so regard-

[73] I have altered Menut's "opinion" to "opinions," since, in his reference to Plato, Oresme says "ses oppinions." See ibid., p. 260, line 150.

[74] Ibid., pp. 261–63.

[75] See Kren, "Oresme: De celo," bk. 2, qu. 13, pp. 667–96. The question is "utrum tota terra semper quiescat."

> ing the other [planets], and yet all phenomena can be saved by this [conception].[76]

In his reply to this *persuasio,* Oresme argues that

> when it is said that it is vain to posit more [operations], etc., this is true, but [the opposed position assuming a moving earth which used this as an argument] cannot save [the phenomena] equally well, as is evident by the arguments which have been offered.[77]

By the time he wrote his French commentary, Oresme's attitude on this question had altered radically. Near the beginning of his lengthy treatment of the possible daily rotation, he reveals his position: "It is possible," he declares, "to embrace the argument and consider with favor ... that the earth rather than the heavens has a diurnal or daily rotation"; moreover, in further support of this position, he insists that "it is impossible to demonstrate from any experience at all that the contrary is true; second, that no argument is conclusive; and third, I shall demonstrate why this is so."[78] After a lengthy series of arguments, in which he demonstrates "why this is so," he concludes that

> one cannot demonstrate by any experience whatever that the heavens move with diurnal motion; whatever the fact may be, assuming that the heavens move and the earth does not or that the earth moves and the heavens do not, to an eye in the heavens which could see the earth clearly, it would appear to move; if the eye were on the earth, the heavens would appear to move."[79]

With the evidence equally balanced toward either of the two alternatives, Oresme opts for the traditional opinion. "Everyone maintains, and I think myself," he explains,

> that the heavens do move and not the earth: For God hath established the world which shall not be moved,[80] in spite of contrary reasons because they are clearly not conclusive persuasions. However, after considering all that has been said, one could then believe that the earth moves and not the heavens, for the opposite is not clearly evident. Nevertheless, at first sight, this seems as much against natural reason as, or more against natural reason than, all or many of the articles of our faith. What I have said by way of diversion or intellectual exercise can in this manner serve as a valuable means of refuting

[76] Ibid., pp. 678–80.

[77] Ibid., pp. 694–96. Here also is the highly abbreviated Latin text: "Ad quartam, cum dicitur quod frustra ponitur plura et cetera, verum est sed non potest salvari eque bene ut patet per rationes factas" (ibid., pp. 693–95).

[78] Menut, *Oresme: Le livre du ciel*, bk. 2, ch. 25, p. 521. For a detailed treatment of Oresme's discussion of the earth's possible axial rotation, see Grant, *Planets, Stars, and Orbs* (n. 22), pp. 642–47.

[79] Menut, *Oresme: Le livre du ciel*, bk. 2, ch. 25, p. 537.

[80] Oresme's reference is to Psalms 92:1, which declares: "Etenim firmavit orbem terrae, qui non commovebitur."

> and checking those who would like to impugn [*impugner*] our faith by argument.[81]

Although, at the end of the discussion, Oresme agrees with Aristotle that the heaven moves with a circular motion, and not the earth, he has nevertheless departed from Aristotle who believed on evidential and metaphysical grounds that the earth lay immobile at the center of the world, whereas Oresme could see no way that reason or experience could resolve the issue—a conclusion that was dramatically different from his reasoned resolution in favor of the earth's immobility in the *Questiones super De celo*.

It was Oresme, in his capacity as a theologian, who resolved this argument in favor of the traditional interpretation, undoubtedly because it had firm biblical support, while lacking sound scientific evidence for either alternative. Although arguments for the earth's rotation appeared more plausible than the alternative, they were not, and could not be, conclusive; hence, the pronouncements and arguments of Scripture and faith had to take precedence. Even if the earth's rotation seemed more probable and plausible than its immobility, rotation had to be assumed false, because Scripture had plainly spoken for an immobile earth at rest in the center of the world. Thus did defense of the faith and theology play a major role in *Le livre du ciel.* In declaring that his arguments might serve as a refutation of "those who would like to impugn [*impugner*] our faith by argument," Oresme was reinforcing an aspect of his treatise that he regarded as of great importance, as is evidenced by his concluding remarks (cited above, see note 23) that his work was, among other things, a defense of the faith against those "who, by chance, would heap calumny upon it or impugn [*impugner*] it." As if to underscore the importance of theology, Oresme concluded his treatise with a chapter that was "concerned especially with the body of Jesus Christ" (bk. 4, ch. 12).

Summary and Conclusion

As I have indicated, the Analytical Table of Contents of *Le livre du ciel et du monde* was comprised of themes that Oresme thought worthy of emphasis. It enabled him to highlight his differences with Aristotle. Many of these topics had counterparts in his earlier *Questiones*, where, however, he made no effort to emphasize his differences with Aristotle. Moreover, themes relevant to religion and theology in *Le livre du ciel* are absent from the earlier *Questiones*, as, for example, "How God exists in the heavens everywhere."[82] Most of the themes in natural philosophy are present in both

---

[81] Menut, *Oresme: Le livre du ciel*, bk. 2, ch. 25, pp. 537–39.

[82] See ibid., pp. 739, 749.

treatises, although, as we saw, Oresme heavily emphasized some themes in *Le livre du ciel* that were given only modest consideration in the *Questiones*. Also, in some instances a theme given extensive attention in the *Questiones* is treated in much briefer compass in *Le livre du ciel*—as, for example, the maximum and minimum powers of things, to which Oresme devoted three questions in his *Questiones*, but only somewhat more than a page in *Le livre du ciel*.[83]

The significant differences between the two treatises should not surprise us given the different circumstances under which they were written. Oresme composed the *Questiones* as a master of arts at the University of Paris early in his career; he wrote the commentary on *Le livre du ciel* near the end of his life, when he had become a distinguished theologian–natural philosopher, and therefore capable of introducing theology anywhere he thought it desirable and effective. The typical medieval questions format was unsuitable for in-depth, sustained examinations of any topic, as Oresme himself recognized.[84] Thus his *Questiones* was much more restrictive than his more free-wheeling *Livre du ciel,* where the commentary format allowed him to pick and choose the passages of the *De caelo* that he would elaborate at great length, and those that he would treat cursorily, or not at all. Once launched into a topic, Oresme was free to introduce theological ideas or bring together concepts and topics that would not have been linked in a questions treatise. In short, Oresme had far greater scope in *Le livre du ciel* than he had had in the *Questiones super De celo*—and he made full use of it.

The ultimate mystery about Oresme's *Livre du ciel et du monde* revolves around the relationship between the comments and discussions he chose to include and the courtly audience for which they were intended. There is no evidence to indicate that Charles V, or anyone else, advised Oresme about achieving some optimum level of intelligibility that would be suitable for courtiers unaccustomed to technical discussions in natural phi-

---

[83] See above, nn. 26, 27.

[84] In his *Questions on Aristotle's De anima*, bk. 3, qu. 3, Oresme says that, "in treating the intellect, we must first see about the substance and nature of it, secondly [we must see about] its accidents and operations. Concerning the nature [of the intellect], the opinions about it must be declared, namely, first the opinion of the Commentator, secondly [the opinion] of Avempace and others, thirdly [the opinions] of Plato and Pythagoras, fourthly [the opinion] of Alexander, [and] fifthly [the opinion] of the faith and of truth. And although it would be better to treat this without questions, yet because of custom, I will do it by means of questions" (my trans. from the Latin text of Marshall, "Nicholas Oresme's "Questiones super libros Aristotelis De anima" [n. 25], pp. 527–28). All told, Oresme devoted five questions to the intellect. In a note (on p. 526) to the paragraph cited, Peter Marshall explains that "what in fact Oresme had in mind here was the inadequacy of the *questiones* format in presenting the topic of the nature of the intellect in an appropriate manner. Oresme would have preferred simply to outline the various opinions, but his actual treatment instead follows the normal format of the other *questiones*."

losophy.[85] In the absence of evidence to the contrary, we ought to assume that Oresme was the sole judge of its content. But this must not be taken to signify that he was uninfluenced by the wishes of Charles V, which he presumably knew.

With this in mind, we can make some plausible conjectures about Oresme's purposes from the assumption that one of his major objectives, encouraged by Charles, was to stimulate debate and argumentation among the courtiers. In this article, I have argued that Oresme sought to achieve this goal by showing (1) that great authorities, especially Aristotle, could be mistaken or misguided or inconclusive about many points in natural philosophy; (2) that conclusions and assumptions about natural philosophy itself were inherently difficult and elusive, as difficult and elusive as the articles of faith; and (3) that there is only one truth: what is true in natural philosophy must also be true in religion and theology, and vice versa.

Although the *Questiones super De celo* and *Le livre du ciel et du monde* were both explications of Aristotle's *De caelo*, they differed vastly in their objectives and methods. Since neither was published until the second half of the twentieth century, we are ignorant about their subsequent influence. But we may assume that the *Questiones* was occasionally read or consulted at the University of Paris. Although Claudia Kren regarded it as an early treatise that reveals awkwardness, confusion, and inarticulateness, Oresme's *Questiones* already displays his keen intelligence and his willingness, even eagerness, to treat unusual ideas and concepts. In the arts faculty of the University of Paris, Oresme's *Questiones* would probably have been regarded as an imaginative work by a bright, young master.

The attitude of the nobles and bureaucrats at the court of Charles V toward *Le livre du ciel et du monde* is, however, an unknown, and will probably remain so. Did the courtiers read the treatise? If so, did they understand it? Were Oresme's attitudes toward natural knowledge and the faith influential? Did *Le livre* actually stir debates at the court? Were those charged with governing able to relate any of Oresme's ideas in *Le livre* to their experiences? If there are answers to such questions, they have yet to be discovered. We can all agree, however, that Charles's request for the French translation of an abstruse treatise in natural philosophy was extraordinary—as was the product of the translator and commentator, Charles's faithful, if not so humble, servant, Nicole Oresme.

[85] It is likely that some courtiers had been educated at the University of Paris, or some other university.

CHARLES V, NICOLE ORESME, AND CHRISTINE DE PIZAN: UNITIES AND USES OF KNOWLEDGE IN FOURTEENTH-CENTURY FRANCE

JOAN CADDEN

One of John Murdoch's most influential contributions to the study of medieval science has been to expand our understanding of the specific intellectual contexts in which medieval ideas about the natural world were located. Long before the concept of intertextuality became fashionable, Murdoch's detailed empirical study of the concerns, arguments, and methods of scholastic sciences demonstrated their ties to, indeed their integration with, the concerns, arguments, and methods of philosophy and theology. Without a grasp of these intellectual contexts, modern scholars are in danger of misconstruing (or missing altogether) the *medieval* meaning and importance of specific developments in scientific theory or approach:

> If it is true, for example, that the new area of importance for fourteenth-century theology was that of justification, grace, and predestination, then the philosophical deliberations over the logic of future contingent propositions surely was a phenomenon of relevance.[1]

The intimate relations among such concerns are an example of what Murdoch calls the "unitary character of late medieval learning"; they have their foundation in the curriculum and milieu of universities, which promoted a complicated conversation within and among the arts, theology, medicine, and law. Although texts are the immediate context in which this unity is grounded, Murdoch has suggested that the institutions that frame it and the relations among the persons who enacted it have a bearing on its particular character.[2]

The work of Nicole Oresme, encompassing (among other things) natural philosophy, epistemology, mathematics, and theology, illustrates Murdoch's account of the fourteenth-century intellectual scene. Seen in dia-

---

[1] John E. Murdoch, "The Analytic Character of Late Medieval Learning: Natural Philosophy without Nature," in *Approaches to Nature in the Middle Ages: Papers of the Tenth Annual Conference of the Center for Medieval and Early Renaissance Studies*, ed. Lawrence D. Roberts, Medieval and Renaissance Studies and Texts 16 (Binghamton, N.Y.: Center for Medieval and Renaissance Studies and Texts, State University of New York at Binghamton, 1982), pp. 171–213, on pp. 173–74.

[2] John E. Murdoch, "From Social to Intellectual Factors: An Aspect of the Unitary Character of Late Medieval Learning," in *The Cultural Context of Medieval Learning: Proceedings of the First International Colloquium on Philosophy, Science, and Theology in the Middle Ages, September 1973*, ed. John Emory Murdoch and Edith Dudley Sylla, Boston Studies in the Philosophy of Science 26; Synthèse Library 76 (Dordrecht and Boston: D. Reidel, 1975), pp. 271–348, on pp. 274–75, 309–10.

logue with other developments—from the science of mechanics to the epistemology of probabilism—it takes on greater significance than it would have were Oresme not engaged in a wider scholarly conversation. Yet his systematic criticism of and polemical attacks upon astrology, particularly in the French *Livre de divinacions*, have suggested the existence of another set of contexts—the intellectual life of the laity, the social dynamics of the court, and the political interests of the monarchy.[3] Historians of science have had limited interest in the connections between the production of learned texts about the natural world and these broader contexts, in part because (unlike much of Oresme's other work) the antiastrological writings have seemed to have existed in intellectual isolation—contesting unspoken views of silent adversaries (the astrologers) and addressing the unyielding position of an equally silent audience (King Charles V of France).

Within the court itself, however, resides evidence of a serious exchange about the nature and role of astral sciences. Furthermore, this conversation reveals cultural and political roles for academic philosophy that extend beyond the simple deployment of horoscopes and elections. Though we have little direct record of their words, Oresme's adversaries were not silent.

[3] At the core of Nicole Oresme's antiastrological corpus are the *Questio contra divinatores horoscopios*, ed. Stefano Caroti, in *Archives d'Histoire Doctrinale et Littéraire du Moyen Age* 43 (51st yr., 1976): 201–310; the *Tractatus contra astrologos iudicarios*, ed. Hubert Pruckner, in *Studien zu den astrologischen Schriften des Heinrich von Langenstein* (Leipzig: Teubner, 1933), pp. 227–45; and *Le livre de divinacions* (hereafter Oresme, *Divinacions*), ed. G. W. Coopland, in *Nicole Oresme and the Astrologers: A Study of His Livre de divinacions* (Liverpool: University of Liverpool Press, 1952), pp. 50–210. The most important related texts are the Ad pauca respicientes, in *De proportionibus proportionum and Ad pauca respicientes*, ed. Edward Grant (Madison: University of Wisconsin Press, 1966), pp. 379–441; and the set of quodlibetal questions edited by Bert Hansen, *Nicole Oresme and the Wonders of Nature: A Study of his De causis mirabilium,* Studies and Texts 68 (Toronto: Pontifical Institute of Mediaeval Studies, 1985). In addition to the studies accompanying the editions, see Charles Jourdain, "Nicolas Oresme et les astrologues de la cour de Charles V," in *Excursions historiques et philosophiques à travers le moyen âge* (Paris: Fermin-Didot, 1888), pp. 559–85; Lynn Thorndike, *History of Magic and Experimental Science* (New York: Columbia University Press, 1934), vol. 3, pp. 398–423; Edward Grant, "Nicole Oresme on Certitude in Science and Pseudo-Science," in *Nicole Oresme: Tradition et innovation chez un intellectuel du XIVe siècle,* ed. P. Souffrin and A. P. Segonds, Science et Humanisme (Paris: Les Belles Lettres, 1988), pp. 31–43; Stefano Caroti, "Éléments pour une reconstruction de la philosophie de la nature dans les *Quodlibeta* de Nicole Oresme," and Max Lejbowicz, "Chronologie des écrits anti-astrologiques de Nicole Oresme: Étude sur un cas de scepticisme dans la deuxième moitié du XIVe siècle," in *Autour de Nicole Oresme: Actes du Colloque Oresme organisé à l'Université de Paris XII*, ed. Jeannine Quillet (Paris: J. Vrin, 1990), pp. 85–118 and 119–74, respectively.

Noted in virtually every scholarly treatment of the antiastrological works, the relationship with the court is addressed more fully in Jourdain, "Nicole Oresme et les astrologues"; Jeannine Quillet, *Charles V, le Roi Lettré: Essai sur la pensée politique d'un règne*, Présence d'Histoire (Paris: Librairie Académique Perrin, 1984), pp. 96–114; Edgar Laird, "Astrology in the Court of Charles V of France, as Reflected in Oxford, St. John's College, MS 164," *Manuscripta* 34 (1990): 167–76. The subject is sometimes addressed under the heading of Oresme's "influence": see, e.g., Coopland, *Nicole Oresme*, pp. 7–13.

Traces of their part in the argument about the foundations and practices of astrology come down to us in the writings of Christine de Pizan, whose father was a prominent physician-astrologer in Charles V's court, and these in turn are confirmed by Oresme's own choices and rhetoric. He was not simply condemning and seeking to dislodge the astrologers, but was engaging in debate with them, thus participating in and contributing to a wider forum for scholarly issues.

The existence of this conversation—which, like Oresme's work as a translator, eroded the boundaries between university and court, Latin and vernacular—also helps to explain the king's silent support of the participants on both sides.[4] Insofar as it represented a royal share in the prestige of learned ideas and practices and insofar as it engaged a range of intellectuals whose careers owed much to his support, the debate itself formed part of a set of political strategies that enhanced the status of the king and the monarchy. Thus, whatever Charles's personal views on the truth and efficacy of judicial astrology, as monarch he saw fit to tolerate, indeed to encourage, the debate, rather than to choose sides and end the discussion. The participants, Oresme included, were not only confronting questions about the structure of the cosmos, the nature of knowledge, and the freedom of the individual will but were thereby also explicitly developing notions of the good ruler and good government. And whatever the utility of astrology to the monarch and the monarchy, the utility of arguing about astrology suggests an expanding set of meanings for the sciences and their proponents.

Just as the university and its curriculum constitute relevant contexts that inform our understanding of the knowledge generated and transformed there, so the court and its culture are contexts that can inform our understanding of the knowledge shaped and engaged there. The environment of the court differed greatly from that of the university with respect to the motives and opportunities for doing science.[5] Thus the divide between these two settings makes the detection of common principles difficult and perhaps unlikely. Nevertheless, in spite of their profound disagreements, the supporters and adversaries of astrology shared certain assumptions, methods, and authorities that bound them together in a loosely unified system of knowledge. And, in spite of the differences in their positions and purposes,

---

[4] On the role of the translations, see especially Serge Lusignan, "Nicole Oresme traducteur et la pensée de la langue française savante," in Souffrin and Segonds, *Nicole Oresme* (n. 3), pp. 93–104; and, more generally, idem, "*Parler vulgairement*": *Les intellectuels et la langue française aux XIIIe et XIVe siècles* (Paris and Montreal: J. Vrin, and Presses de l'Université de Montréal, 1986).

[5] Cf. Guy Beaujouan, "Motives and Opportunities for Science in the Medieval Universities," in *Scientific Change: Historical Studies in the Intellectual, Social, and Technical Conditions for Scientific Discovery and Technical Invention from Antiquity to the Present*, ed. A. C. Crombie (New York: Basic Books, 1963), pp. 219–36, on p. 235.

they participated together in the social and political processes of Charles's court.

CHARLES LE SAGE AND ASTROLOGICAL LEARNING

Charles V is best known for his management of the French nobility, his effective system of taxation, and his successful prosecution of one segment of the Hundred Years War. He became regent in 1356 when his father, Jean II, was a prisoner of the English, northern and western France were under English control, and the city of Paris was in revolt. When he died in 1380, the English crown had been expelled from all but a few footholds, the free companies of English and French soldiers that had plagued the land had been brought under control, and France was (temporarily) restored to order. The successes of his reign no doubt contributed to his being called "Charles the Wise," but that title was due more to his carefully cultivated reputation as a prudent and bookish man and his activities as an informed patron of the arts, literature, and the sciences within his court.[6]

By the end of his reign in 1380 Charles's library at the Louvre palace contained more than eleven hundred volumes, a collection comparable in size to that of the Sorbonne. It was the first in France to be constituted as a permanent, public institution, with a specially arranged space and a permanent librarian.[7] The library was one layer of Charles's patronage, which encompassed not only ownership of the texts but also support of those who produced them and the encouragement of learned discussion at court. About 450 books—more than a third of the collection—were on science and medicine, and several bear specific signs of the king's interest in the stars and the judicial sciences.[8] One book contains a miniature of the king himself conversing with nine astrological authorities;[9] two (a French geo-

[6] Claire Richter Sherman, "Representations of Charles V of France as a Wise Ruler," *Medievalia et Humanistica*, n.s., 2 (1971): 83–96; Quillet, *Charles V* (n. 3), pp. 46–60; Françoise Autrand, *Charles V: Le Sage* ([Paris]: Fayard, 1994), pp. 714–19; Donald Byrne, "Rex imago dei: Charles V of France and the 'Livre des propriétés des choses,'" *Journal of Medieval History* 7 (1981): 97–108, on pp. 102–3.

[7] Léopold Delisle, *Recherches sur la Librairie de Charles V*, 2 vols. and atlas of plates (Paris: H. Champion, 1907), identified 1,239 books from early catalogues (about a hundred of which came to the collections after Charles's death), the provenance of surviving manuscripts, and other evidence. There may have been other items in the collection as well. On the character of the institution, see vol. 1, p. 2. See also Autrand, *Charles V* (n. 6), pp. 719–20.

[8] The legal works are Delisle, *Librairie de Charles V*, #380–#436. His heading "arts and sciences," #437–#870, includes about eighty volumes on moral philosophy, political theory, chivalry, etc.; the rest are on science and medicine. The remaining items, #871–#1,239, are mainly history and literature.

[9] Claire Richter Sherman, *The Portraits of Charles V of France (1338–1380)*, Monographs on Archaeology and the Fine Arts Sponsored by the Archaeological Institute of America and the College Art Association, 20 (New York: New York University Press for The College Art Association of America, 1969), pp. 18–19 and fig. 1.

mancy and a set of the Alfonsine tables) bear Charles's own mark which, following his father's practice, he inscribed in his favorite books.[10] Works on astronomy and astrology dominated, many bearing the familiar names of Muslim and Jewish authors. They occupied more than 170 volumes, or about 40 percent of the scientific and medical titles, and not a few of them were present in French as well as Latin. In quality and depth, as well as in number, they represented a highly serious body of material, shying away neither from the technical difficulty nor from the Oriental associations of the celestial sciences.[11]

Although the king's library was deeply indebted to the learning of university scholars, at its center were many texts that were marginal in the official course of study, some of them even suspect from the point of view of orthodox natural philosophy or theology.[12] The off-center subjects such as astrology and geomancy and their situation at the heart of the kingdom—not only in the Louvre but also in the interests of the king himself—gave rise to a sense of urgency and an atmosphere of debate that lent significance to academic opinions like those of Oresme.

Deeply entwined with and often not fully distinguished from astronomy, astrology could lay claim to a place among the liberal arts.[13] At Oxford, where the disciplines of the quadrivium flourished, and in southern Europe, where secular institutions and medical education were stronger, the study of the heavens found a secure place in the developing curricula of the thirteenth and fourteenth centuries. At Paris, a lack of inclination and tradition was reinforced at once by theological suspicion of astrology and, toward the end of the thirteenth century, by a general vigilance concerning the doctrinal health of the arts faculty. By the middle of the fourteenth century, however, teaching and research on the motions and influences of the heavens had begun to accumulate around the edges of the academic program at Paris—in private homes and on feast days.[14] This interest covered not

---

[10] Delisle, *Librairie de Charles V*, vol. 1, pp. 3–5; vol. 2, ##592, 595, 752; and *Cabinet des manuscrits de la Bibliothèque Impériale*, 3 vols., Histoire Générale de Paris, Collection de Documents (Paris: Imprimerie Impériale, 1868–81; repr. New York: Lenox Hill, 1973), vol. 1, p. 19.

[11] There were works attributed to Ptolemy (including a French *Tetrabiblos*), Albumasar, Alchabitius, Alkindi, and Abraham ibn Ezra, as well as Sacrobosco and John of Saxony. Cf. Richard Lemay, "The True Place of Astrology in Medieval Science and Philosophy: Towards a Definition," in *Astrology, Science and Society: Historical Essays*, ed. Patrick Curry, (Woodbridge, Suffolk: Boydell Press, 1987), pp. 57-73.

[12] See Beaujouan, "Motives and Opportunities" (n. 5), p. 235.

[13] The study of celestial motions and the study of their influences could be and sometimes were distinguished, but these endeavors were inextricably associated with each other. See Lemay, "True Place of Astrology" (n. 11) and Theodore Otto Wedel, *The Medieval Attitude toward Astrology, Particularly in England*, Yale Studies in English 60 (New Haven: Yale University Press, 1920).

[14] On the teaching of mathematical sciences at Paris, see Richard Lemay, "The Teaching of Astronomy in Medieval Universities, Principally at Paris in the Fourteenth Century,"

merely the motions of the heavens but also their use in formulating judgments and choices in human affairs.[15] Students and teachers of the arts faculty assimilated astrological traditions into a strong intellectual framework, including the substance of Aristotelian cosmology and the methods of scholastic investigation.[16] Thus, not only was the court of Charles V indebted to the university for its access to many of the texts and ideas represented in the king's library, but the condemnations of astrology associated with the court had roots in the deep concerns of university philosophers and theologians with problems of free will and determinism, certainty and contingency.

Starting in the late 1350s, astrology became the focus of intense intellectual interest in the household of the Dauphin, Charles, Duke of Normandy. He drew heavily upon university men's expertise and access to texts as he built the scientific section of his library. Astrological works were among the first books he commissioned for his library, and the French translation of Ptolemy's *Quadripartitum* with a gloss by ʻ Alī ibn Riḍwān dates from this period; long and filled with technical language, it was most likely the work of a university master.[17] In 1361 Robert de Godefroy made a French translation of the *Liber novem judicum*, a work on judicial astrology that advertised its connection to Moslem science, and to which were appended horoscopes of Charles and his four children. Robert described himself as "master of arts and *astronomien*" of the Dauphin—a dual identity as academic and courtier.[18]

Yet advanced university training was not required for participation in the astrological boom. The Franklyn of Chaucer's *Canterbury Tales*, invented only a generation later, was a bourgeois who made a great show of his command of astrological jargon. In 1361–62, at Charles's request, Pèlerin de Prusse produced a number of French tracts that he announced as useful for the astrology of elections—the choice of auspicious and the

Manuscripta 20 (1976): 196–217, on pp. 200–4 and Guy Beaujouan, "L'enseignement de l'arithmétique élémentaire à l'Université de Paris aux XIIIe et XIVe siècles: De l'abaque à l'algorisme," in *Homenaje a Millás-Vallicrosa*, 2 vols. (Barcelona: Consejo Superior de Investigaciones Cientificas, 1954), vol. 1, pp. 93–124, on pp. 97–104.

[15] Lemay, "Teaching of Astronomy," p. 200. The chronology is summarized in Beaujouan, "L'enseignment de l'arithmétique," pp. 100–101.

[16] Murdoch, "Analytical Character" (n. 1).

[17] MS Paris BN fr. 1348 is apparently the original: Bibliothèque Nationale, *La librairie de Charles V* (Paris: Bibliothèques Nationale, 1968), #198. Max Lejbowicz, "Guillaume Oresme, traducteur de la Tetrabible de Claude Ptolémée," *Pallas* 30 (1983): 107–33, gives the fullest review of the question of Nicole Oresme's responsibility for this translation, concluding that he was not; Claire Richter Sherman has recently offered art historical evidence on the other side in *Imaging Aristotle: Verbal and Visual Representation in Fourteenth-Century France* (Berkeley: University of California Press, 1995), pp. 18–20, esp. n. 34.

[18] Charles Victor Langlois, *La vie en France au moyen âge du XIIe au XIVe siècle*, vol. 3, *La connaissance de la nature et du monde d'après des écrits français à l'usage des laïcs* (Paris: Hachette, 1927), p. 207; Delisle, *Librairie de Charles V* (n. 7), #200, p. 116.

avoidance of inauspicious times for specific actions. Though he mentioned himself a number of times, the author did not qualify himself as "master."[19] No doubt a student at some time, he represented the environment of the court more than that of the university.[20]

Like the initial arrival of celestial sciences at the university, the Dauphin's early acquisitions were serious but not ambitious. Just as the first required course in the mathematical sciences instituted at the arts faculty of Paris in 1366 probably involved the study of a single elementary work, such as the *De spera*, Pèlerin assured his patron that his tract on the astrolabe would introduce the use of that instrument "without involving strange operations and requirements in the form of calculation, so that the operations would be simple and comprehensible to anyone."[21] Although much of the material Charles acquired in the years before his coronation was of a general nature, there are signs of a strongly pragmatic side to his interest, including the horoscopes cast for Charles and his children.[22] In his *Traité des élections universelles des douze maisons*, written for Charles in 1361, Pèlerin de Prusse included such sample interrogations as "Whether a man will have a kingdom."[23] Such bluntly judicial uses of the science represented its most objectionable dimension: they reflected a confidence (not in evidence at the university) in the set course of events and its knowability.

Although astrology never fully escaped its marginal status at the arts faculty of Paris, in the years that followed the death of Jean II and the coronation of Charles in 1364 it retained and even enhanced its status in Charles's court. With the help of his librarian, Gilles Malet, a lay bibliophile who worked his way up through the ranks of the king's household,[24] Charles acquired much of his vast and excellent astronomical and astrological corpus—treatises, tables, glosses—in Latin and French. During this period, the members of Charles's entourage most closely associated with the arts faculty were not the source of support for his interest in astrology, as they seem to have been in the early years; rather, it was physicians among Charles's retainers who advanced the cause of the celestial arts.

---

[19] Delisle, *Cabinet* (n. 10), vol. 3, pp. 336–37 n. 1.

[20] What little is known of Pèlerin is summarized in the Introduction to Edgar Laird and Robert Fischer, *Pèlerin de Prusse on the Astrolabe: Text and Translation of His Practique de astralabe*, Medieval and Renaissance Texts and Studies 127 (Binghamton, N.Y.: Center for Medieval and Early Renaissance Studies, 1995), pp. 1–5.

[21] Delisle, *Cabinet*, vol. 3, p. 337, n. 1: "sans meller ouvrages et besoingnes estrainges par guise de calculacion, afin que les ouvrages soient simples et de chascune personne entendables."

[22] Oxford, St. John's, MS. 164, fol. 158 in Sherman, *Portraits* (n. 9), p. 22 and fig. 6.

[23] Autrand, *Charles V* (n. 6), p. 745, "Interrogation d'un homme s'il aurait un royaume."

[24] Delisle, *Librairie de Charles V*, vol. 1, pp. 10–20. Gilles himself does not appear to have had a particular interest in astrology: of the two dozen books he is known to have presented to Charles, only a bestiary and a "Marvels of the World" approach a scientific subject (p. 15).

The link between medicine and astrology, in matters concerning the general welfare as well as in matters of individual health, was familiar to French intellectuals and the lay public. In 1348 the medical faculty had identified the configuration of heavenly bodies as a general cause of the plague, and in 1365 John of Burgundy's medical treatise on the plague opened:

> Everything below the moon, the elements and the things compounded of the elements, is ruled by the things above, and the highest bodies are believed to give being, nature, substance, growth and death to everything below their spheres. It was, therefore, by the influence of the heavenly bodies that the air was corrupted and made pestilential.[25]

Thinking of this kind was no doubt greatly enhanced at the university by the influence of scholars from the south, like Pietro d'Abano, whose work was widely copied and referred to in Paris. But the faculty of medicine at Paris (unlike its sisters at Montpellier, Padua, and Bologna) was institutionally and intellectually weak compared to the other faculties. Henry of Langenstein signaled his disdain when he commented that, in terms of certainty of knowledge, astrology was last in line—after medicine and meteorology.[26] Buffeted by contingencies and unknowns, medicine nevertheless possessed status as a practical art linked to both authority and reason, and thus lent credibility to astrology.[27]

Men learned in medicine moved more easily between university and court than did masters of arts, and their social role involved not only the mediation of theory and practice but also the job of explaining and prognosticating on matters of health and disease to a lay audience. Most, if not all, of the physicians who served at court would have had some training in astrology. Certainly those trained at the university of Montpellier, like Jean Jacme and Jean de Tournemire (for whose expenses and services the king paid the impressive sum of 400 gold francs in 1378),[28] were proficient in that science. The two most prominent physicians attached to Charles's court were explicitly associated with astrology as well: Gervais Chrétien and Tommaso da Pizzano. Gervais, who studied arts and medicine at Paris, founded a medical college at the university to which Charles donated astro-

[25] Rosemary Horrox, ed. and trans., *The Black Death* (Manchester: Manchester University Press, 1994), #62, pp. 184–85.

[26] Nicolas Steneck, *Science and Creation in the Middle Ages: Henry of Langenstein (d. 1397) on Genesis* (Notre Dame, Ind.: University of Notre Dame Press, 1976), p. 100.

[27] See Michael R. McVaugh, "The Nature and Limits of Medical Certitude at Early Fourteenth-Century Montpellier," *Osiris*, 2d ser., 6 (1990): 62–84. The problem of medical knowledge played itself out in ways distinct from but related to the problem of natural knowledge. Cf. Grant, "Nicole Oresme on Certitude" (n. 3).

[28] Léopold Delisle, *Mandements et actes divers de Charles V (1364–1380) recuillis dans les collections de la Bibliothèque Nationale*, Collection de Documents Inédits d'Histoire de France, 1st ser., Histoire Politique (Paris: Imprimerie Nationale, 1874), #1722.

nomical instruments, astrological books, and an endowment for the study of mathematical disciplines.[29] Tommaso, a doctor of medicine from Bologna, is referred to in royal documents as "our beloved *astronomien*" (a term used indifferently for "astronomer" and "astrologer"), "our beloved and faithful physician," and, toward the end of Charles's life, "our beloved and faithful councilor and physician." He served Charles for more than ten years, during which he was called on to apply his astrological skill to more than medical uses: a hostile witness, Philippe de Mézières, complained later, "O, how often was Thomas of Bologna, with his weak judgment, mistaken and deceived!"[30]

The "Debate" About Astrology

Philippe's attack is an abbreviated and personal echo of the philosophical program and extended polemic that Nicole Oresme launched against astrology during Charles's years as dauphin, regent, and king. In the face of such adamance, the books and instruments in Charles's possession, and the visibility of astrologers at court have given rise to the impression that a conflict about astrology occurred during Charles's reign. Thus, for example, Charles Jourdain, in his foundational essay on the subject, declared that "judicial astrology had captivated not only the fourteenth century but also the court of Charles V," and Edward Grant asserts that Oresme's mathematical arguments against astrology were not "a merely scholastic exercise," but that he wanted to "strike a blow at the astrologers, who had aroused his deep concern by their considerable influence on the king of France."[31]

These circumstances alone, however, are insufficient to illuminate the nature and outcome of the exchange—indeed, to establish the existence of

[29] S. Le Paulmier, "Notice sur Gervais Chrétien, fondateur du Collège de Maître Gervais, à Paris, et sur Guy Chrétien, trésorier du roi," [pt. 1], *Revue Catholique du Normandie* (1897): 40–60, on pp. 49–50.

[30] Christine de Pizan, *Le livre des fais et bonnes meurs du sage roy Charles V* (hereafter Christine, *Charles V*), ed. S. Solente, Société de l'Histoire de France, Série antérieur à 1789, 2 vols. (Paris: Librairie Ancienne Honoré Champion, 1936–40), vol. 1, pp. i–xv: the descriptive phrases "nostre amé astronomien," "nostre amé et feal physicien," and "nostre amé et feal conseiller et physicien" are on p. iv; Philippe de Mézières cited, pp. x–xi ("O quantes foiz Thomas de Bouloigne à cestui petit jugement failly et fu deceu"). On Thomas's career, see Charity Cannon Willard, *Christine de Pizan: Her Life and Work* (New York: Persea, 1984), and the Introduction to Christine, *Charles V*. Although medieval authors occasionally attempt to make a terminological distinction between *astronomia* and *astrologia*, the two terms are usually used interchangeably.

[31] Jourdain, "Oresme et les astrologues" (n. 3), p. 582; Edward Grant, ed. and trans., *Nicole Oresme and the Kinematics of Circular Motion: Tractatus de commensurabilitate vel incommensurabilitate motuum celi* (Madison: University of Wisconsin Press, 1971), p. 125. See also Grant's introduction to Nicole Oresme, *De proportionibus proportionum* (n. 3), p. 64.

an exchange at all. When Jourdain cites astrology's claim to the status of a science, or Stefano Caroti speaks of "astrology's" answer to Oresme, the voice in question is strangely disembodied.[32] It inheres in one sense in the texts that multiplied in the king's library and in the objections that Oresme raises within his own arguments, but it hardly constitutes an interlocutor in a real debate.

Only the history of astrology at the university and the court establishes a set of conditions that make plausible a two-sided conversation. More specifically, the strong relationship between astrology and medicine—present at the University of Paris, enhanced by university-educated personnel from the south, and developed in the practices of the court—formed a basis for the defense of astrology. The role of men like Gervais Chrétien and Tommaso da Pizzano not only lent respectability to the enterprise of reading the heavens for practical purposes, but also focused attention on some of the most widely accepted and philosophically respectable dimensions of astrology. In addition, the respect for learning and the high level of intellectual discussion in Charles's entourage created conditions under which an educated defense of astrological beliefs and practices was possible. Finally, like Oresme's French version of Aristotle's treatise on the heavens discussed by Edward Grant in this volume, the controversy surrounding astrology reveals its full historical significance only when seen in the context of theoretical and practical concerns of the monarchy.

While we know something of the careers of Charles's physician-astrologers and the writings of some have come down to us, none has left us a defense of astrology. One figure associated with the court has, however, transmitted the outlines of such a defense. Christine de Pizan is regularly cited to support the contention that Charles was enamored of the science of astrology and held it in the highest regard.[33] Brought by her father Tommaso to Paris in 1368 at the age of four and raised close enough to Charles's court to make plausible claims of affiliation, Christine was in a position to know something about the king's tastes and convictions. As the daughter of one of the physician-astrologers she would have had cause to exaggerate the status of the arts upon which her father's position depended. After the deaths of the king, her father, and her husband, Christine, with

---

[32] Jourdain, "Oresme et les astrologues," p. 567; Stefano Caroti, "Nicole Oresme's Polemic against Astrology in His 'Quodlibeta,'" in Curry, *Astrology* (n. 11), pp. 75–93, on p. 88: "astrology's reply," etc.

[33] Christine's testimony is far more relevant and reliable than that of Simon de Phares, who wrote nearly a hundred years after Charles V's death: Symon de Phares, *Recueil des plus célèbres astrologues et quelques hommes doctes, faict par Symon de Phares du temps de Charles VIII, pub. d'après le manuscrit unique de la Bibliothèque Nationale*, ed. Ernest Wickersheimer (Paris: Honoré Champion, 1929). See also Jean Patrice Boudet, *Lire le Ciel: La bibliothèque de Simon de Phares, astrologue du XVe siècle*, Publications de Scriptorium 9 (Brussels: Centre d'Etude des Manuscrits, 1994).

two young children and few resources, deployed her strong education and her high connections to make a career as a writer. As an aspiring client of those highly placed in the world of Charles VI, she had incentives to speak to the interests of her patrons, as well as to uphold the reputation of her father—for his sake and her own. Her *Livre des fais et bonnes meurs du sage roy Charles V*, written in 1404 for Philippe, Duke of Burgundy, brother of the late king, is a glowing monument in praise of that monarch and contributed significantly to the perpetuation of his reputation as a wise king.[34]

Christine's works have been underexploited as a source for the practices, positions, and arguments that constituted the second side of the astrological controversy. On the one hand, historians of science have not taken her knowledge of astrology seriously, while, on the other hand, her biographer Charity Cannon Willard has not appreciated the complexity of contemporary objections to astrology. Willard surveys the clues to Christine's cosmological and astrological views and suggests that she is answering Tommaso's critics, but finds Christine "more secular" and thus "more forward looking" than Oresme and others, whose "strictures on the use of astrology [were] bounded above all by the religious concepts of their day."[35] Two of Christine's works that formed part of the political discourse of Charles VI's reign contain substantial discussion of the foundations and uses of astrology, in addition to a few valuable details about Tommaso and several (probably exaggerated) claims about Charles V's (undoubtedly real) interest in astrology: the *Livre des fais et bonnes meurs* and the more general *Livre du corps de policie* (completed by 1407). Christine nowhere gives a systematic defense of the science, but, in the course of these works on good rulers, she does articulate many arguments directly related to points raised by Oresme in his antiastrological writings. She was not only broadly educated, but also closely attuned to the cultural valences of texts and authorieis and highly skilled at putting them to use.[36]

Christine herself cannot be considered the voice of opposition to those works, some of which were written before her birth, but she was in an excellent position to be familiar with the substance of what astrology's proponents had to offer. These proponents included not only the astrologers

[34] For a detailed account, see Willard, *Christine de Pizan* (n. 30).

[35] Charity Cannon Willard, "Christine de Pizan: The Astrologer's Daughter," in *Mélanges à la mémoire de Franco Simone: France et Italie dans la culture européene*, vol. 1, *Moyen âge et renaissance*, Centre d'Études Franco-Italien, Universités de Turin et de Savoie, Bibliothèque Franco Simone 4 (Geneva: Slatkine, 1980), pp. 95–111, on p. 108.

[36] See Monica H. Green, "'Traittié tout de mençonges': The Secrés des dames, 'Trotula,' and Attitudes towards Women's Medicine in 14th- and Early 15th-Century France," in *Christine de Pizan and the Categories of Difference*, ed. Marilynn Desmond (Minneapolis: University of Minnesota Press, forthcoming). I am grateful to the author for access to a prepublication copy.

themselves but also a wider circle of men and women associated with the court who had or affected an interest in intellectual problems in general, or astrology in particular. While Oresme's concerns about the susceptibility of princes to the dangerous advice of astrologers are fairly conventional, the challenge he offers in the prologue to his *Livre de divinacions* suggests the existence of not only an *audience* in need of enlightenment but also of *opponents*, even enemies, whose medium is talk rather than written texts and whose language for discussing these lofty and urgent matters is French rather than Latin, "I compose this little book in French so that lay people can understand it ... and if anyone wants to refute what I shall say as to my basic understanding, let him confront it openly and rationally—not by insults—and write against it."[37] That his lay detractors are speaking, not writing, both underscores the problem of sources for documenting the other side of the debate and confirms the qualifications of someone in Christine's position to report on it. At the same time, the call to conduct the debate openly and rationally may resonate with some of its political dimensions to be explored below.

Oresme's arguments against astrology have been the subject of numerous and increasingly nuanced interpretations;[38] the positions of his largely invisible opponents lend significance to some of them, leaving others unanswered. Christine does not convey familiarity with Oresme's mathematical arguments as they bear on the knowability of celestial motions, but she does address the dignity and utility of astrology, the dangers of charlatanry, the association of astrology with superstition, the problems of necessity and free will, and the role of the science in affairs of state, all of which figure in Oresme's works. Thus, although she was not directly involved in whatever exchanges may have occurred from the late 1350s to the 1370s, and although her exposition of these subjects is likely to have been affected by intervening circumstances, she gives us access to views otherwise beyond our reach.

Taken together, Oresme's works and Christine's discussions create the impression of a dialogue different from that suggested by the objections and responses internal to Oresme's own work. In particular, a comparison between the pro-astrological arguments presented in the two works of Christine's just mentioned and in Oresme's *Livre de divinations* reveals substantive and rhetorical concessions on both sides. More important, it suggests the extent to which Oresme and his adversaries, though they did not cease to disagree profoundly, were operating within a shared intellectual framework that encompassed assumptions about nature, theology, and

[37] Oresme, *Divinacions*, Prologue, p. 50: "Ja compose ce livret en francois affin que gens lays le puissent entendre ... et se aucun veult reprouver ce que je diray quant a ma principal entencion, si le face en apper, et par raison, nom pas en detraction et escripse encontre."

[38] See above, n. 3.

political discourse, as well as a gallery of standard authorities. This evidence suggests that many of Oresme's arguments were indeed heard and understood, and that, in spite of his polemical tone, he was aware of and responded seriously to the positions of his lay adversaries. That Oresme was engaged in a real conversation rather than just haranguing an indifferent audience expands our understanding of him beyond his role as a conduit carrying academic learning to the laity: it shows him also to be an active participant in court culture.[39] Furthermore, the exchange will help to explain the posture of the king, who has previously appeared to have been impervious to the arguments of one of his most favored protégés.

Because he wrote it in French and because it addressed directly the unwritten discussion of (or attacks on) his objections to astrology, the *Livre de divinacions* best represents Oresme's side of the debate with the court astrologers, though his other works may shed some light on it as well. Like his other works in French, this one was neither a simple translation nor a watered-down popularization.[40] Although Oresme's interlocutors no doubt included individuals who knew only French, those bearing the titles "astrologer" and "physician" knew Latin as well, and, as Bert Hansen has pointed out, the division in Oresme's work on this and related subjects is as much between the scholastic and the polemical as between Latin and French.[41]

Christine herself knew enough Latin to be a pivotal player in a public exchange of Latin and French letters concerning *The Romance of the Rose*, which involved Jean Gerson, chancellor of the University of Paris, and others.[42] The conduct of that debate (which took place in 1401), the principal participants, its tone, and the attention it received all help to put the confrontation about astrology in perspective. A public affair characterized by heated arguments and elegant rhetoric, it attests the permeability of the line between university and court culture. Like the astrology that most concerned Oresme, the vernacular romance belonged to lay culture but engaged issues of concern to the clerical elite; like the authorities to whom Christine appealed in her defense of astrology, her learned arguments about the nature of women and love derived from scholarly sources, but resonated with courtly values. In the process of that exchange, Christine became learned and Gerson and others became courtly, so that in spite of the heated language and serious disagreements, the participants created and occupied the same sort of common ground that Oresme and Christine's sources ap-

---

[39] Sherman's *Imaging Aristotle* (n. 17) gives detailed evidence of the close collaboration between Oresme and the king in the production of the French Aristotle.

[40] See Lusignan, "Oresme traducteur" (n. 4).

[41] Hansen, *Oresme and the Marvels of Nature* (n. 3), pp. 22–25.

[42] Eric Hicks, ed., *Le débat sur le Roman de la rose*, Bibliothèque du XVe siècle (Paris: Champion, 1977).

pear to have shared. Indeed, the deep and earnest divisions were a necessary feature of a social and intellectual exercise that bound together different sectors of a newly forming elite.

Oresme's *Livre de divinacions* is thus a serious work. Indeed, it represents a significant step in the development of his ideas about the relationship between the pristine study of mathematical astronomy and the contaminated practices of judicial astrology. Unlike his earlier *Tractatus contra astrologos*, the *Livre*, which introduces six levels of celestial science, presents a continuum rather than a dichotomy between those two endeavors. In addition and as a consequence, Oresme shifts the center of his attack on judicial astrology from the enterprise of astrology per se to the specific rules and practices by which it is conducted. He may have had philosophical reasons for these adjustments, but the existence of an ongoing exchange with court intellectuals who favored some forms of judicial astrology, and perhaps even their responses to his attacks, would have constituted additional incentives for him to reconsider his earlier position.[43] It certainly opened up for discussion the middle ground between the clearly pure and the clearly contaminated, and in turn created the opportunity for the astrologers' party to claim that the territory in which they operated was within the boundaries of reason and orthodoxy.

The Goodness of Astrology

Christine's arguments tend to seek such uncontested territory. Her praise of astrology is often framed in general Aristotelian and Christian terms. In the context of extolling the value of *sagesse* in a prince, she distinguishes between the knowledge of lower causes and the knowledge of higher causes, and she elevates the speculative, including metaphysics and astrology, above the practical.[44] Here, what she seems to mean by "astrology" is "the art of knowing the movements of the celestial spheres and planets"—the theoretical, astronomical aspects of the science.[45] Oresme is willing to concede this much and more, describing as "speculative and mathematical, a very noble and very excellent science," the part of astrology that deals not only with celestial movements but also with measurements and tables.[46]

[43] Lejbowicz, "Chronologie" (n. 3). The chronology of these works has been disputed. I follow Lejbowicz, who places the *Tractatus* and the *De configurationibus* before the *Livre*, and the *Livre de divinacions* before the *Questio contra divinatores*. As Grant's analysis of the *Livre du ciel et du monde* (in this volume) indicates, however, Oresme has not abandoned the incommensurability of celestial motions.

[44] Christine, *Charles V*, pt. 3, chs. 2–3, vol. 2, pp. 10–12, and chap. 4, pp. 18–19. See Willard, "Astrologer's Daughter" (n. 35), pp. 98–99, 105–6.

[45] Ibid., chap. 11, vol. 2, p. 34: "l'art de connoistre les mouvemens des celestielz esperes et planetes."

[46] Oresme, *Divinacions*, chap. 2, p. 54, and chap. 1, p. 52: "speculative et mathematique, tres noble et tres excellente science." Oresme also includes knowledge of "signes," which

Both Christine and Oresme praise the elevated subject matter of astrology and its role as a stepping-stone to knowledge of the Creator. She cites Ptolemy; he cites Aristotle.[47] But, whereas Oresme is quick to distinguish this sort of astrology from its more questionable judicial parts, Christine uses similar rhetoric to color the whole enterprise with respectability and high-mindedness.

The same strategy applies in reverse where the relationship of astrology to various (other) forms of divination is concerned. The title of Oresme's French polemic evokes the language of scriptural and patristic condemnations of sorcery and related abominations, which he specifically invokes in the body of the work;[48] indeed, he indicates that he has chosen it with these connections (as well as with Cicero) in mind.[49] Christine is clearly sensitive to such charges and takes pains to separate the true science from the forbidden arts. For practical as well as for religious reasons, she warns princes to eschew and banish those who use astrology improperly as a mask for darker practices:

> But it must always be understood that no prince must call upon anyone who employs, under cover of these [ancient authors], an evil art forbidden by the church, for therein lies too great a danger.... For I mean only pure and perfect astrologers [whose] astronomical learning, [which] is so exalted and so exceeds in subtlety all the other [arts], is pure, natural, and without evil art or invocations of any kind.[50]

Christine is clearly resisting the sort of association implied by Oresme when, in the opening sentence of his *Livre de divinacions*, he lumps together astrology, geomancy, nigromancy, and "whatever such arts (if one may call them arts)," as not only evil in and of themselves but also—as Christine suggests—"temporally dangerous."[51]

---

might be taken as a concession to the occult aspects of astrology that are the focus of his attack in this work, but may also simply refer to elements on the map of the heavens.

[47] Christine, *Charles V*, pt. 3, chap. 4, vol. 2, p. 17; Oresme, *Divinacions*, chap. 15, p. 112; cf. *Ad pauca respicientes* (n. 3), pt. 2, p. 426.

[48] Oresme, *Divinacions*, chap. 9, pp. 74, 76, 78. See also his quodlibetal question *Contra divinatores horoscopios* (n. 3), which is not so named in all manuscripts, pp. 209–15, 216, n. a. Lejbowicz discusses the rhetoric and scriptural allusions of Oresme's attack in "Chronologie" (n. 3), pp. 121–26.

[49] Oresme, *Divinacions*, chap. 11, p. 92.

[50] Christine de Pizan, *Le livre du corps de policie* (hereafter Christine, *Policie*), ed. Robert H. Lucas, Textes Littéraires Français (Geneva: Droz, 1967), pt. 1, chap. 25, pp. 78–79: "Mais totesfoys n'est mie a entendre que soubz umbre d'iceulx nuls prince doie appeler nul qui use de mauvaise art defendue de l'eglise, car en ce gist trop grant peril.... Car je n'entens que des purs et parfais astrologiens, laquele science d'astrologie tant soit treshaulte et tant que toutes les aultres passe en subtilité est pure, naturel et sans mauvais art ne invocacions quelconques."

[51] Oresme, *Divinacions*, Prologue, p. 50: "quelxconques tielx ars, se on les doit appeller ars"; "perilleuses temporelment." Later he confirms that these should not be regarded as sciences: chap. 1. p. 54. Henry of Langenstein similarly presses the identification of astrolo-

Less grave but equally persistant was the charge that interest in astrology was a form of idle curiosity and thus a form of intellectual temptation or fascination, in the etymological sense. Oresme characterizes the desire for knowledge of the future as "gluttony" and argues that the first science of a prince must be politics, so he should not allow himself to give in to curiosity and neglect the public good by studying such distractions as planets and astrolabes.[52] More particularly, Oresme argues (citing Avicenna) that "a number of the rules of astrology are based on poetry and rhetoric, that is to say, on fables and beliefs which are not acceptable in natural science."[53] It is specifically against this objection that such study is concerned with useless fables that Christine mounts a defense of appropriately placed curiosity in the form of a long paraphrase of Aquinas's commentary on Aristotle's *Metaphysics*, treating the role of wonder—and particularly of such marvels as lunar phases, solar eclipses, and the movements of the heavens in general—in arriving at knowledge of the causes of things. Fables, like philosophy, arise out of the the same wonder and doubt that lead in turn to the investigation of causes; thus, "Those ancients who were moved by great wonders to be philosophers could be called poets."[54]

The Orthodoxy of Astrology

As Christine's allusion to the problem of Church-forbidden arts suggests, she felt called upon to defend astrology on theological grounds. On the negative side, she dissociated it from sorcery and superstition; on the positive side, she asserted that it was "pure and natural." She knew about the execution of the astrologer Cecco d'Ascoli, and was aware that the premises and practices of astrology were fraught with the dangers of determinism, even without the intervention of demons.[55] Christine did not address the full range of natural philosophical and theological objections to judicial

gers with "geomanciam, auspicia, ornumanciam et huiusmodi": *Contra astrologos coniunctionistas* in Pruckner, *Studien* (n. 3), pt. 2, chap. 8, p. 192.

[52] Oresme, *Divinacions*, chap. 9, p. 78 ("humain lignage est trop glout de savoir les choses d'avenir"); chap. 13, p. 104. Cf. Giles of Rome, *Li livres du gouvernment des rois: A Thirteenth-Century French Version of Egidio Colonna's Treatise "De regimine principium" Now First Published from the Kerr Manuscript*, ed. Samuel Paul Molenaer, Columbia Studies in Romance Philology and Literature (New York: Macmillan for Columbia University Press, 1899), bk. 2, pt. 2, chap. 8, pp. 201–2.

[53] Oresme, *Divinacions*, chap. 11, p. 86: "plusieurs des regles d'astrologie sont fondees sur poetire [*sic*] et sus rethorique, c'est a dire, sus fables et sus persuasions qui ne sont pas a recevoir en naturelle science."

[54] Christine, *Charles V*, pt. 3, chap. 66, vol. 2, p. 172: "Yceulz anciens, qui furent meus par grans merveilles à estre philosophes, pevent estre diz poetes." On the contested place of wonder in natural philosophy, see Katharine Park and Lorraine Daston, *Wonder and the Order of Nature, 1150-1750* (New York: Zone, forthcoming).

[55] On Christine's awareness of Cecco, Willard, "Astrologer's Daughter" (n. 35), p. 103; on demons, Henry of Langenstein, *Contra astrologios coniunctionistas* (n. 49), pt. 2, chap. 8, p. 192.

astrology, but she did launch a conventional defense against the standard charge that doctrines of astrological necessity denied free will. Oresme must have been familiar with the kind of response Christine was to offer, since he insisted, concerning judicial astrologers, that "it is their opinion that everything happens by necessity—however much they sometimes assert the contrary."[56]

Christine tells the story of the great Archimedes, whose mathematical and technological accomplishments she praises. A practitioner of astrology and geomancy (clearly a bad sign, in the context of Christine's insistence on their distinctness), he predicted his own death and thought his fate inescapable: "He said that the movement of the heavens held him so firmly that he could not free himself." She marvels that so great a scholar could be so deceived. First she deploys theological language—free and at liberty, the soul can choose—then, citing Aristotle's *Ethics*, the language of reason:

> In what is subject to the soul, namely deliberation of the will, the influences of the heavens have no sovereignty, notwithstanding that it may be true that the action of the heavens gives the individual a number of inclinations, such as to some people joyfulness, lust, or other natural motions. But this notwithstanding, the individual can rein them in by reason and resist with respect to deeds all such inclinations.

The great astrologer Ptolemy agreed, she asserts, and the life of Aristotle himself (who was "very badly conditioned by nature") attests to this truth.[57]

Astrology and Celestial Causation

For Christine, as for many of her contemporaries, the basis for the claims of astrology can be found in natural philosophy, according to which the truths of astrology derive from a chain of causal influences by which heavenly motions affect earthly events. She cites both Aristotle and Ptolemy as her authorities,[58] and asserts that knowledge "of the movements of the heavens, how they are perpetuated, the size of the spheres, their diversity, and their influences" leads to knowledge of "much of the generation of things here below, and one almost arrives at knowledge of the causes of this world."[59]

---

[56] Oresme, *Divinacions*, chap. 11, p. 90. See also *Ad pauca respicientes* (n. 3), pt. 2, p. 424: "que c'est leur oppinion que tout avienne de necessite, combien qu'ilz dient aucunes fois le contraire."

[57] Christine, *Policie*, pt. 1, chap. 24, pp. 76–77: "Il disoit que le mouvement du ciel le tenoit si fermement qu'il ne se pouoit partir.... Mais en ce qui est subgect a l'ame, c'est assçavoir deliberacion de voulenté les influences du ciel n'ont point de seignuerie, non obstant qu'il peut estre vray que l'action du ciel donne a l'omme pluseurs inclinacions, si comme aux aucuns joliveté, luxure, ou autres mouvemens naturelz. Mais non obstant ce, l'omme y peut mettre frain par raison et resister quant au fait a toutes teles inclinacions." Aristotle was "tresmal condicioné par nature."

[58] Ibid., chap. 25, p. 78.

[59] Christine, *Charles V*, pt. 3, chap. 67, pp. 170–71: "des mouvemens du ciel, comment il se soustient, des grandeurs des esperes, de la diversité d'elles et de leurs influences ... grant

Christine's "much" and "almost" would appear to be hedges against all manner of objections. For his part, Oresme makes only grudging concessions at the level of causation: "sometimes" people are moved by the inclinations they have received from the heavens and nature, but these causes are not only subject to preemption by the will but also weak and distant in their influence.[60]

Christine's works reflect no awareness of Oresme's more complex arguments about the intricate permutations of possible causal relations and the related difficulty of determining causes.[61] Her causal approach is consistent with academic natural philosophy, however, and represents a commitment to a strong form of astrological naturalism. It contrasts with the response of Pèlerin de Prusse, who, in spite of his greater status as a man of science and his aggressive claims about the powers of judicial astrology, had evaded both the question of causation and the problem of determinism by declaring that the stars were not causes but merely signs of God's will.[62] Oresme, who did not reject the existence of omens and prodigies, nevertheless emphatically rejected the idea that astrology could be saved by invoking what he regarded as a false notion of signs.[63] To the extent that they agreed that causation was central to a natural philosophical understanding of celestial actions, Oresme and Christine were engaged in the same conversation.

Although clearly speaking of natural processes, Christine is not much concerned with explaining the specific means by which the stars exert their powers. Nevertheless, in the context of refuting charges of determinism, she does give an account consistent with a medical model of celestial influence. According to her formulation, notwithstanding "some inclination and influence of the heavens," the soul is free; the body, however, cannot escape:

> But as for the body, it is quite true that in some things the individual is subject by birth to the actions of the body of the heavens, just as to the alterations of the course of the heavens and also of the four primary qualities, namely heat,

partie de la naiscence des choses de ça bas, et à pou près on parvient aux cognoiscences des causes de ce monde." On the role of Aristotelian natural philosophy, see Lemay, "True Place of Astrology" (n. 11).

60 Oresme, *Divinacions*, chap. 12, p. 98: "aucune fois."

61 Oresme, *Contra divinatores horoscopios* (n. 3), esp. pp. 268–86.

62 Laird, "Astrology in the Court of Charles V" (n. 3), who calls Pèlerin's claims about elections "shocking" (p. 169). The preemption of God's will was not much of a problem, since there was agreement that the stars were God's instruments: see John D. North, "Medieval Concepts of Celestial Influence: A Survey," in Curry, *Astrology* (n. 11), pp. 5–16.

63 Oresme, *De mirabilibus*; *Contra divinatores horoscopios*, pp. 296–97. The central theme of his *De mirabilibus* runs counter to Pèlerin's view of signs.

> cold, dryness, and moistness. For it is never within one's power to help being hot in summer and so on for the other [seasons].[64]

Some sort of causal conduit was widely accepted, even by the opponents of astrology. Operating through usual and familiar natural principles—the four primary qualities and the seasons associated with them—Christine's version of celestial influence was among the most plausible available in the fourteenth-century.[65] Indeed, according to Henry of Langenstein in his *Contra astrologos*, the effects of the heavens on the sphere of the elements are the proper subject of astrology.[66] Henry's position that the stars cause war only by affecting people's animal passions, such as greed, might have seemed perfectly reasonable to Christine, though he meant it to represent a severe limitation of the validity and utility of astrology.[67]

In his *Livre de divinacions*, Oresme is willing to allow that the heavens influence the weather and that astrology is, in principle, a source of knowledge about atmospheric conditions. In practice, however, sailors and peasants make better predictions than astrologers. Similarly, casting nativities and reading palms may indeed yield a grain of truth—especially regarding a person's complexion and inclination—but the information is useless for determining specific outcomes.[68] Objections such as these were closely linked in Oresme's other works to larger questions about the complexity of causation, and to a host of mathematical and epistemological problems that represent the most advanced and sophisticated elements in Oresme's philosophy in general.[69] Thus, from the perspective of fourteenth-century philosophy, much more was at stake than the idle chatter or even dangerous predictions of deluded courtiers or even malicious advisors.

These more general and fundamental questions appear to have been largely invisible to Christine, and are thus unlikely to have been a significant dimension of the conversations of the court intelligentsia.[70] However optimistic Oresme may have been about the sophistication (and Latin) of his lay audience, in the work most clearly addressed to nonspecialists, the French *Livres de divinacions*, he did not see fit to elaborate upon his most

---

[64] Christine, *Policie*, pt. 1, chap. 24, pp. 76–77: "quelque inclination et influence du ciel.... Mais quant au corps, bien est vray que en aucunes choses l'omme est subjet par naissance es actions du corps du ciel, si comme es alteracions du cours du ciel et aussi des quatre premieres qualités, c'est assçavoir de chaleur, froideur, seicheresse et moisteur. Car il n'est mie en la puissance de l'omme que en esté il n'ait chault et ainsi des autres."

[65] North, "Medieval Concepts of Celestial Influence" (n. 60); Lemay, "True Place of Astrology" (n. 11).

[66] Henry of Langenstein, *Contra astrologos coniunctionistas* (n. 51), pt. 2, chap. 7, p. 190.

[67] Ibid., pt. 2, chap. 1, p. 171.

[68] Oresme, *Divinacions*, chap. 2, p. 56; chap. 3, p. 58. See also chap. 11, p. 90.

[69] See, for example, Oresme, *Contra divinatores horoscopios*, pp. 168–75, 282–84, et passim.

[70] See Edward Grant, "Nicole Oresme, Aristotle's *On the Heavens*, and the Court of Charles V," in this volume.

original arguments.[71] He may well have judged those approaches too technical to have the desired effect. He may also have had more substantive reasons for backing away from arguments based on the incommensurability of celestial motions, since they would apply to the good as well as to the bad aspects of the science.[72] Whatever his reasons, in the French treatise Oresme chose to emphasize not the impossibility of astrological predictions and judgments, but the inadequacy of the specific rules that practitioners applied in their work. Thus he says of weather prediction, "The specific rules written for this part [of astrology] are apparently false," and of chiromancy:

> And the rules written for this are almost all false. Of the *experimens* that are written in a number of books, it is certain that most are not true and are just lies and tricks, and this is obvious to whoever wishes to test them. And it is also thus for a number of rules of judicial astrology. But there are some people so foolish [*gens si simples*] that they believe that everything they find in writing is the truth.[73]

By locating the false rules of judicial practices in books, Oresme allows the implication that he is targeting the views of more learned practitioners like Tommaso da Pizzano as "disreputable and suspect."[74] Furthermore, the expression "gens simples" would ordinarily refer to illiterates, but Oresme uses it here to refer to the same opponents whom he challenges at the beginning of the work to *write down* a reasoned argument against him. This would certainly be felt as an insult by members of the aristocracy and of the king's household who had little or no training in the scholastic art of questioning everything they read, yet at the same time it functions as an exhortation to this audience to engage in the rational and even empirical evaluation of propositions.

Unlike his more radical, more technical, arguments made elsewhere against the possibility of developing adequate rules under any circumstances, Oresme's formulation here leaves a loophole large enough for Christine de Pizan to drive a siege engine through. Just as she has taken pains to dissociate respectable astrology from practices forbidden by the Church, she is pleased to distinguish the learned and skilled astrological

[71] On the accessibility of and potential audience for Oresme's Latin works, see Hansen, *Oresme and the Marvels of Nature* (n. 3), pp. 24–25.

[72] Lejbowicz, "Chronologie," pp. 141–43.

[73] Oresme, *Divinacions*, chap. 2, p. 56; chap. 3, p. 58: "Et samblablement sont faulses les regles speciaux qui sont escriptes pour ceste partie.... Et pour ce les regles qui sont escriptes sont pres que toutes fausses. Des experimens qui sont escrips en plusieurs livres est il certain que le plus ne sont pas vrayes et ne sont que mensonges et tromperies, et appert magnifestement a qui les veult esprouver. Et ainsi est il de plusieurs regles d'astronomie judicative, mais il est aucunes gens si simples qu'ilz cuident que tout ce qu'ilz treuvent en escript soit verite."

[74] Oresme, *Divinacions*, chap. 10, p. 80: "diffamees et suspectes."

elite from the misguided and deceptive charlatans of Oresme's polemic. Astrology is, she warns, accessible to few and misunderstood by many:

> Because of its difficulty, that is, its great subtlety, there are few who are well grounded in it. And, for that reason, many have criticized it, because they do not grasp or know how to understand it—just as it is said that usually those who are ignorant of things readily criticize them.[75]

Pèlerin de Prusse may have been suggesting a similar defense when he opened his French book on astrological elections with the Hippocratic caveat that life is short and the art is long.[76] Thus, according to Christine, "pure and perfect astrologers," applying valid methods (the possibility of which Oresme had allowed), may be suspect in the eyes of the ignorant, but would be untainted by the kinds of errors, evasions, and deceptions catalogued by Oresme under the heading of arguments from experience.[77] Indeed, she has positioned herself so as not to be outdone in the condemnation of "abusers who, under the cover of science, pretend to know yet know nothing, and often deceive and trick princes and lay people by their sneaky craft."[78]

Astrology and the Prince

The question of error and fraud was critical for both Oresme and Christine because of the danger that it would put the prince and the public good in jeopardy. For Christine, the role of astrology in government was an important basis for its claims to legitimacy and the keystone of its claims to utility; for Oresme, the political arena was the sphere in which the dangers of astrology, demonstrated by reason and experience, were most clearly manifest. The relation of astrology to politics had special rhetorical weight since, as both authors demonstrated, it gave access not only to a vast array of classical, biblical, and modern exempla, but also to the resonances of the "advice to princes" genre, which had wide Latin and vernacular currency. Once again, the two are addressing the same questions in the same terms, if from different positions.

Having been delivered the high ground, Christine's pure and perfect astrologers, practicing a form of astrology that is "natural and without evil art or invocations of any kind," do not look so different from the (rare) man of

---

[75] Christine, *Policie*, pt. 1, chap. 25, p. 79: "Pour la difficulté d'elle, c'est assavoir que pour la grande subtilité qui est, pou sont qui bien y soient fondés. Et pour celle cause pluseurs l'ont blasmee pour ce qu'ilz ne la sentoient ne sçavoient comprendre, si comme il est dit que communement ceulx qui ignorent les chose les blasment voulentiers."

[76] Laird, "Astrology" (n. 3), p. 169.

[77] Christine, *Policie*, pt. 1, chap. 25, p. 79; Oresme, *Divinacions*, chap. 8; also chap. 12.

[78] Christine, *Policie*, pt. 1, chap. 25, p. 78: "abuseurs lesquelz soubz umbre de science se faignent sçavoir et rien ne sceuent, et souvent deséoivent et trompent les princes et les layes gens par leur industrie cauteleuse."

learning—in some respects a representation of himself—whom Oresme holds up as worthy of royal patronage, namely:

> Those who make observations and who take pains to scrutinize the rules of [astrological] judgments (or to find new ones) and who look at and reflect on other arts or sciences—good or bad—dealt with in the first chapter without misusing [them], and who know how to reflect on the nature of things and to condemn rationally what ought to be condemned.[79]

Christine's case, however, rests not just on the benign general study of astrology but specifically on its political utility, "although some people repudiate the science for the advice of princes"; she distinguishes clearly, as does Oresme, between the prince's personal and political deployment of the art, placing the latter at the center of her case.[80]

On this point Oresme and Christine both argue not only from authority but also from history. Each tells a story about Thales; one gives examples of bad kings, the other of good ones. Venturing into the territory of Oresme, who translated it into French, Christine cites the authority of Aristotle's *Politics*.[81] She invokes the name of King Ptolemy to support the positive role of astrology; to what must have been a familiar point, Oresme responds that Ptolemy the king and Ptolemy the astrologer were not the same man.[82]

Much of Christine's evidence is curiously irrelevant to the most common forms of judicial astrology, such as nativities or elections, though her cautionary tales include Caesar's indifference to a warning about the ides of March.[83] She tells the story of Supplicius Gallus, who, by virtue of his study of astrology, saved his people in battle against the king of Persia. He was able to prevent panic among his troops by explaining that the eclipse of the moon, which they took to be a bad omen, was, in fact, "a natural thing."[84]

The use of astrology to maintain order in the ranks added an important dimension to Christine's advocacy of astrology as "very profitable for the state."[85] One of the most serious charges leveled against practitioners was

---

[79] Oresme, *Divinacions*, chap. 13, p. 104: "Ceulx qui font observacions et qui mettent painne a examiner les regles des jugemens, ou trouver nouvelles, et qui regardent et considerent les autres ars ou sciences touchies ou premier chappitre, bonnes ou males, sans abuser, et qui scevent considerer la nature des choses et reprouver par raison ce qui fait a reprouver."

[80] Christine, *Policie*, pt. 1, chap. 25, p. 78: "quoy que aucuns repugnent la science en conseil des princes." Here she mentions ordering his personal affairs; Oresme, *Divinacions*, chap. 14, p. 108, entertains the utility of astrology in governing regimen.

[81] Thales: Christine, *Policie*, pt. 1, chap. 15, p. 79; Oresme, *Divinacions*, chap. 8, pp. 70, 72. Aristotle: Christine, *Policie*, pt. i, chap. 25, p. 79.

[82] Christiine, *Policie*, pt. 1, chap. 24, p. 77 and chap. 25, p. 78; Oresme, *Divinacions*, chap. 14, p. 106.

[83] Christine, *Policie*, pt. 1, chap. 25, p. 80.

[84] Ibid., chap. 26, p. 81: "chose naturele."

[85] Ibid., p. 80: "tresprofitable a la chose publique."

that they stirred up the rabble.[86] To the extent that Christine could represent the most learned and least volatile forms of astrology in the service of public order, she could sustain the distinction between dangerous, ignorant pseudo-astrologers and the sages worthy to counsel princes.

The Outcome of the Controversy

Judging from Charles V's continued support for physician-astrologers, as well as from his possession of books and instruments, historians have declared Oresme the loser in his battle against his largely silent opponents. Jourdain admired the risk he took in persistently opposing himself to the king's presumed position; Albert Menut observed that "Oresme's arguments appear to have had no effect on Charles"; and Grant concluded that Oresme "failed dismally" in his attempt to weaken the king's allegiance to his astrologers.[87] While Charles was indeed in some sense his audience, reading Oresme's work in this way produces a static picture that is inconsistent with the place of these arguments in his larger philosophical program, and thus inconsistent with the "unity of knowledge" at the university of Paris and the French court: the invisibility of Oresme's interlocutors has occluded the relationship between the debate about astrology and contemporaneous controversies about political theory.

Through his translations of and commentaries on Aristotle's *Politics*, Oresme both raised the intellectual level of those at the center of power and acted as a participant in the political scene. Virtue and rational judgment for the benefit of the common good, rather than force and arbitary power, were central to good government. Thus, while several canon law scholars in the court circle played up the divine origins of royal power, several philosophers, including Oresme, proposed a model influenced both by Aristotle and by the claims of the estates, which acknowledged in a limited way the right of the community and the public good. The differences here were serious, but the terms of the debate excluded the most radical claims of both the nobility and the third estate.[88] Similarly, Oresme's *Livre de divinacions* was both a serious intellectual project and a contribution to urgent questions about the foundations of good government, in which both nature and

---

[86] Henry of Langenstein, *Contra astrologos coniunctionistas* (n. 51), pt. 2, chap. 5, p. 180. See also Thorndike, *History of Magic* (n. 3), 3: 422.

[87] Jourdain, "Oresme et les astrologues" (n. 3), p. 26; Albert Menut, introduction to Nicole Oresme, *Le livre du ciel et du monde*, ed. and trans. Albert D. Menut and Alexander J. Denomy (Madison: University of Wisconsin Press, 1968), p. 6; Grant, introduction to *De proportionibus* (n. 3), p. 64.

[88] See, for example, Jean de Golein's *Traité du sacre* in Marc Bloch, *Les rois thaumaturges: Etude sur le caractère surnaturel attribué à la puissance royale particulièrement en France et en Angleterre* (Paris: A. Colin, 1961), pp. 478–87 and Nicole Oresme's *Le livre de politiques d'Aristote*, ed. Albert D. Menut, *Transactions of the American Philosophical Society*, n.s., 60/6 (1970). See also Autrand, *Charles V* (n. 6), pp. 728–42.

reason played a critical role. He agreed with Christine about the importance of a wise and educated king, if not about the particular content of the wisdom and education. The lines of connection between his positions and those of Christine suggest that he both heard and was heard by those for whom the authority of nature (as read by men without clear identification with church, university, or aristocracy) was a credible basis for understanding the experiences and shaping the decisions of political governance and everyday life.

In the *Livre de divinacions* Oresme's ideas are not merely transmitted but shaped for a nonuniversity audience, and their significance is played out in relation to the environment of Tommaso da Pizzano and his daughter. Once Oresme had reframed his position to concede many claims of judicial astrology in principle had and presented his challenge at the level of practice, he entered into a specific "astrological culture," to borrow Max Lejbowicz's term, in which the question is not whether to make astrological judgments, but how to make good ones.[89] These concessions, such as locating astrology's failure in the rules of application rather than in the foundational tenets, d0 not represent an abandonment of his strong opposition, but they do represent a considerable shift in its terms. As such, the *Livre* is, like his French Aristotelian works, a part of Oresme's intellectual biography as well as an element in the history of Charles V's reign. In conjunction with the serious defenders of judicial astrology whose views are echoed however faintly in Christine de Pizan's voice, Oresme has taken a step toward bridging what Richard Lemay has called the "quasi-biological gap" between the pursuit of astrology for its own sake or a higher good, and the use of astrology to make a difference in the world.[90]

Different as they were, the careers of Oresme and Christine both depended on their ability to occupy, indeed to help create, a social and intellectual space between the university and the court, between Latin and vernacular, between the literary and the practical. For neither Charles, nor Christine, nor Oresme was the development of this space in general, or the discussion of astrology in particular, an end in itself. For each it had a different use and a different meaning, but for all it related in some way to the principles of good government.

---

[89] Lejbowicz, "Chronologie" (n. 3), p. 131. See also Laird, "Astrology" (n. 3), p. 168.

[90] Lemay, "True Place of Astrology" (n. 11), p. 60. Thus construed, the exchange implied here would suggest modification of Autrand's view (*Charles V*, p. 728) that Charles kept the thinkers and the doers separate in his court.

Astrology, Astrologers, and the Government of the Realm

Astrology was by no means the only, or even the most important, subject of intellectual debate at court. Arguments from philosophy, theology, and canon law fueled a series of political debates in which the stakes—the fate of the papacy, the relations among the estates—were much higher than those concerning celestial causation. Charles's entourage included university-trained men who undertook considerable programs of political philosophy and propaganda.[91] As with astrology, so with political philosophy: a broad, flexible, and even contradictory discussion suited Charles's purposes—assertive without being overtly aggressive, and aimed at various audiences. Oresme himself explicated the relationship between lively discussion and a healthy polity in the representations of good government that he devised for his French version of the *Politics* made with the particular circumstances of Charles's reign in mind: there positive and constructive exchange contrasts with the silent cruelty of bad government, and at the same time is easily distinguishable from dangerous forms of speech, such as the whispers of conspiracy and the demagoguery of rebellion.[92]

If astrology was not central to the political concerns of Charles's reign, neither was it irrelevant. Both the elements of the debate that addressed the conduct of the good prince and the circumstances of the debate that loosely tied the sides together were related to Charles's attempt to assert his legitimacy, support his authority, and enhance his power as a ruler. The circumstances of his political career limited his resources, and he skillfully deployed what instruments he had access to—ranging from the imagery of astrology as powerful knowledge, to the political assets created by patronage, and the advantages of cultivating limited debate.

Charles was not yet twenty when the English took his father prisoner at the battle of Poitiers. His position as regent was far from secure. On the one hand, a small number of powerful noble dynasties, not all of which were resistant to English rule, were prepared to seize upon the opportunity of a monarchy in distress to reestablish regional autonomy and aristocratic influence against the Capetian agenda of centralization and royal authority. On the other hand, the city of Paris (and to a lesser extent other towns), from its position of increasing economic and strategic importance, sought to gain advantages for the third estate—if not by eliminating the privileges and exemptions of the Church and nobility, at least by securing some of its own and by playing a more active role in the affairs of the kingdom. Skeptical of the ability of the young prince to protect and govern the realm, and united in their resistance to the taxes that had been raised for the unsuccess-

---

[91] See Quillet, *Charles V* (n. 3).

[92] Sherman, *Imaging Aristotle* (n. 17), plates 7, 8, 10; pp. 127, 135–36.

ful war against England, the three estates briefly claimed effective control of the country, going so far as to issue their own coins, though the alliance soon disintegrated. The insecurity of Charles's position is underscored by the success of Charles II, king of Navarre and count of Evreux, the regent's brother-in-law and himself of royal lineage, whom the leadership of Paris and some members of the nobility for a time regarded as more capable of ruling the country and more susceptible (if not more sympathetic) to their influence.

During this period, the young prince's willingness both to negotiate with and to besiege Paris, both to hold captive and to make peace with Charles of Navarre, constitutes the beginning of an eclectic and ad hoc style of rule, arising at least in part from necessity and weakness and resulting in an effective if delicate balance. From about 1359 to 1375, during the period when the ransomed Jean II resumed the reins of government and during the first decade of his own rule, Charles dealt similarly with the challenges of competing interests and a precarious power base—struggling especially to gain control of the administration of the country at the expense of the nobility, on whom he continued to depend for fiscal support, for maintaining peace in the countryside, and for prosecuting the war against the English. A homogeneous group of barons and prelates (many from the same families), led especially by the brothers Guillaume and Jean de Melun—the archbishop of Sens and King Jean's chamberlain, respectively—placed narrow constraints on Charles's autonomy, effecting what has been called a revolution by prying the government of the realm apart from the government of the royal household and estates. This process involved keeping tight control of Charles's advisors, both before and after his coronation, and elevating the role of the General Council, whose membership they dominated. Charles was able to resist the full force of this control, as the history of taxation in this period illustrates. His gradual ascent to personal rule in the latter part of his reign was accelerated by the demise of the Melun brothers, but depended substantially on his ability to take control of and deploy effectively the general councillors and the members of his personal staff.[93]

Whatever stability and order Charles achieved were thus not the result of a straightforward assertion of royal power. Although a number of his predecessors, including Philippe IV (d. 1314), had developed both the model and the mechanisms for monarchical authority and centralization, Charles was not in a position to make direct ideological claims or to exercise direct political control, whatever his inclinations might have been. His political skill consisted in mustering and coordinating a complicated array of symbolic, administrative, personal, and other resources. Among these,

[93] Raymond Cazelles, *Société politique: Noblesse et couronne sous Jean le Bon et Charles V*, Mémoires et Documents (Société de l'École des Chartes) 28 (Geneva: Droz, 1982).

astrologers, astrology, and the controversy associated with them took their place, significant both in themselves and as examples of his strategy of rule. In particular, Charles (1) borrowed and adorned himself with the legitimacy associated with learning; (2) enhanced his status by turning the borrowed authority into a good to be bestowed by patronage; and (3) cultivated and deployed a staff whose status derived from these associations.

Borrowing Authority

The King's political moderation, his *prudence*, whether arising from his personality or resulting from his constrained circumstances, was central to his reputation as Charles "le Sage." Taking advantage of the semantic range of *sagesse*, he strove successfully to elide that quality with an attribution of *wisdom* as well. Pictorial and textual representations of Charles produced for him, often in commissioned works, depict him among students, among books, among authorities on judicial astrology, and with an armillary sphere.[94]

Taking her cue from this theme, Christine de Pizan characterized the king's "perfect *sagesse*, namely wisdom, knowledge, understanding, prudence, and art"; in particular, "he knew how to answer and speak well and handsomely about everything, including about the exalted matters of philosophy, like [those] of astrology, on which [he was] an expert."[95] Charles cultivated his association with the prototype of the wise ruler, Solomon, whose deep understanding extended to the natural world in general and to astral divination in particular.[96] Again emphasizing in particular Charles's love of books and command of astrology, Christine advanced the comparison with "King Ptolemy."[97] Prudence and books, wisdom in general and knowledge of the heavens in particular, link Charles specifically to ancient royal authority through a medium distinct from (but not incompatible with) divine, constitutional, military, or other sources of power.

These carefully cultivated images of the king extended beyond a static representation of authority. Like Ptolemy a collector of books, Charles made the Louvre library a visible and active intellectual site. In the first miniature of Oresme's *Traite de la sphere*, with an open book on his lap, Charles appears learned; conversing with astrological worthies in the miniature at the opening of the *Livres des neuf juges*, he becomes one of the

---

[94] Sherman, "Representations" (n. 6); Laird, "Astrology" (n. 3), pp. 168–69.

[95] Christine, *Charles V*, pt. 3, chap. 4; vol. 2, p. 16: "parfaite sagesse, cest assavoir, sapience, science, entendement, prudence et art"; "de toutes bien et bel sceut respondre et parler, et encore des haultes choses de philosophie comme d'astrologie, tres expert en ycelle."

[96] Autrand, *Charles V* (n. 6), pp. 714–19; Byrne, "Rex imago Dei" (n. 6), pp. 101–3; 1 Kings 4:33. The Book of Wisdom (e.g., 7:17–18) greatly enhanced Solomon's association with astral divination.

[97] Christine, *Charles V*, pt. 3, chap. 12; vol. 2, p. 43:"Moult fu sage cellui roy Ptolomée, et moult sceut de la science d'astronomie, et mesura la rondeur de la terre."

authorites. Knowing how to "answer and speak well and handsomely" on scholarly subjects, he formed a bridge between the esoteric and technical knowledge of books and science, on the one hand, and the courtly medium of speech in which a king persuades, commands, judges, and educates his subjects, on the other. This function established a unique and independent role for the king, appropriating a function, authority, and influence normally reserved for clerks.[98]

Here astrology had a special role to play, in part by virtue of its marginal status. Whatever Charles's degree of learning in subjects such as law or theology, those subjects were so overtly weighty and controversial and so fully owned by credentialed individuals invested by powerful ecclesiastic and academic institutions that any overt encroachment on their prerogatives would have been difficult to effect and, above all, politically imprudent—far from *sage*. A miniature in the presentation manuscript of Oresme's French version of Aristotle's *Ethics* shows a king and his councillors listening to a cleric lecture. Charles was more inclined to leave debates about moral, social, and political issues in the hands of credentialed clerics (including Oresme) tied to one university faculty or another.[99] Although astrology was clearly not uncontroversial or inconsequential, it was territory not unequivocally owned by others, a place to stake a claim and build a reputation. As Edgar Laird points out, Charles V's involvement in the production of astrological texts enhanced his independence from the experts. Oresme's very resistance to such independence confirms it: "It is decent and honorable for kings and princes to know some general things about these subjects not by laborious investigation, but through the accounts of others."[100]

The king's particular appropriation of astrological learning served another purpose beyond the dignity of his wisdom and the independence of his authority. His subjects, allies, and enemies gave at least some degree of credence to the efficacy of judicial astrology. A century earlier, in a vernacular guide to the education of princes, Giles of Rome had singled out the aristocracy unwilling "to undertake any project or a battle without knowing

---

[98] Laird, "Astrology" (n. 3), pp. 172–73.

[99] For example, at Charles's behest, Jean Golein undertook the task of explicating the ritual of coronation in such a way as to assert the king's divine powers without overstepping ecclesiastical boundaries, and to assert the king's independence from the nobility within the relatively benign framework of patriotism. See Bloch, *Les rois thaumaturges* (n. 86), pp. 139–40; Jean Golein, *Traité du sacre*, in ibid., pp. 478–87. On Oresme's deployment of the *Ethics* and *Politics*, see Sherman, *Imaging Aristotle* (n. 17).

[100] Oresme, *Contra judiciarios astronomos*, ed. Coopland, in Oresme: *Divinacions* (n. 3), chap. 6, p. 135: "Dico ergo primo quod decens est et honorabile regibus et principibus quedam generalia de istis non laboriosa investigacione sed per narracionem aliorum cognoscere." See Laird, "Astrology" (n. 3), pp. 172–73; and Jourdain, "Oresme et les astrologues" (n. 3), pp. 580–81.

the course and dispositions of the stars,"[101] and we know that Tommaso da Pizzano prepared an election for the marriage of the duke of Savoy on the occasion of his son's marriage.[102] Kings did not necessarily rely on astrological predictions, but the specifically royal and political associations were persistent.[103] Thus astrology served as one of many sources of power, practical as well as symbolic.

Perhaps equally important, Charles was aware of the widespread popular belief in astrology and in other means by which natural or supernatural forces might be divined and managed, or even controlled. Indeed, the credulousness of the common folk was one of the dangers cited by the opponents of astrology. Henry of Langenstein, writing in Paris in 1373, warned particularly against "certain semi-astrologers ... who stir up and disquiet the minds of the people with false rumors of future events."[104] Rabble rousing and panic mongering were certainly not in the king's interest at a time when heavy taxes, recurrences of plague, and bands of unemployed soldiers threatened the stability of town and country alike, but the faith of the populace in divination and magic could be put to royal use. The very ambiguities about which Oresme complained[105] left plenty of room for the king and his circle to make of them what they wanted: rather than placing deterministic constraints upon royal actions, the results of consultations with astrologers would generally have lent support to whatever action or position was in Charles's best interest. Furthermore, like the beneficent healing power of the king's touch, his access to the secrets of the heavens connected the realm and its ordinary subjects with higher powers through Charles's person.[106]

Distributing Benefits

Once Charles exercised his figurative ownership of prudent and powerful learning and his literal ownership of the books in which it was encoded,

---

[101] Giles de Rome, *Li livres du gouvernement des rois* (n. 52), referring to astronomy: "la quele les gentiz hommes par aventure aprennoient ça en arrieres por cen que il estoient trop curieus de savoir le jugement des estoiles quer il ne voloient onques commencier euvres ne batailles jusqu'a tant quer il seüssent tout le cours et les dispositions des esteiles, et cen siet hon par astronomie, par quoi il l'aprennoient volontiers" (p. 200).

[102] Solente, Introduction to Christine, *Charles V*, vol. 1, p. vii.

[103] Hilary M. Carey, "Astrology at the English Court in the Later Middle Ages," in Curry, *Astrology* (n. 11), pp. 41–56.

[104] Henry of Langenstein, *Contra astrologos coniunctionistas* (n. 51), pt. 2, chap. 5, p. 180: "quibusdam semiastrologis, ymmo in arte illa indoctis, qui vituperium sibi prognosticant, artem diffamant, falsis futorum rumoribus populorum mentes turbant et inquietant." The dating is conjectural: Steneck, *Science and Creation* (n. 26), pp. 17–18.

[105] Oresme, *Divinacions*, chap. 12, p. 94: "Item, leurs paroles sont aucune fois doubles, amphiboliques, a deux visages." Cf. Robert Bartlett, *Trial by Fire and Water: The Medieval Judicial Ordeal* (Oxford: Oxford University Press, 1986).

[106] Bloch, *Les rois thaumaturges* (n. 88), pp. 134–40, 232–36; Hansen, *Oresme and the Marvels of Nature* (n. 3), pp. 22–24.

themselves enshrined in the secular temple of his library, he was in a position to command and dispense them as a patron. Charles's patronage was not limited to the world of books and scholarship, nor within that world was it limited to natural science, but the history of his commissions and gifts, like the history of his interests and image, affords astrology a privileged place. The support of scholarship was, at least in principle, a long tradition of the French monarchy, of late extended most visibly to the University of Paris.[107] On a mundane level, Charles's relationship with the university had more to do with balancing the legal and fiscal claims of Church and city and with maintaining order in the Latin Quarter than with images and ideologies.[108] At a loftier level, the king supported impecunious students, including a number from his home base of Normandy, at the Collège de Navarre. Here dynastic politics no doubt had some influence: the house of Navarre was in chronic opposition to Jean II and Charles V, and in frequent alliance with the English. Charles's scholarships not only diluted the influence of Charles le Mauvais of Navarre, but also provided the king with loyal supporters.

Members of Charles's immediate circle supported and influenced the University of Paris by founding new colleges that illustrate the multifaceted relationship between learning and royal power. Natural philosophy was not the only, nor even the most prominent, avenue of operation. Jean de Dorman, from one of the great families that held Charles's power in check, having entered his service in 1358, had Charles's support for the foundation of the Collège de Beauvais and the endowment of twelve scholarships in 1370. Devoted to the study of civil and canon law and dedicated to the service of the king and public good, the college produced many professionals employed by the royal court and thus represented an investment in the promotion of useful knowledge.[109]

Similar goals are more easily detectable in the foundation of the Collège de Notre Dame by Gervais Chrétien, physician first to Jean II and then to Charles. From a family of the lesser nobility in Basse-Normandie with some record of royal service, Gervais was nevertheless poor enough to have received a royal scholarship at the Collège de Navarre. He achieved some status at the university, as dean of the medical faculty and regent, and he held a series of clerical offices, especially in Normandy.[110] The establishment of the college closes the astrological circle, connecting university and

---

[107] Speaking of Charles's generosity toward that institution, Christine de Pizan repeats the legend that Charlemagne was its founder and first protector: Christine, *Charles V*, pt. 3, chap. 13; vol. 2, pp. 46–49.

[108] Pearl Kibre, *Scholarly Privileges in the Middle Ages: The Rights, Privileges, and Immunities of Scholars and Universities at Bologna, Paris, and Oxford*, Mediaeval Academy of America, Publication 72 (Cambridge: Mediaeval Academy of America, 1962), pp. 155–66.

[109] See Autrand, *Charles V* (n. 6), pp. 706–8, on the political significance of the college.

[110] Le Paulmier, "Notice sur Gervais Chrétien" (n. 29).

court. Master Gervais presented his sovereign with a copy of Avicenna's *Canon of Medicine,*[111] and the king presented Gervais with medical and astrological books, as well as astrological instruments for the college.[112]

Medical learning had been one of the strongest sources of astrological knowledge at court, and the court, in turn, strengthened astrology at the university by providing it with a more secure and hospitable home than the arts faculty. Charles's support placed the royal seal of approval on the middle ground that medical astrology could claim to occupy in the astrological debate. Concerned with the body, not the will, and embedded in an art that acknowledged the difficulty of its claims about causation and prognosis, medical astrology was best positioned to parry, if not to answer, objections like those of Oresme. The latter had, after all, conceded that both Galen and the *Secretum secretorum* regarded the advice of an astrologer as appropriate in determing regimen.[113] In what is the closest we can come to Charles's mature position on astrology, the royal act of 1377 that establishes an endowment for the college specifies that it will support two students "who shall live in the house of our beloved and faithful physician, Master Gervais Chrétien … and read the mathematical sciences [which are] legal and permissible to read at our university of Paris."[114]

The king's direct role in the foundation of the Collège de Notre Dame, often cited as evidence of his commitment to astrology, thus involved more than the generic benefits of patronage—ingratiation with the client, a reputation for largesse. He placed his weight and his resources behind a specific, substantive curricular development consistent with his broad development of astrology. Like astrology itself, the institutional niche represented by the college was a place where Charles could assert and display his authority without direct confrontation with the competing centers of power. The college, its curriculum, and the particular books presented by the king were points at which theory met utility, the heavens met human well-being, and the monarch, through the secrets of the macrocosm and the power of healing, presided over the public good.

Given the widespread confidence in the wisdom and work of astrologers, the protests of Oresme, Philippe de Mézières, and others, even if dis-

---

[111] Delisle, *Librairie de Charles V* (n. 7), vol. 2, #810, pp. 131–32.

[112] Symon Phares is the source of this information: *Célèbres astrologues* (n. 33), p. 4. It is made plausible by circumstantial evidence. Charles not only possessed many such books but also quite a few instruments: Jules Labarte, *Inventaire du mobilier de Charles V, roi de France*, Collection de Documents Inédits sur l'Histoire de France, 3rd ser., Archéologie (Paris: Ministre de l'Instruction Publique, Imprimerie Nationale, 1879), ##1190, 1983 (?), 2072, 2216, 2268, 2270, 2427, 2714, 2817, 3119, 3121.

[113] Oresme, *Divinacions*, chap. 14, p. 108.

[114] Léopold Delisle, *Mandements* (n. 28), #1497: "qui demourent en l'ostel de nostre amé et feal phisicien maistre Gervais Chrestien, lesquels nous avons ordené que ils liront les sciences mathematiques licites et permises à lire en nostre université de Paris."

seminated in French, were not likely to affect the status of the science—and there are no signs that they did. Thus astrology's political utility did not depend on its being unopposed. Indeed, the very existence of the debate, insofar as anyone noticed it, may even have enhanced its aura of legitimacy, both by associating it (in yet another way) with learned discourse and by providing a pretext for legitimizing arguments such as those of Christine de Pizan.

The King's Men

In the context of Charles's reign, Master Gervais Chrétien and Master Nicole Oresme had more in common than in conflict. Natives of Charles's most secure political base, Normandy, beneficiaries of his aid as scholarship students, participants in the construction and use of his library, royal servants and clerical beneficiaries who owed their advancement in no small part to the king, and active participants in the intellectual life and administration of the university, these men were even less opponents on the social and political planes than they were on the intellectual plane.[115] Both were part of a fluid and diverse group of retainers and counselors with whose help Charles tried, with varying degrees of success, to exercise personal control over the kingdom and enhance the authority of the crown. Although they had in common their loyalty to the king, the members of Charles's intellectual circle disagreed about astrology, political theory, and other matters. The overlapping circles of household and government of necessity contained and balanced contradictions, and Charles was better off promoting an uneasy accommodation than throwing in his lot with one party or another.

Given the political environment of all but the last few years of his reign, Charles valued and advanced loyal and independent civil servants. From the famous and flamboyant Bertrand du Guesclin, whose military exploits considerably strengthened Charles's hand not only against the English armies and free companies but also against the French nobility, to the less conspicuous Gilles Malet whose post as librarian positioned him to serve in a variety of useful roles, Charles found it expedient, when possible, to rely on men who had come from nowhere. When, in the mid-1370s, he gained control of the Council General, he placed there men like the Abbot of Vaast (later Bishop of Chartres) Jean Le Fèvre, a doctor of law, and Jean Le Mercier, who administered royal taxes. Both were of bourgeois origins, as their names suggest, and served as useful counterweights to the members of

[115] François Neveux, "Nicole Oresme et le clergé normand du XIVe siècle," in Quillet, *Autour de Nicole Oresme* (n. 3), pp. 9–36; Le Paulmier, "Notice sur Gervais Chrétien" (n. 29).

the high nobility who remained in positions of power.[116] The glosses and illustrations of Oresme's French version of Aristotle's *Politics* gave abstract and symbolic voice to the critical role of the middle classes as the embodiment of the Aristotelian mean, and as players on the immediate political scene.[117] The former scholarship recipient Nicole Oresme and the immigrant Tommaso da Pizzano were both Charles's *conseillers*. Throughout his political career, first as Dauphin and duke of Normandy and then as king, Charles made particular use of his household staff and an overlapping group of intellectual companions, including an inner circle, referred to by one historian as the "club du roi," with whom he was fond of discussing books at the end of his workday.[118]

Although perhaps of more modest origins than the more prominent members of his Council, the intellectuals in Charles's entourage were not completely without worldly concerns—Oresme did not disdain the office of bishop that he attained, and Tommaso da Pizzano was undoubtedly wise in the ways of court society. Nevertheless, they were (on the whole) more disinterested than Charles's noble advisors and, at least as important, they could be tied to a set of conventional images that represented the truly learned as detached from social and political ties, interests, or indebtedness—as wise, honest, and without guile. Christine de Pizan was drawing upon these cultural associations (and perhaps attempting to adorn her father in them) when she told the following story to illustrate Charles V's life of learning (as well as to honor her father, who also wrote a tract on alchemy):

> [Charles] heard that near Avignon there was a speculative clerk who led the life of a philosopher and worked very subtly in the art of alchemy.... And this clerk had been a disciple of Master Arnald of Villanova.... The king, who liked to see all subtle things, wrote to him that he wanted [the alchemist] to come there and his trouble would be well and amply rewarded. By letters dictated in very fine Latin, the clerk humbly thanked the king for the honor of which he was not worthy, but [said] that truly he was a solitary, speculative man with strange habits; he was not suited for court; he did not know how to mouth the flattering blandishments appropriate to lords. Thus he was much more at ease and at rest in a poor life, eating cabbages and turnips, while contemplating philosophy, since he did not covet other riches.[119]

---

[116] Raymond Cazelles calls them "une équipe peu nombreuse et polyvalente, chaqu'un d'eux pouvant être attaché à des tâches diverses" (*Jean le Bon et Charles V* [n. 91], p. 549).

[117] Sherman, *Imaging Aristotle* (n. 17), plate 10; pp. 227–30.

[118] Autrand, *Charles V* (n. 6), pp. 728–31.

[119] Christine, *Charles V*, pt. 3, chap. 22, pp. 64–65: "Charles ... entendit que vers Avignon avoit un speculatif clerc, qui tenoit vie de philosophe, et moult soubtilment ouvroit en l'art d'arquemie ... et avoit esté ce dit clerc disciple de maistre Arnault de Villeneuve.... Le roy, qui toutes soubtilles choses desiroit à veoir, lui escripst que il voulsist venir par deça, et bien et grandement lui seroit sa peine meritée. Le clerc par ces lettres ditté[e]s en tres beau latin mercia le roy humblement de l'onneur qu'à lui non digne faisoit, mais que vrayement, comme il fust homme solitaire, speculatif, et d'estranges manieres, n'estoit abte à court, ne en

The king replied that he understood perfectly—that, had God not given him the burdens of lordship, he would have preferred to search for truth and virtue rather than to hear dishonest flattery. Seeing the king's good intentions, the alchemist went to Paris, stayed awhile, and returned with many beautiful gifts. This figure embodies the oxymoronic ideal of the honest and unambitious courtier and contrasts it with the caricature of the self-interested and disingenuous courtier. Thus, while Charles's support of and interactions with men of learning were concrete, even practical, they also served a larger symbolic political purpose.

Among the intellectuals in Charles's circle physicians and astrologers were not the only nor even the most significant players, but they were among the most independent of the forces that Charles needed to balance. Given the modest political influence of the medical faculty, even a man like Gervais Chrétien, who occupied important positions at the university over a long period, supported no positions and undertook no endeavors that in any way compromised the king's interests. This group of clients (Gervais, Pèlerin de Prusse, and Tommaso, as well as Gilles and others) was particularly beholden to and dependent on the king. Compared to canon and civil lawyers, their profession afforded them little public status; their properties, benefices, and access to other forms of patronage often derived from their connections with the king. The close analogy between astrology and medicine with respect to epistemology and applications only reinforced these men's strong ties to the person of the king: however much their knowledge was grounded in general science with implications for the public good, its applications would always be particular and circumstantial.

One further dimension of this group's background reinforced its close ties to the king and independence from competing centers of power: its foreign connections. Although the Norman origins of Oresme, Gervais, the physician Evrard de Conty,[120] and others were one source of allegiance and dependence, others came from a greater geographical and intellectual distance. Jean Jacme, whom Charles ennobled in 1359, established his medical credentials at the University of Montpellier and was at one time physician to the King of Aragon.[121] His adversary at Montpellier, Jean de Tournemire—like Jacme an author, and for a time physician to Pope Gregory XI at Avignon—also rendered service to Charles.[122] No less outsiders than those whom Charles brought up from the lower orders, these men, along with

bouche ne sceust avoir les blandices flateuses, qu'il convenoit à seigneurs; si estoit trop plus aise et à repos à pouvre vie, mengant choulz et rabes en speculant philosophie, comme il ne fust convoiteux d'autre richece." Cf. Christine de Pizan, *Lavision-Christine*, ed. Mary Louise Towner (Washington, D.C.: Catholic University of America, 1932), pt. 2, pp. 137–40.

[120] Ernest Wickersheimer, *Dictionnaire biographique des médecins en France au Moyen Age*, 2 vols. (Paris: Droz, 1936), s.v. "Evrard."

[121] Ibid., s.v. "Jean Jacme."

[122] Ibid., s.v. "Jean de Tournemire."

Tommaso, brought to the court the substance and aura of sciences and practices associated with southern Europe. Oresme is no doubt alluding to the alien nature of astrology when he mentions that, according to Ptolemy, southerners are better at it than northerners.[123] In Spain, Italy, and southern France, medicine and the celestial sciences (both separately and in combination) had a special prominence and luster; they were more secular than the learning of the north, and more associated with Arabic and Jewish traditions. It was not just the persons of these men but the expertise they offered that placed them at some distance from the University of Paris. Besides being free from the political entanglements of Paris, they came from an environment in which the theological and philosophical objections to astrology raised by Oresme had little impact. Under the king's aegis their distinct intellectual outlook participated in the general coherence of the debates at court, while retaining some of its alien tone.

Charles may also have been consciously engaged in some mild orientalizing. His status as "very Christian king" was never in question and had its own political uses, so Charles could not afford to have the hidden powers of the exotic overwhelm his reputation for humble orthodoxy, but his status could be enhanced by small doses. If the connection of academic astrology at Paris with the authorities of the Islamic world had become masked by the fourteenth century,[124] it was prominently featured in the books of Charles's library, including the manuscript depiction of the king among the heathen dignitaries of the art.[125] Christine de Pizan defended judicial astrology by citing its use by Abraham and the Chaldeans, while Nicole Oresme was not alone in suspecting Mohammet of false revelations obtained by means of magic or astrology.[126] The associations were real as well as symbolic: Charles put his Hebrew books at the disposal of Jewish scholars and in general assumed a posture of toleration, in spite of strong warnings within his own circle about Jewish practice of sorcery, astrology, and illicit medicine.[127] Indeed, one way in which its opponents stigmatised astrology was to emphasize its foreign associations. In 1387 Philippe de Mézières depicted Superstition, the allegorical figure who defends astrology, as adorned with geomantic signs and with Greek, Hebrew, and Chaldean letters.[128] The same sorts of signs probably appeared on the talismanic ring that the Dauphin

[123] Oresme, *Divinacions*, chap. 11, p. 86.

[124] Lemay, "True Place of Astrology" (n. 11), pp. 58–59, n. 1; 71–72.

[125] Edgar Laird ("Astrology" [n. 3], p. 171) observes that the figure of Alchabitius, unlike the Christian figures, is given dark hair in one manuscript.

[126] Christine, *Mutacion de* fortune, ll. 7717–20, quoted in Willard, "Astrologer's Daughter" (n. 35), p. 110, n. 22; Oresme, *Divinacions*, chap. 12, p. 100.

[127] Autrand, *Charles V* (n. 6), pp. 741–42.

[128] Jeannine Quillet, *La philosophie politique du Songe du vergier (1378): Sources doctrinales*, L'Eglise et l'Etat au Moyen Age 15 (Paris: Librairie Philosophique J. Vrin, 1977), pp. 118–19.

Charles had presented to his father upon the latter's return from captivity.[129] The effect of these associations was twofold. First, they connected the king with quasi-occult knowledge and thus with privilege and power. Second, they linked him with monarchs of the East like Alexander and Solomon, legendary for their command of esoteric learning and their unquestionable political authority.

CONCLUSION

The diverse, flexible, and complex strategies that Charles V employed in the course of his reign reflect something about the unity of knowledge in the fourteenth century, as well as about his political style and personal inclinations. Like other European monarchs, from Charlemagne to Henry II of England to Pope Innocent III, he called upon the content and the personnel associated with formal, elite, Latin learning in the service of political administration. By the late fourteenth century, however, Charles was able to go beyond simply using such learning to appropriating and integrating significant portions of it. He presided over the creation of a forum, a body of knowledge, and a method of discourse that brought together (however imperfectly) the two cultures of university and court.

The conditions of exchange in the court, the household, and the library constituted the forum where individuals as different as Oresme, Gervais, Tommaso, Christine, Philippe de Mézières, and Charles himself could witness and participate in discussions of common concerns in a shared language. The body of knowledge produced under those conditions drew heavily upon a variety of learned Latin traditions, which were profoundly transformed not only in the process of translation and adaptation, but also in the course of their appropriation by diverse members of the king's entourage. The method or practice of vernacular intellectual debate, which might have owed some debt to (and derived some legitimation from) university disputations, was related to literary traditions and conventions of the court as well. The exchanges concerning astrology, partially reconstructed here, both depended upon and contributed to these features of the court environment.

Insofar as these developments reflect a loose unity of knowledge at university and court, they represent much more than a simple transfer from the former to the latter. Like his French version of Aristotle's *Politics*, Oresme's vernacular intervention against astrology was a product of—indeed, a conscious contribution to—Charles's program of governance. Just as our understanding of debates about natural philosophy at the University

---

[129] Autrand, *Charles V* (n. 6), pp. 380–83.

of Paris would be incomplete without an account of its relationship to theological issues, so our grasp of controversy about astrology at the court depends on locating it in relation to political questions. Furthermore, these domains of inquiry were no more isolated from each other than were the participants in the exchanges treated here. The world of Christine de Pizan, Nicole Oresme, and Charles V was diverse and complicated, but it was not divided. In fourteenth-century France, contexts overlapped and texts spoke to each other across the dissolving boundaries of language, culture, gender, order, tradition, genre, and discipline. Too uneven and piecemeal to constitute a strongly coherent form of the unity of learning, these exchanges nevertheless reveal commonality beneath controversy.

ACADEMIC CONSULTING IN FIFTEENTH-CENTURY VIENNA: THE CASE OF ASTROLOGY

MICHAEL H. SHANK*

Present-day state legislatures and the historiography of early modern science have at least one thing in common: a low opinion of the universities in their territories. Although not universal, this estimate is very widespread, cutting across the many ideological differences and methodological divides that have emerged since it was first proposed several centuries ago. Despite its venerable age, this interpretation still has wide currency in the most recent historiography of science, which has—if anything—reinforced its standing.[1] To put the matter in stark terms: whereas medieval science is university science, early modern science is extrauniversity science. Medieval university science is speculative and theoretical, whereas nonuniversity early modern science is characterized by its practice, most notably its experimental practice. Each of the preceding propositions can be held separately, with a noncommittal attitude toward the other. More often than not, however, they go hand in hand, and provide a set of powerful, mutually reinforcing generalizations about the history of science in the half-millennium after 1200.

These generalizations have drawn aid and comfort from the stereotype of the ivory tower, an enduring hyperbole of university isolation that will live long on account of its apposite character in a limited number of cases. To transform this simplistic picture, one need only point to the weighty counterexamples of law and medicine, which are the most important and the least surprising.[2] This study brings together fragmentary and scattered evidence about the involvement of university masters in astrological prac-

---

* Abridged versions of this paper benefited from criticisms, comments, and suggestions offered by the participants in the joint HSS/AHA session in honor of John Murdoch (Washington, D.C., December 1992), the Science Studies colloquium at the University of California–San Diego (November 1994), and the Notre Dame Program in History and Philosophy of Science (September 1996). In addition to the editors, I thank in particular Joan Cadden, Ralph Drayton, Richard Kremer, David Lindberg, and Robert Westman.

[1] For overviews of the historiography, see John Gascoigne, "A Reappraisal of the Role of the Universities in the Scientific Revolution," in *Reappraisals of the Scientific Revolution*, ed. David C. Lindberg and Robert S. Westman (Cambridge: Cambridge University Press, 1990), pp. 207–60; and the preface to Mordechai Feingold, *The Mathematicians' Apprenticeship: Science, Universities and Society in England, 1560–1640* (Cambridge: Cambridge University Press, 1984), pp. 1–22.

[2] See in particular the introduction to Michael McVaugh and Nancy Siraisi, eds., *Renaissance Medical Learning: Evolution of a Tradition*, *Osiris*, 2nd ser., 6 (1990): 7–15, as well as the contribution of Danielle Jacquart, "Theory, Everyday Practice, and Three Fifteenth-Century Physicians," pp. 140–60.

tice near Vienna that will complicate the picture further. The alternative sketched below illustrates the perceived value of university expertise beyond the confines of the lecture hall, even among members of the faculty of arts. The case of astrology is particularly instructive as one of the most notable loci of interaction between the political sphere and science or natural philosophy, broadly understood.[3] Secondarily, this paper traces an intriguing pattern of change in the background of the astrological practitioners who worked near the Habsburg court—a perceptible shift from medical astrologers in the early fifteenth century, through mathematical astronomers/astrologers from the 1430s to the 1460s, and finally to the figure of the prophet-astrologer in the late fifteenth century.

In the background of this study lies the question of the university's relation to the larger society, a problem with which I have been concerned ever since John Murdoch, more than twenty years ago, first persuaded me that the medieval university is a fascinating institution. Since then my conviction has grown even firmer: the importance of the university for the history of medieval science, natural philosophy, medicine, learning, and culture is difficult to underestimate. John's scholarship has done much to illuminate "unities" among various aspects of medieval learning that have suffered from excessive compartmentalization at the hands of historians—most notably, across the boundaries between science and philosophy, and to a lesser extent between philosophy and theology.[4] In the present state of our ignorance, the medieval university remains the primary context and spring-

---

[3] Lynn Thorndike, *History of Magic and Experimental Science*, 8 vols. (New York: Columbia University Press, 1923–58); Otto Neugebauer, "The Study of Wretched Subjects," *Isis* 42 (1951): 111; David Pingree, "Hellenophilia vs. the History of Science," *Isis* 83 (1992): 554–63; J. D. North, *Horoscopes and History*, Warburg Institute Surveys and Texts 13 (London: Warburg Institute, University of London, 1986); Hilary Carey, *Courting Disaster: Astrology at the English Court and University in the Later Middle Ages* (New York: St. Martin's Press, 1992); Brian Copenhaver, "Astrology and Magic," in *The Cambridge History of Renaissance Philosophy*, ed. Charles B. Schmitt et al. (Cambridge: Cambridge University Press, 1988), pp. 264–300; Stefano Caroti, "La critica contro l'astrologia di Nicole Oresme e la sua influenza sul Medioevo e nel Rinascimento," *Atti dell' Accademia Nazionale dei Lincei*, Classe di Scienze morali, storiche et filologiche, ser. 8, *Memorie* 23 (1979): 545–685; and the specific literature cited below.

For the Viennese environment in particular, see Helmuth Grössing and Franz Stuhlhofer, "Versuch einer Deutung der Rolle der Astrologie in den persönlichen und politischen Entscheidungen einiger Habsburger des Spätmittelalters," *Anzeiger der Österreichischen Akademie der Wissenschaften*, phil.-hist. Kl., 117 (1980): 267–83; Franz Stuhlhofer, "Georg Tanstetter (Collimitius): Astronom, Astrologe und Leibarzt bei Maximilian I. und Ferdinand I.," *Jahrbuch des Vereins für Geschichte der Stadt Wien* 37 (1981): 7–49; Helmuth Grössing, *Humanistische Naturwissenschaft: Zur Geschichte der Wiener mathematischen Schule des 15. und 16. Jahrhunderts* (Baden-Baden: Koerner, 1983), pp. 47–52, 121–26.

[4] John E. Murdoch, "From Social into Intellectual Factors: The Unitary Character of Medieval Learning," in *The Cultural Context of Medieval Learning: Proceedings of the First International Colloquium on Philosophy, Science, and Theology in the Middle Ages*, ed. John E. Murdoch and Edith D. Sylla, Boston Studies in Philosophy of Science 26 (Dordrecht: Reidel, 1975), pp. 271–339.

board for most of the writings and activities amalgamated under the rubric "medieval science," including its ties to, and inclusion in, philosophy.

Indeed one can, without being unduly circular, learn much about the practice of medieval science without looking far beyond the walls of the university. This is not surprising, for two reasons. First, the universities fought vigorously for relative autonomy from their local jurisdictions. Their frequent success is illustrated not only by their extensive financial and legal privileges,[5] but also by the fact that some of the most famous landmarks in medieval university history (such as the condemnations of 1277) are associated with local infringements on academic autonomy. Second, no other late-medieval institution comes close to competing with the university as a base for natural philosophy, medicine, and the mathematical sciences. For these reasons, an insular treatment of *university* science offers an excellent first approximation to an understanding of *medieval* science.[6] To justify such an insular approach, one need only note that the university was—and largely remains to this day—a peculiar subculture with its own language and vocabulary; its own systems of decision-making, rewards, and promotion; its own patterns of administration and leadership; and even its own rituals. Indeed, the cultural peculiarity of the medieval university, to a greater extent than its early-modern counterpart, finds expression in a scale of values that sometimes subverts the traditional hierarchies of the surrounding culture. Like few other medieval institutions, the university made it possible for poor or lower-caste intelligent youngsters—if they were male and Christian—to make good, and even rise to the papacy.[7]

As a second approximation, however, an isolationist approach to the university is inadequate. The medieval universities were far from being ivory towers: they interacted, sometimes vigorously, with their local surroundings. If *secundum imaginationem* we could cut academic institutions loose from their moorings and push them out to sea, they would not survive long. Intellectual, social, and economic life within the university depended crucially on the permeability of its walls to ideas and texts; to masters and students; and to money, paper, food, and drink. This paper focuses on one

[5] Typical academic privileges included safe-conduct during travel, freedom from taxation, price controls on many services and goods (bread, wine, writing supplies), and the right to be tried by an ecclesiastical rather than a civil court.

[6] As Edward Grant pithily noted, at no time other than the Middle Ages did the university, in its nearly millennial history, devote so much of its energy and attention to natural philosophy—what we now call science: Edward Grant, "Science and the Medieval University," in *Rebirth, Reform, and Resilience: Universities in Transition, 1300–1700*, ed. James M. Kittelson and Pamela J. Transue (Columbus: Ohio State University Press, 1984), pp. 68–69.

[7] For an exceptional university woman (in disguise), see Michael H. Shank, "A Female University Student in Late Medieval Kraków," *Signs* 12 (1987): 373–80; reprinted in *Sisters and Workers in the Middle Ages*, ed. J. M. Bennett, E. A. Clark, Jean F. O'Barr, B. A. Vilen, S. Westphal-Wihl (Chicago: University of Chicago Press, 1989), pp. 190–97.

aspect of that intellectual permeability, namely, the interaction of university masters with the city and the court.

Then as now, the leading university masters did more than teach and write; they also constituted a resource and pool of talent that could be consulted as the need arose. The demand for academic consulting demonstrates with clarity the recognition by political authorities that university learning, esoteric though it seems to uninitiated late-twentieth-century onlookers, was neither seen, nor treated, as useless, and could provide a framework for political decisions. The expression "academic consulting" is, of course, anachronistic, but fruitfully so. It serves, first, to draw attention to the substantive contributions of university expertise to extra-academic political life, rather than merely to matters of prestige enhancement or social climbing, for example. As Larry Scanlon has recently pointed out about the medieval *Secretum secretorum*, allegedly written by Aristotle for Alexander the Great: "This belief—that antiquity's greatest philosopher would have devoted himself to producing a *Fürstenspiegel* for its greatest conqueror—suggests the conviction that royal power not only desired secular wisdom, but needed it."[8] Their clerical status nothwithstanding, academics were the leading masters of that desirable secular wisdom. Second, the expression connotes an activity associated with ongoing academic responsibilities. The qualification is important, for in the late Middle Ages university consultants were outnumbered by university-trained individuals, especially lawyers, who had left the university to assume high positions of political and administrative responsibility. By dint of their skills and numbers, these former students were changing the social composition of European courts.[9] The importance of the alumni should not obscure entirely the extra-academic involvements of the university masters. The ongoing academic relationship is crucial in several ways. In contrast to the artist in the employ of a court or city, the academic consultant belonged to an institution whose power gave him more autonomy than that of a guild member.

Although these questions might be studied with respect to several universities, the case of Vienna between the late fourteenth and the mid-fifteenth centuries is particularly fascinating on account of both its acknowledged significance in the history of astronomy and its proximity to several courts that were growing in importance throughout the period. After sketching the background of the issue in Vienna and hinting at the wide-

---

[8] Larry Scanlon, *Narrative, Authority, and Power: The Medieval Exemplum and the Chaucerian Tradition* (Cambridge: Cambridge University Press, 1994), p. 279.

[9] This is particularly the case for lawyers. See Peter Moraw, "The Court of the German Kings and the Emperor at the End of the Middle Ages, 1440–1513," in *Princes, Patronage and the Nobility: The Court at the Beginning of the Modern Age, c. 1450–1650*, ed. Ronald G. Asch and Adolf M. Birke (London: The German Historical Institute/Oxford University Press, 1991), p. 113.

spread character of consulting, I shall turn specifically to the case of astrology.

FROM PARIS TO VIENNA

As has long been known, the universities of Paris and Vienna share a close genetic connection. After the University of Vienna was founded in 1365, the Parisian master Albert of Saxony served briefly as its first rector. During the next twenty years, the university led a precarious half-existence. When it revived ca. 1384, the academic and administrative impetus once again came from Paris-trained German masters whom the Great Schism had recently forced to return home, most notably Henry of Langenstein (d. 1397).[10]

Not surprisingly, the institutional model of Paris loomed large at Vienna, extending from the inspiration derived from the Parisian statutes to the ideals for the university library. The two universities also shared a notable contextual similarity: unlike Oxford, for example, Paris and Vienna were cities in which courts played a leading role, opening up many opportunities for interaction between ruler, town, and gown—from friction to mutual assistance, and from consultation to flattery. In Vienna, however, the university had from the outset a closer relationship with the local court than was the case in Paris. Whereas the French crown could claim little credit for the existence of the University of Paris, the Habsburg dukes Rudolph IV (d. 1365) and Albert III (d. 1395) were directly responsible for, respectively, the birth of the university in 1365 and its revival in the 1380s.[11] After bringing Langenstein and his colleagues to Vienna in 1384, Albert III remained a patron of the university; he also endowed a *collegium ducale* that provided the most prominent masters with stipends and lodgings, and built an *aula magna* in which the university held its major gatherings. That Langenstein appealed directly to the duke when he needed funds to expand the holdings of the university library testifies to his expectation of support.[12] Conversely, as founders and patrons of the university, the Habsburgs had a strong claim on its attention and, as we shall see, they did

---

[10] Michael H. Shank, *"Unless You Believe, You Shall Not Understand": Logic, University, and Society in Late Medieval Vienna* (Princeton: Princeton University Press, 1988), chap. 1.

[11] According to one chronicle, it was Albert III's chancellor, Berthold of Wehingen, who convinced the duke to bring Langenstein and Oyta to Vienna, and to support them comfortably. Berthold had been a master of arts at Vienna during the early days of the first foundation; see the "Wiener Annalen, 1349–1404," *Monumenta Germaniae Historica, Deutsche Chroniken* (Hanover and Leipzig, 1933), 6: 233.

[12] Shank, *"Unless You Believe"* (n. 10), p. 20; Gustav Sommerfeldt, "Zur Zeit der Begründung der Universität Wien," *Mitteilungen des Instituts für Österreichische Geschichtsforschung* 29 (1908): 291–322, esp. 306.

not hesitate to draw upon the expertise of its masters. The range of academic consulting can be illustrated only briefly here. But it is important to emphasize that this practice was restricted neither to Vienna, nor to natural philosophical, medical, or astrological issues; nor were the dukes the exclusive beneficiaries.

One of the most unusual instances of academic consulting involves a request from the city council of Vienna, which solicited advice from the university about matters of taxation, interest, and property law in the early 1390s. This case spawned separate treatises on legal contracts by three masters (Henry of Langenstein, Henry of Oyta, and Johannes Reuter), as well as an abject apology from Langenstein for his tardiness in answering the city council's request. In this instance, the city hoped to solve a specific problem by appealing to general principles of law. For help with this task, it called upon the expertise of academic theologians and lawyers. The case was politically sensitive: it occurred in the wake of the pope's excommunication of Duke Albert III for taxing the clergy, and the duke's subsequent reinstatement after he rescinded the tax; it also involved a clash between a flexible older precedent and a rigid law enacted by Rudolph IV (d. 1365), the founder of the university and brother of Albert III.[13] As its request demonstrates, the city council was not seeking academic legitimation. Rather, it evidently believed, perhaps with prodding from the nobility, that university learning could help advance its cause—perhaps by helping to exact more concessions from the duke. The city council received not a unanimous, authoritative report, but a typically academic contribution: three different opinions that, to the council, may have looked like an unresolved university disputation.[14] Clearly, the university masters were not making policy but offering arguments and advice.

Later, toward the end of the Great Schism (1378–1417), the Habsburg dukes themselves drew heavily on the expertise of their university masters, particularly theologians, for both legal and political advice. This phenomenon was not merely Viennese, but European in scope. Around 1415, when the Council of Constance convened to end the papal schism, academics impressed their *modi operandi* upon the council, organizing their representatives into "nations" like those of their faculties of arts. Indeed, the Council of Constance arguably represents the political "coming of age" of the European university community. To be sure, the importance of academics at the Council derived in part from the influence that alumni who now held high political office exerted on the proceedings. But another part of that significance was that the universities participated in the Council as in-

[13] Winfried Trusen, *Spätmittelalterliche Jurisprudenz und Wirtschaftspolitik dargestellt an Wiener Gutachten des 14. Jahrhunderts,* Vierteljahrschrift für Sozial- und Wirtschaftsgeschichte, Beiheft 43 (Wiesbaden: Franz Steiner Verlag, 1961), esp. pp. 3–7, 146–55.

[14] Ibid., p. 20.

stitutions in their own right, with representatives invited by Emperor Sigismund and the competing popes. Eventually, once their members and alumni had constructed a theory of ecclesiastical power around which the contending factions could unite, the universities, through their representatives, joined the cardinals in settling the schism by electing a new pope.[15]

IN THE WAKE OF HENRY OF LANGENSTEIN

That the reliance on academic learning outside the university was widespread should be a truism in the cases of law, theology, and medicine. It is far from being obvious in the specific case of astrology in Vienna, however, for several reasons. First, since the study of ancient and medieval astrology has traditionally been considered a topic of questionable worthiness, it is only in recent decades that it has begun to receive the scholarly scrutiny it deserves. In the specific case of the Viennese environment, the historical study of astrology has suffered from an additional handicap: the perception that the university and its masters had remained untainted by astrology.[16] Since Henry of Langenstein, the leading light of the early university, was a prominent critic of judicial astrology, his opposition was presumed to have shaped his context decisively.

Langenstein's late-fourteenth-century arrival in Vienna must be seen, however, against a background in which royal and princely courts increasingly were seeking astrological guidance. The house of Habsburg, with its early imperial aspirations, was not exempt. Already in the early thirteenth century, astrologers at the court of Emperor Frederick II (d. 1250) had reportedly predicted that the young Rudolph of Habsburg (1218–1291), who was staying at the imperial court, would later become Frederick's successor.[17] After Rudolph in fact became the first Habsburg to head the Holy Roman Empire as German king in 1273, astrology continued to play an

[15] Between the late 1370s and the early decades of the fifteenth century, the Great Schism crippled the power and moral authority of the papacy. These divisions gave academics an opportunity to move into the breach. Although the latter arguably did not exploit the opportunity as fully as they might have, their participation proved extensive and decisive—from the early academic proposals to settle the Schism in the late 1370s right on down to the various alternatives that led to the Council of Constance, and indeed during the Council itself. See R. N. Swanson, *Universities, Academics, and the Great Schism* (Cambridge: Cambridge University Press, 1979), passim, e.g., pp. 15–18, 175–78, 193–98, 204–7; Shank, *"Unless You Believe"* (n. 10), pp. 178–79, 185.

[16] Rudolf Klug, "Johann von Gmunden, der Begründer der Himmelskunde auf deutschem Boden," *Sitzungsberichte der Akademie der Wissenschaften—Wien*, phil.-hist. Kl., 222 (1943): 9, 58–63.

[17] See Alphons Lhotsky, "Apis Colonna: Fabeln und Theorien über die Abkunft der Habsburger. Ein Exkurs zur *Cronica Austriae* des Thomas Ebendorfers," in Lhotsky, *Aufsätze und Vorträge*, ed. Hans Wagner and Heinrich Koller, 5 vols. (Munich: Oldenbourg, 1971), 2: 15–16.

important role at his court. Franz Stuhlhofer has noted that three significant events in Rudolph's life occurred on astrologically significant dates: his coronation (Mars and Sun in sextile), his marriage (Venus and Mercury in conjunction), and his decisive battle with Ottokar (Mars and Sun in conjunction).[18]

From an astrological point of view, then, the environment into which Langenstein was stepping was probably not neutral, even though the university in the late fourteenth century did little to encourage the association between judicial astrology and the court. Langenstein's skeptical outlook on this score goes back at least to his years as a Parisian master, when King Charles V of France asked him to discuss the significance of the comet of 1368: although the French king took astrology very seriously, Langenstein concluded that it was futile to read in the comet signs about princes and plagues—thereby showing that consultants do not always tell their powerful clients what they want to hear.[19]

Like most of his contemporaries, Langenstein took it for granted that the planets naturally caused effects upon sublunar physical bodies. But his natural philosophical principles (notably his theory of causation, and his belief in the continuous action of planets) provided the basis for doubting that these regular effects were accentuated during astrologically propitious times. For example, he pointed out the problems involved in defining planetary conjunctions in terms of proximity in longitude only, while neglecting sometimes significant differences in latitude, or in assuming that the effects of a comet could be delayed for months or even years.[20] In works written in Paris and Vienna, Langenstein also criticized the validity of astrology on epistemological grounds: in the hierarchy of certainty, he classified astrology last, behind medicine and meteorology.[21] For all of

[18] Grössing and Stuhlhofer, "Versuch einer Deutung" (n. 3), p. 276. Although the thirteenth-century astrological writer Leopold of Austria called himself the "son of the Duchy of Austria" (filius ducatus Austriae) in his *Compilatio*, it is not clear what his relation to the Habsburgs was; see Pierre Duhem, *Le système du monde: Histoire des doctrines cosmologiques de Platon à Copernic*, 10 vols. (Paris: Hermann, 1913–59), 3: 289–90.

[19] Herbert Pruckner, *Studien zu den astrologischen Schriften des Heinrich von Langenstein* (Leipzig: Teubner, 1933), e.g., p. 89. For a concise, up-to-date summary of the contentious atmosphere and the astrological personnel at the French court, see Carey, *Courting Disaster* (n. 3), pp. 106–11; and Edgar Laird and Robert Fischer, *Pélerin de Prusse on the Astrolabe: Text and Translation of His Pratique de l'astrolabe*, Medieval and Renaissance Texts and Studies, 127 (Binghamton, N.Y.: State University of New York, 1995), pp. 1–8.

[20] See, e.g., Pruckner, *Studien zu den astrologischen Schriften* (n. 19), pp. 114, 152–55, 169–71; Nicholas Steneck, "A Late Medieval Arbor Scientiarum," *Speculum* 50 (1975): 264ff; John D. North, "Celestial Influence—The Major Premiss of Astrology," in North, *Stars, Mind and Fate: Essays in Ancient and Medieval Cosmology* (London: Hambledon Press, 1989), pp. 291–96.

[21] Caroti, "La critica" (n. 3), pp. 616–17; Nicholas Steneck, *Science and Creation in the Middle Ages: Henry of Langenstein (1397) on Genesis* (Notre Dame: University of Notre Dame Press, 1976), p. 100. Epistemologically, he thus appears to have been even less tolerant of medical astrology than Oresme.

these reasons, he joined Oresme in believing that astrology aided neither princes nor the commonwealth, and indeed was politically noxious.[22]

Against such a background, the reader will not be surprised to learn that we have at present no evidence of Langenstein's involvement with the Habsburg court on astrological matters during his Viennese years. But Langenstein did not hold the power of veto over all such activities at the university during the next two generations: in the fall semester of 1390, during his tenure there, a course on Alkabitius (presumably on his textbook, the *Introductorius* or *Isagoge*) appears on the list of ordinary lectures, for the first and last time in the Acts of the Faculty of Arts.[23]

Beyond this fleeting astrological incursion into the curriculum, several medical astrologers were associated with both the court and the university during Langenstein's career. We know almost nothing about them, save that they had astrological interests, which came to the fore when a comet appeared in the late winter of 1402. Frederick of Drosendorf (d. 1404) was a master of arts from Prague who settled in Vienna and taught at the university in the 1390s while he pursued a medical degree. Later on, one source designates him as "great astrologer," another as "astrologer of Austria," strongly suggesting an official association with the court. He was also a canon of the cathedral of St. Stephen's, to the chapter of which he left a small library, including several astronomical and astrological books.[24] Five years after Langenstein's death, the comet of 1402 spurred him into action. On 12 March 1402, he petitioned the faculty of arts for permission to determine two questions, one of which is probably the question on the comet

[22] Pruckner, *Studien zu den astrologischen Schriften* (n. 19), p. 200; Caroti, "La critica contro l'astrologia di Nicole Oresme" (n. 3), p. 626.

[23] *Acta Facultatis Artium Universitatis Vindobonensis, 1385–1416*, ed. Paul Uiblein (Vienna: Böhlau Verlag, 1968), p. 54 [hereafter cited as *AFA* I]. The lecturer was Benedikt de Makra. A former student at the University of Prague, he had matriculated at Vienna in 1387 but left the university four years later (*AFA* I, 67, lines 19–21).

[24] Alphons Lhotsky, *Die Wiener Artistenfakultät, 1365–1497*, Österreichische Akademie der Wissenschaften, phil.-hist. Kl., 247 (Vienna, 1965), pp. 115–16. Arriving from the University of Prague, Drosendorf matriculated at Vienna in 1391. He died in 1404, the year in which he received his medical degree. The "Wiener Annalen" calls him *astrologus Austriae*, an official-sounding designation echoed by the *magnus astrologus* in the necrology of the cathedral chapter; see Harry Kühnel, *Mittelalterliche Heilkunde in Wien* (Graz-Köln: Böhlau, 1965), pp. 21–22 (whose numbers and references are not always correct). Drosendorf left twenty-five books to the cathedral chapter, of which sixteen were medical and five astronomical/astrological, including "in astronomica glosam Almagesti," the Alphonsine tables, the "introducciones Zahelis in iudicia Arabum," an "arithmetic of astronomy," and—most intriguing of all—"tres sexternos de theorica nova, que tabularum terminis dimissis realem sperarum et motuum habitudinem declarat": Theodor Gottlieb, *Mittelalterliche Bibliothekskataloge Österreichs* (Vienna: A. Holzhausen, 1915), 1: 281–82; see also *AFA* I: 511. Curiously, the title of this "theorica nova" is almost identical to that of Georg Peuerbach's *Theoricae novae*, as recorded by Regiomontanus in his 1454 copy (MS Vienna, ÖNB lat. 5203, fol. 2r) but omitted in the version he printed (Nürnberg, c. 1472?); see E. J. Aiton, "Peurbach's *Theoricae novae planetarum*: A Translation with Commentary," *Osiris*, 2nd ser., 3 (1987): 8 n. 14.

preserved in a Graz manuscript. According to the colophon, one of the respondents was Berthold of Wehingen (d. 1410), bishop of Freising, former rector of the university, chancellor to Dukes Albert III and Leopold IV (d. 1411), and one of the figures responsible for bringing Langenstein to Vienna.[25] The comet also stimulated another medical astrologer to write a treatise on the subject. The son of an apothecary, Jacob Engel of Ulm had been trained in the arts and medicine at Paris in the 1380s and matriculated at Vienna in 1391. It is not clear whether Engel taught in the faculty of arts, but he did serve as personal physician to the Habsburg dukes Albert III and Leopold IV.[26] His treatise on the comet of 1402 devotes the first nine chapters to various aspects of comets, including their signification. His review of the literature ends with Albertus Magnus in the thirteenth century, without a word about either Langenstein's lengthy treatise on comets, or the reservations about astrology common in the Parisian natural philosophical tradition that Engel had left a decade earlier. Engel concludes his treatise by claiming that "faithful catholics must not pay attention to the significations of either this or other stars"—intriguing advice after spending most of the treatise doing the contrary.[27]

Engel and Drosendorf were but two in a series of academically trained physicians associated with both the university and the Habsburg court in the late fourteenth and early fifteenth centuries. Several others had had recent contact with the University of Padua, a prominent center of medical astrology.[28] It is perhaps one of these individuals who deserves the credit

[25] *AFA* I: 203, 500; MS Graz, Universitätsbibliothek 443, fols. 105v–107r: Utrum cometa sit exalatio calida et sicca futurorum eventuum presignificativa. For Berthold of Wehingen's role in bringing Langenstein to Vienna, see the "Wiener Annalen, 1349–1404" (n. 11), 6: 233.

[26] *AFA* I: 191, 415, 523. A facsimile of the late-fifteenth-century incunabulum edition of "De cometis" appears in Jane Jervis, *Cometary Theory in Fifteenth-Century Europe*, Studia Copernicana, vol. 26 (Dordrecht/Boston: Kluwer, 1985), pp. 157, 161. From Engel's treatise, we learn that he was in Paris in 1382, presumably as a student. After translating a treatise on phlebotomy into German for the court of Albert III in 1390, he matriculated at the University of Vienna the following year, as a master of arts and *licenciatus in medicina*. By 1402, he had been associated with Vienna for more than ten years. Kühnel, *Mittelalterliche Heilkunde* (n. 24), pp. 91–92.

[27] "Capitulum ultimum in quo ostenditur quod fideles catholici non debent advertere nec illam nec aliarum stellarum significationes" (Jervis, *Cometary Theory*, p. 160). See also Thorndike, *History of Magic and Experimental Science* (n. 3), 4: 80–87.

[28] On Padua in the fifteenth century, see Nancy Siraisi, *Medieval and Early Renaissance Medicine: An Introduction to Knowledge and Practice* (Chicago: University of Chicago Press, 1990), pp. 64, 68. To cite only three examples: before Jakob Engel, Nicholas of Hebersdorf (d. 1419) was physician to Duke Albert III from 1378 onward. He was among the first students to matriculate at the newly revived University of Vienna in 1382. At some point thereafter, he moved to Padua until March 1403. A few months after his return, he was elected dean of the Viennese medical faculty, serving occasionally as consulting physician to Albert V and Leopold IV. Harry Kühnel, "Die Leibärtzte der Habsburger bis zum Tode Kaiser Friedrichs III.," *Mitteilungen des Österreichischen Staatsarchiv* 111 (1958): 10–11; Kühnel, *Mittelalterliche Heilkunde* (n. 24), pp. 60–61.

for computing almanacs for 1403–1417, arranged in a novel format that allowed the practitioner to compare the positions of different planets at a glance.[29]

During the early years of the revived university, most hints of astrological activity pertain to physicians who usually had been trained elsewhere and appear to have joined the university pro forma. Since the Viennese medical faculty initially was very small, one guesses that it was the ducal physicians who lent support and prestige to the faculty, rather than the other way around.

JOHN OF GMUNDEN AND THE HABSBURG COURTS

Several clues suggest that between the 1420s and 1440s this early pattern of court physicians as leading astrologers was beginning to shift toward greater involvement in astrology by masters of arts. Not coincidentally, these were the years following the end of the Great Schism. In the later years of the Schism, the influential theologian and cardinal Pierre d'Ailly turned to aspects of astrology that he had rejected during his student years in Paris under Nicole Oresme, delved deeply into his astrological vision of history, and wrote extensively about it.[30] The works of Pierre d'Ailly found readers in the circle of John of Gmunden, the leading astronomer at Vienna from the 1420s until his death in 1442.[31] Gmunden himself was a master of

---

A "Franciscus medicus" (not mentioned by Kühnel but identified by Uiblein as "Franciscus de Motonibus de Tervisio") had matriculated at Vienna in 1385 but was in Padua by 1410 at least. He completed his medical studies and joined the Padua faculty in 1423. A gift-manuscript mentions him as "artium lib. et med. doctorem tunc ducum Austrie physicum;" his titles and career track suggest involvement in medical astrology: *AFA* I: 511, [607]; Tiziana Pesenti, *Professori e promotori di medicina nello Studio di Padova dal 1405 al 1509: Repertorio bio-bibliografico*, Contributi alla Storia dell'Università di Padova 16 (Padua: Edizioni Lint, 1984), p. 158.

Galeazzo de Sancta Sophia (d. 1427) had also arrived in Vienna from Padua in 1394, and was a leading member of the university and of the Habsburg court, as ducal physician, until his return to Padua sometime after 1405: Kühnel, *Mittelalterliche Heilkunde* (n. 24), p. 39; Kühnel, "Leibärzte" (n. 28), pp. 14–15; Pesenti, *Professori e promotori*, pp. 182-86. See also Johann Schroff and Johann Aygel, discussed in n. 35 below. Colophons provide additional evidence of Vienna-Padua connections by later scholars: Pesenti, *Professori e promotori*, pp. 32, 184.

[29] The manuscript is Vienna, ÖNB lat. 2406; see Ernst Zinner, *Leben und Wirken des Johannes Müller von Königsberg genannt Regiomontanus*, 2nd ed. (Osnabrück: Otto Zeller Verlag, 1968), pp. 185–86 (hereafter cited as Zinner, *Regiomontanus*).

[30] See the excellent book by Laura Ackerman Smoller, *History, Prophecy, and the Stars: The Christian Astrology of Pierre d'Ailly* (Princeton: Princeton University Press, 1994), esp. pp. 51–54.

[31] MS Vienna, ÖNB lat. 5266, contains several of Pierre d'Ailly's most important works, together with other astronomical works. The scribes, who were copying in 1434, are associated with the University of Vienna and the nearby Klosterneuburg monastery, at the time headed by Georg Müstinger, a good friend of Gmunden's who was devoted to the science of the stars. See Smoller, *History*, p. 135; and Michael H. Shank, "The 'Notes on al-Biṭrūjī'

arts who had begun, but never completed, his theological degree, instead teaching exclusively astronomy, mathematics, and related topics. So far as we know, in his unusual lifelong *official* teaching career he did not take up astrology, although he introduced to the curriculum new lectures on the astrolabe and the albion.[32] At his death, however, he willed to the faculty of arts several astronomical instruments and a library that included a serious collection of astrological books. These volumes testify to a more-than-passing interest in a discipline that had no official academic outlet in Vienna. His will also specified that in their new home the astrological books were to be kept in a locked cabinet.[33]

The older historiography on Gmunden has portrayed him as an opponent of astrology because his writings contain an attack—in German—on James of Clusa's astrological prediction of a conjunction of all the planets in 1432. But Gmunden had much more to say about the astrologer's incompetence and foolishness than he did about opposition to astrology per se.[34] Indeed, he seems to have had more than a private interest in astrology. He was well known at both of the Habsburg courts of his mature years. By 1429 Duke Albert V had appointed him "superintendent of the university," a long-term position of liaison between the duke and the university.[35] In

---

Attributed to Regiomontanus: Second Thoughts," *Journal for the History of Astronomy* 23 (1992): 15–30, esp. 18–19, and the literature cited therein.

[32] Claudia Kren, "Astronomical Teaching at the Late Medieval University of Vienna," *History of Universities* 3 (1983): 15–30, esp. 19–22. Contrary to Kren, it was Gmunden who lectured on the albion (p. 20) and Johann Angerer on *De sensu et sensato* (p. 19), not vice versa.

[33] In contrast to his theological books and Boethius's *Musica*, which were to be chained (for normal use on the premises), the following works were specifically to be locked up: Guido Bonati's *Summa astrologiae*, John of Eschenden's *Summa iudiciorum*, Hali ibn Ragel's *Summa iudiciorum* and his commentary on Ptolemy's *Tetrabiblos*, Alkabitius's *Introductorius*, and Leopold of Austria's *Summa iudiciorum* [= *Compilatio*]. The Faculty of Arts discussed Gmunden's proposals for his books and instruments twice before his death, first in tentative fashion in 1435, and again in 1440, before eventually probating his will in 1443. See Paul Uiblein, "Johannes von Gmunden: Seine Tätigkeit an der Wiener Universität," in *Der Weg der Naturwissenschaft von Johannes von Gmunden zu Johannes Kepler*, ed. G. Hamann and H. Grössing, Österreichische Akademie der Wissenschaften, Sitzungsberichte, phil.-hist. Kl., 497 (Vienna, 1988), p. 49; the text of the will appears on pp. 59–62.

[34] "Waz aber got will verhenken in obuerschriben dinger daz hat er im nicht chunt than er hat es auch nicht aus der kunst es ist auch nicht versehentleich daz du vorigen geschrift ausge von ainem gelerten und weysen, sunder daz dy in ainer narrenweys oder zu ainer petriegnus des volkchs hat geticht. Geschribn zu Win des andern tags des augusts anno domini 1432" (Klug, "Johann von Gmunden" [n. 16], pp. 62–63). See also Hans Kaiser, "Johannes von Gmunden und seine mathematischen Leistungen," in Hamann and Grössing, *Der Weg der Naturwissenschaft*, (n. 33), p. 89. James of Erfurt is also known as Jacobus de Clusa.

[35] "Superintendens ex parte principis" (MS Vienna, Universitätsarchiv, Ph7 [hereafter cited as *AFA* II], fol. 99r). See Konradin Ferrari d'Ochieppo and Paul Uiblein, "Der 'Tractatus Cylindri' des Johannes von Gmunden," in *Beiträge zur Kopernikus-Forschung*, Katalog des Oberösterreichischen Landesmuseum, 86 (Linz, 1973), pp. 33, 38 n.14; and Uiblein, "Johannes von Gmunden" (n. 33), p. 46. Some earlier holders of the office of superintendent had been personal physicians to the Habsburgs, notably Johannes Schroff (d. 1417)

this kind of relationship with the court, more typical of the medical astrologer than the master of arts, Gmunden's astrological books probably found uses beyond self-education.[36] This presumption is strengthened by a manuscript from 1438, at the end of Albert V's rule, which contains several of Gmunden's works copied by one of his students, and concludes with a table of planetary aspects in relation to the moon. The audience is clearly highly placed, for the table tells the user which days are to be preferred or avoided for such nonbourgeois activities as undertaking joint ventures with princes (to be avoided when the moon and Saturn are in a quartile aspect).[37] Whether the manuscript was initially intended for Albert is not certain.

What is certain is that the very next year, in 1439, Duke Albert V died, apparently without an heir, and Lower Austria fell to Duke Frederick V, from the Styrian branch of the Habsburgs. Soon thereafter Frederick V was elected German king, the prelude to his eventual coronation as Emperor Frederick III in 1452.[38] As we shall see, with the Third Frederick were associated great expectations that add a new dimension to our understanding of the emperor's astrological interests.

By 1440 the Gmunden manuscript with its "Table of aspects" was definitely in Frederick III's hands. Onto the back flyleaf, he copied his personal trademark, the vowel sequence AEIOU and the motto explicating it, just above a notice recording an event of momentous political significance: Albert V had an heir after all, for his widow had given birth to a posthumous son, Ladislaus.[39] If Ladislaus lived, Frederick's rule over a consoli-

---

and Johann Aygel (d. 1436), both of whom had also trained in Padua: see Kühnel, "Die Leibärtzte" (n. 28), p. 15; Kühnel, *Mittelalterliche Heilkunde* (n. 24), pp. 36–37, 63–64; *AFA* I, 524, 537–38; and *AFA* II, fol. 46v: "[22 December 1421] secundo proposuit superintendens universitatis videlicet magister Johannes Aygl."

[36] A hint of the displacement of the medical astrologer by the astronomer appears in a letter from June 1456, in which Georg Peuerbach implies that he observed the comet of 1456 ("Halley's") with the former Habsburg Duke Albert's physician, whose astronomical knowledge he evidently did not admire; see Albin Czerny, "Aus dem Briefwechsel des grossen Astronomen Georg von Peuerbach," *Archiv für Österreichische Geschichte* 72 (1888): 299. Peuerbach does not name this contemporary of Gmunden's, but the chronology seems to point to Johannes Zeller of Augsburg: Kühnel, "Die Leibärtzte" (n. 28), p. 94.

[37] "Tabula aspectus planetarum ad lunam," MS London, British Library, Add. 24071, fol. 75v. The splendid illuminated initial that opens the manuscript (showing a cowled figure—probably Grosseteste—holding up an astrolabe) strengthens the presumption of a nonacademic provenance.

[38] With this title, some historians designate him as Frederick IV, however (e.g., Joseph Chmel, n. 59 below). His election as German king effectively made him emperor-to-be, and he was therefore treated as emperor well before his coronation by the pope.

[39] The explicit signification Frederick gave to this vowel sequence in the manuscript is "En, amor electis iniustis ordinor ultor / Sic Fridiricus ego mea iura rego," that is, "I am loved by the elect, I am appointed punisher of the unrighteous; Thus I Frederick manage my laws." There were, however, many other versions; see Alphons Lhotsky, "AEIOV: Die 'Devise' Kaiser Friedrichs III. und sein Notizbuch," and "Die Bibliothek Kaiser Friedrichs III.," in Lhotsky, *Aufsätze und Vorträge* (n. 17), 2: 181–82, 227, 231; Gerhard Ladner, "The Middle Ages in Austrian Tradition: Problems of an Imperial and Paternalistic Ideology," *Viator* 3

dated Habsburg territory that included Hungary and Bohemia would be temporary. The presence of this notice in a book from the imperial library containing planetary aspects is suggestive indeed. A tantalizing counterpart to the Gmunden manuscript is an ivory quadrant constructed for Frederick in 1438—the very date of the manuscript. It too is inscribed with the vowels *aeiou*. The backside of the quadrant is devoted to lunar positions, with special markings for the trine, quartile, and sextile aspects described in the manuscript of the same year.[40] The resonance between the artifact and the manuscript suggests that the latter may have been intended for Frederick in the first place, before his election as German king.

Since Frederick's involvement with astrology late in life is documented, these early clues give substance to the notion that he had a lifelong concern with the subject. We are told that the emperor relied upon astrology "to cast the horoscopes of his contemporaries as a means of gauging their characters, fates, and deaths."[41] A first-hand account by his sixteenth-century biographer reports that astrology was his chief preoccupation in old age.[42]

Indeed, circumstances beyond his control may have given Frederick more than the usual number of reasons to delve into astrology. By associating his name with the number three, his new position placed Frederick III at the center of millenarian expectations that had been simmering since the thirteenth century. In the 1240s, quasi-messianic speculations had focused on the person of Emperor Frederick II, whom supporters and opponents alike had seen as a chastiser of ecclesiastical corruption, even during his lifetime. According to one commentary on the book of Jeremiah, Frederick was to overthrow the corrupt church in 1260. His death in 1250 merely

(1972): 433–68, esp. 445–47. On the manuscript (MS London, British Library, Add. 24071), see also Dana B. Durand, *The Vienna-Klosterneuburg Map Corpus* (Leiden: Brill, 1952), p. 60.

[40] The quadrant is preserved in the Kunsthistorisches Museum. Photographs appear in Klug, "Johann von Gmunden" (n. 16), p. 26; and in Ernst Zinner, *Deutsche und niederländische astronomische Instrumente des 11.–18. Jahrhunderts*, 2nd ed. (Munich: C. H. Beck, 1967), plate 46, description on p. 322. According to Firneis, the quadrant also matches the data in MS Vienna, ÖNB lat. 5418, a Gmunden manuscript of 1434: see Maria G. Firneis, "Johannes von Gmunden—Der Astronom," in Hamann and Grössing, *Der Weg der Naturwissenschaft* (n. 33), p. 79. See also Zinner, *Regiomontanus* (n. 29), p. 23.

[41] Alphons Lhotsky, "Kaiser Friedrich III.: Sein Leben und seine Persönlichkeit," in Lhotsky, *Aufsätze und Vorträge* (n. 17), 2: 156;. See also Grössing, *Humanistische Naturwissenschaft* (n. 3), p. 88. In addition to MS London, British Library, Add. 24071, Frederick's library contained at least two other manuscripts with astrological and/or astronomical contents, although it does not appear that he made special efforts to acquire these: MSS Vienna, ÖNB lat. 2271 and ÖNB lat. 2352 (an astronomical miscellany) came from the library of King Wenceslaus. See Alphons Lhotsky, "Die Bibliothek Kaiser Friedrichs III.," in Lhotsky, *Aufsätze und Vorträge*, 2: 227; and H. J. Hermann, "Eine unbeachtete Wenzelhandschrift in der Wiener Hofbibliothek," *Mitteilungen des Instituts für Österreichische Geschichtsforschung* 21 (1900): 162–65.

[42] Joseph Grünpeck, *Die Geschichte Friedrichs III. und Maximilian I.*, trans. Th. Ilgen (Leipzig: Verlag der Dyk'schen Buchhandlung, n.d.), p. 20.

postponed these expectations, leading to speculation that a new Frederick would carry out the tasks his namesake had left unfinished.[43]

Such expectations were widespread in the late-medieval Holy Roman Empire. In late-fourteenth-century Vienna, Henry of Langenstein knew about a prophecy stating that the "third" Frederick would arise upon the occasion of a Jupiter-Saturn conjunction.[44] In the early fifteenth century, similar Frederick legends lived on in Thuringia, and also in Vienna, where they are reported by Thomas Ebendorfer of Haselbach, a prominent theologian at the university.[45] Great hopes for good as well as fears for evil were thus associated with the next Frederick: these intensified when a Habsburg named Frederick became German king, and later emperor. The obvious questions, for him as well as his contemporaries, were: Has the time come? Is this Frederick the expected one? Lending such speculations even greater eschatological urgency was the Ottoman threat, with which Pseudo-Methodius's legend of the last Roman emperor resonated. According to this seventh-century prophecy, to which Pierre d'Ailly had recently given renewed currency, the last Roman Emperor would first vanquish the enemies of religion and then abdicate, returning the imperial insignia to God as a prelude to the Second Coming.[46] In the context of this larger prophetic literature, Frederick's consultations with astrologers take on a dimension that is at once very personal and highly political.

We know neither how extensively Gmunden catered to Frederick's astrological interests, nor how specific the latter were in the 1440s.[47] Duke

[43] Norman Cohn, *The Pursuit of the Millennium*, rev. ed. (New York: Oxford University Press, 1970), pp. 108–26; David Abulafia, *Frederick II: A Medieval Emperor* (London: Allen Lane, 1988), pp. 428–35; Marie Tanner, *The Last Descendant of Aeneas: The Hapsburgs and the Mythic Image of the Emperor* (New Haven: Yale University Press, 1993), chap. 6, esp. pp. 123–26.

[44] Dietrich Kurze, *Johannes Lichtenberger († 1503): Eine Studie zur Geschichte der Prophetie und Astrologie*, Historische Studien, Heft 379 (Lübeck: Matthiesen Verlag, 1960), p. 42 n. 252.

[45] See the chronicle of Theodor Engelhusius, reported in Georg Voigt, "Die deutsche Kaisersage," *Historische Zeitschrift* 26 (1871): 131–87, esp. 160–61; and Marjorie Reeves, *The Influence of Prophecy in the Later Middle Ages: A Study of Joachimism* (Oxford: Clarendon Press, 1969), pp. 333–34.

[46] For the Byzantine and trans-Byzantine contexts of this legend (seventh through nineteenth centuries), see Paul J. Alexander, "Byzantium and the Migration of Literary Works and Motifs: The Legend of the Last Roman Emperor," *Medievalia et Humanistica*, n.s., 2 (1971): 47–68, esp. 48–49. For its association with the Third Frederick in the Viennese context, see Reeves, *Influence of Prophecy* (n. 45), p. 334. On pseudo-Methodius and Pierre d'Ailly, see Jean-Patrice Boudet, "Simon de Phares et les rapports entre astrologie et prophétie à la fin du moyen âge," *Mélanges de l'École Française de Rome—Moyen Âge* 102 (1990): 638–40; and Smoller, *History* (n. 30), pp. 106–7.

[47] The long list of gifts that members of Frederick's court received from the Frankfurt city council and the accounts of that visit in 1442 do not seem to mention an astrologer or medical astrologer, although two barber-surgeons are recorded. Heinig nevertheless notes that "there were at all times 3 or 4 physicians, barbers, and astronomers at the court" (Paul-Joachim

Albert V had ruled from the Hofburg, a convenient ten-minute walk away from the university in Vienna. But Frederick, from the Styrian branch of the Habsburgs, was not very welcome in Vienna.[48] He preferred to hold court in Wiener Neustadt, a long day away on foot. In any case, the relationship between Frederick and Gmunden did not last for more than a few years: the latter died in 1442, two years after Frederick's election as German king.

THE IMPERIAL COURT, 1440S–1470S

It was also in 1442 that Enea Silvio Piccolomini left the service of antipope Felix V to join Frederick III's court; he became a major figure behind the emperor's policies in the decade that followed. For many years before becoming cardinal of Siena and later Pope Pius II (d. 1464), Piccolomini served Frederick III as leading advisor, secretary, and ambassador.[49] His imperial ideology was, unlike Frederick's, well articulated.[50] It seems not to have required support from astrology, a topic on which Piccolomini was almost reticent. He did not discuss his views on the subject on the one occasion when he had a chance to do so systematically, namely, in his own *Fürstenspiegel*. Piccolomini wrote the *De liberorum educatione* (1450) as an educational treatise for the ten-year-old Ladislaus Posthumus, Albert V's heir to the thrones of Hungary, Bohemia, and Lower Austria, who was spending his minority at the court of his imperial relative Frederick. Near the end of his work, Piccolomini discussed the mathematical sciences. Although he encouraged instruction in both geometry and arithmetic, Piccolomini warned Ladislaus about spending too much time on mathematics, which could be noxious to those who tarried in the subject. Turning to the stars, he tersely noted that "the prince should not be denied a moderate reading of astronomy, which discloses the heavens and interprets the secrets of matters on high to mortals; the knowledge of this has often been of help to, and a source of glory for, great leaders."[51] Piccolomini's examples,

Heinig, "How Large Was the Court of Emperor Frederick III?" in Asch and Birke, *Princes* [n. 9], p. 148; see also pp. 143–45).

[48] Frederick escaped a lynching in the Viennese Augustinerkirche in 1441, the year before Gmunden's death. The incident, following an inflammatory speech by a nobleman, underscores the volatility of the messianic mode: the rowdy crowd reportedly called Frederick "King of the Jews" and shouted "Crucify him!": Lhotsky, "Kaiser Friedrich III" (n. 41), pp. 135–36.

[49] C. Ugurgieri della Berardenga, *Pio II Piccolomini, con notizie su Pio III e altri membri della famiglia*, Biblioteca dell'Archivio Storico Italiano, 18 (Florence: Leo S. Olschki, 1973), pp. 113–96.

[50] John B. Toews, "The View of Empire in Aeneas Sylvius Piccolomini (Pius II)," *Traditio* 24 (1968): 471–87.

[51] "Nec astronomie moderata lectio regis puero negari debet, que celos ostendit et arcana superum mortalibus pandit. hujus notitia magnis sepe ducibus adjumento decorique fuit"

drawn from Quintilian, scarcely constitute an endorsement of prognostication; on the contrary, they stress the demystifying power of astronomical knowledge, using a *locus classicus* in which the astronomical explanation of a lunar eclipse calms superstitious soldiers who had seen the event as an omen. More significant, perhaps, his brief discussion of mathematics and astronomy contrasts with page upon page devoted to eloquence, proper punctuation, and Latin grammar and orthography.

Piccolomini's advice to Ladislaus on the subject of astrology contrasts sharply with that of the widely diffused pseudo-Aristotelian *Secretum secretorum*, in which Aristotle allegedly urged Alexander the Great to do nothing without the council of an astrologer.[52] Yet even if Piccolomini had doubts about the assiduous study of mathematics and astronomy, his restraint about astrology in his educational manual should not be confused with utter skepticism: assorted remarks in his correspondence suggest that he took celestial influence on some events for granted (see, for example, his genuine distress at dissension at the Council of Basel, which he blamed on the influence of Mars in 1437).[53] Nor did his reticence about astrology in formal contexts stand in the way of his social or epistolary relations. We know from his correspondence in the mid-1440s that he had developed a literary relationship with Johannes Schindel, a leading astronomer in Prague and the former astrologer to Emperor Sigismund of Luxemburg (d.

(Joel Stanislaus Nelson, *Aeneae Silvii De Liberorum Educatione: A Translation with an Introduction,* Catholic University of America, Studies in Medieval and Renaissance Language and Literature, 12 (Washington, D.C.: Catholic University of America Press, 1940), pp. 210, 212; the passage quoted is from p. 212 (my translation).

[52] M. A. Manzalaoui, ed., *Secretum secretorum: Nine English Versions,* Early English Text Society, 276 (Oxford: Oxford University Press, 1977), pp. 46–47. Note, however, that at least the French translation of Giles of Rome's *De regimine principum,* a text largely inspired by the *Secretum,* recommended that princes not become involved in astrology: Carey, *Courting Disaster* (n. 3), p. 173.

[53] See Thorndike, *History of Magic* (n. 3), 4: 393. Piccolomini also mentioned astrology in passing as part of his festive quodlibetal disputation at the University of Vienna in 1445. Concluding a short passage on the relation of knowledge to action, he seems to suggest a contrast between *scientia* and *prudentia* paralleling that between *astronomia* and *astrologia*: "Scientia enim solum intelligit, quid est bonum, prudentia vero, postquam novit, etiam operatur, ut ante dixi. Astronomia de celi conversione stellarumque cursibus notitiam prebet; idem facit astrologia. Quia hec tamen superstitiosum quid habet, dum siderum cursu necessitates hominum et mores predicare conatur, aliud astronomia et aliud astrologiam dicimus esse. Et hec ad primam petitionem respondisse sufficiat" (Lhotsky, *Wiener Artistenfakultät* [n. 24], pp. 265–66). That this festive disputation was held before Frederick III (and his ward, Duke Sigismund of Tirol) may help to explain the ambiguity of the passage: on the one hand, Piccolomini seems to draw a parallel between *astrologia* and *prudentia* (perhaps the emperor's position?); on the other, he distances himself from it as *superstitiosum* (which sounds very much like the traditional position of the university).

Finally, Piccolomini was not averse to helping astrologers find their way to practitioners of magic in the mountains; see his letter to Gorro Lolli (15 January 1444, from St. Veit in Carinthia), sent to Italy with the servant of the medical astrologer and personal physician of the Duke of Saxony: *Selected Letters of Aeneas Silvius Piccolomini,* ed and trans. Albert R. Baca (Northridge, Calif.: San Fernando Valley State College, 1969), pp. 23, 94.

1437).[54] Later, Piccolomini was on excellent terms with Frederick's main court astrologer during the early 1450s, Johann Nihili (Nicka) of Bohemia—not suprisingly, perhaps, since the emperor and his associates took astrology seriously, and valued Nihili in particular. The clues are once again scanty, but the outlines of the picture are clear.[55]

Nihili's astrological advice undoubtedly played a role in several of Frederick III's crucial decisions. The "meister Hans Astronomus" who accompanied Frederick to Rome in 1452 for the imperial coronation and his marriage to Eleonor of Portugal was undoubtedly Nihili.[56] The imperial astrologer must have been involved in, or overseen the preparations for, those events by recommending suitable dates and casting the relevant horoscopes (analogous to the lengthy anonymous one prepared ca. 1450/51 for the bride-to-be).[57] Franz Stuhlhofer has suggested a plausible astrology-

[54] Rudolph Wolkan, ed., *Der Briefwechsel Eneas Silvius Piccolomini*, Abteilung II: Briefe als Priester und als Bischof von Trient, (1447–1450) Fontes Rerum Austriacarum, Abt. 2: Diplomataria et acta, vol. 67 (Vienna: Hölder, 1912), *Briefwechsel*, Abt. I, vol. 1, pp. 582–84.

[55] Beyond the circle of the immediate imperial family, the astrologer Johannes von Lübeck dedicated a "Prognosticon super Antechristi adventu Iudaeorumque Messiae" to Johannes Hinderbach, bishop of Trento and Frederick's chancellor; see Katherine Walsh, "Deutschsprachige Korrespondenz der Kaiserin Leonora von Portugal: Bausteine zu einem geistigen Profil der Gemahlin Kaiser Friedrichs III. und zur Erziehung des jungen Maximilian," in *Kaiser Friedrich III. (1440–1493) in seiner Zeit: Studien anlässlich des 500. Todestages am 19. August 1493/1993*, ed. Paul-Joachim Heinig, Forschungen zur Kaiser- und Papstgeschichte des Mittelalters, 12 (Cologne: Böhlau Verlag, 1993), p. 424.

[56] Although this inference runs counter to claims by Dietrich Kurze (Kurze, *Johannes Lichtenberger* [n. 44], p. 8 n. 21), it seems to be the most plausible one in light of what we know about the careers, ages, and relationships of the principals. Kurze speculated that "meister Hans Astronomus" was Regiomontanus because the Eleonor of Portugal horoscope was in Regiomontanus's hand. But the latter circumstance does not make Regiomontanus the author, as Grössing has argued convincingly (see n. 57 below). Regiomontanus's young age (fifteen) and his habitual deference to Georg Peuerbach, who a few years later had not yet entered the service of a court (see below), make both Regiomontanus's authorship of this horoscope and his identity with "meister Hans Astronomus" most unlikely. On the positive side, Enea Silvio Piccolomini often refers to Nihili as "Johannes astronomus" in his letters during this period; see Wolkan, *Der Briefwechsel Eneas Silvius Piccolomini* (n. 54), 1: 172, 349.

[57] The horoscope survives in only two known copies, both associated with Hartmann Schedel, the oldest of which (MS Munich, Bayerische Staatsbibliothek, Clm 453, fols. 78r–85v) is, according to Zinner (*Regiomontanus* [n. 29], p. 47–48), in Regiomontanus's hand). It is reproduced in facsimile and transcribed in Felix Schmeidler, ed., *Joannis Regiomontani Opera Collectanea: Faksimiledrucke von neun Schriften Regiomontans und einer von ihm gedruckten Schrift seines Lehrers Purbach* [*sic*] (Osnabrück: O. Zeller Verlag, 1972), pp. xi–xiii, 2–33. Its authorship remains a problem. Since the recipient is addressed as "magnifice vir" and "Eques insignis" in the opening and closing lines, respectively, he cannot be Frederick himself. The level of explanation, together with the absence of references to the recipient's own knowledge of the subject, seems to eliminate Nihili (and other astrologers) as recipient. Was the horoscope perhaps commissioned by another member of Frederick's court, as a "second opinion" and a check on that prepared by Nihili? Although such a circuitous hypothesis seems to revive the candidacy of Regiomontanus, Grössing has argued convincingly that Peuerbach is more likely to be the author than the fifteen-year-old Regiomontanus, who was not yet a bachelor of arts: see Grössing and Stuhlhofer, "Versuch einer Deutung" (n.

laden scenario to explain curious aspects of the imperial marriage. The wedding itself (on 16 March 1452) occurred at the time of a Mercury-Venus conjunction. Stuhlhofer argues that this date was the next best alternative to the most propitious Venus-Sun conjunction on 22 January 1452, which had to be scuttled when the bride failed to appear in Italy before 2 February.[58] Once Eleonor had arrived, the astrologer's participation would have been crucial in setting a revised schedule that maximized the planetary benefits for events of such significance for the future of the Holy Roman Empire. Frederick eventually appointed Nihili a member of his council.[59]

Piccolomini's cordial relations with the imperial astrologer transpire in his correspondence. In 1453–54, he wrote to Johann Nihili as a friend ("amice optime"). Although an astrologer, Nihili did not trust his welfare to the stars alone; he also appealed to his friend's connections with high dignitaries. Piccolomini proved very accommodating, using his considerable influence with the papal court to recommend the preferment of Nihili to church benefices, and composing several letters of introduction for him ("pro viro mihi amicissimo"). In giving political news from Frederick's court, Piccolomini describes Nihili as an excellent astronomer ("astronomum insignem"), and mentions his judgments (e.g., the prediction ca. June 1453 that the Bohemian heresy would increase during the following year on account of the evil conjunction).[60] At the same time, he also uses a strikingly nondeterminist appeal to the divine will in closing a letter to Nihili.[61]

Although Nihili held a bachelor of arts degree from Leipzig,[62] no one has yet uncovered a formal association between him and the University of Vienna. As political events would soon prove, however, he had at least one

3), esp. pp. 268–75; Grössing, *Humanistische Naturwissenschaft* (n. 3), pp. 89–90. Nihili was approximately a generation older than Peuerbach, having received his bachelor's degree from Leipzig in 1431: see Pavel Spunar, ed., *Repertorium auctorum bohemorum provectum idearum post Universitatem Pragensem conditam illustrans*, Studia Copernicana, 25 (Wroclaw: Ossolineum, 1985), p. 140.

[58] Stuhlhofer has also suggested an astrological explanation for the delay in consummating the union: see Grössing and Stuhlhofer, "Versuch einer Deutung" (n. 3), p. 280.

[59] Wolkan tantalizingly dates the appointment to 5 February 1452, a few days after Eleonor's arrival in Italy: *Der Briefwechsel Eneas Silvius Piccolomini* (n. 54), p. 97 n. But the correct year appears to have been 1454; see Joseph Chmel, *Regesta Chronologico-Diplomatica Friderici IV. Romanorum Regis (Imperatoris III.)*, Abt. I (1440–1442) (Vienna, 1838; repr. Hildesheim: Georg Olms, 1962), #3165, p. 319. The emperor's letter of recommendation for Nihili in January 1454, when the latter returned to Bohemia to settle an inheritance, confirms his favor: Wolkan, ed., *Briefwechsel*, Abt. III, vol. 1 (1450–1454), Fontes Rerum Austriacarum, Abt. 2: Diplomataria et acta, vol. 68 (Vienna: Hölder, 1918), p. 612.

[60] Wolkan, ed., *Briefwechsel*, Abt. III, vol. 1, pp. 378, 420–21, 425, 440, 461, 491–92. The reference to the prediction is on p. 172.

[61] Ibid., p. 174: "deus bene omnia pro suo arbitrio diriget, quamvis nobis eius consilii ignote sunt cause." Note that Piccolomini had also been in correspondence with the Bohemian astronomer Johannes Schindel.

[62] See Spunar, *Repertorium* (n. 57), p. 140.

friend at the university, a candidate for the master's degree named Georg Peuerbach. To understand Peuerbach's role, a brief digression into political history is necessary. Throughout the 1440s and 1450s, the nobility of Lower Austria evidently did not view Frederick III through messianically tinted lenses; indeed, they resented his direct regency over their territory during Ladislaus Posthumus's minority. They even tried to kidnap Ladislaus in Italy on the occasion of the imperial coronation and wedding in 1452. When this plot failed, they laid siege to Wiener Neustadt upon the emperor's return, until the latter agreed to release Ladislaus from his tutelage. Ladislaus Posthumus thus became duke of Lower Austria and king of Hungary and Bohemia at the tender age of thirteen.[63]

At this time, Georg Peuerbach had recently returned from several years in Padua, and was completing his master of arts degree at Vienna. A letter survives from this period, in which Johannes Nihili urges him to become astrologer to King Ladislaus's court, a post that Peuerbach held during the next several years.[64] Indeed, in early December 1453 Peuerbach petitioned the faculty of arts for a key to the secure section ("cancellum") of its library—that is, the section that housed John of Gmunden's astrological books.[65] Peuerbach's correspondence with Nihili makes it clear that financial considerations loomed large in his decision to seek the position with Ladislaus. During these years, Peuerbach spent only a part of his time at court. He continued to teach at the University of Vienna, and appears to have worked for several courts—Frederick's imperial court in Wiener Neustadt, Ladislaus's roving court in Lower Austria, Bohemia, and Hungary, and perhaps even the court of Sigismund of Tyrol.[66]

When he stayed in Wiener Neustadt, Peuerbach was in contact with several individuals whom Helmuth Grössing has dubbed the "humanist circle of Wiener Neustadt."[67] What is striking about the group is the fact that most

[63] *The Commentaries of Pius II*, trans. Florence Gragg, Smith College Studies in History, 22 (Northampton, Mass., 1937), bk 1, pp. 64ff; Erich Zöllner, *Geschichte Österreichs von den Anfangen bis zur Gegenwart* (Munich: Oldenbourg Verlag, 1961), p. 147.

[64] "De stipendio impetrando apud dominum regem Ladislaum consulo omnino, ut per medium amicorum totis viribus pro eodem laboretis. Sint 24 librae, de quibus sitis contentis ad presens" (letter from Johannes Nihili [Bohemus] to Peuerbach written sometime after Ladislaus's accession to the throne—between 1452 and 1454: see Czerny, "Aus dem Briefwechsel" [n. 36], p. 289).

[65] "In die (Sancti Nicolai ... Magister Georgius de Pewrbach petivit clavem ad cancellum librarie et fuit exauditus, et fuerunt sibi lecta 4^{or} puncta que huiusmodi petentibus leguntur secundum quandam conclusionem" (Vienna, Universitätsarchiv, Ph8, fol. 69r).

[66] See the allusion to Sigismund's gift to Peuerbach in Zinner, *Regiomontanus* (n. 29), p. 33. Note that Frederick III was for a time the guardian of both Ladislaus and Sigismund.

[67] Grössing, *Humanistische Naturwissenschaft* (n. 3), p. 126. The members of this group appear together in a dialogue, *De remedio amoris*, by Johann Tröster, who earlier had had close connections with the imperial court. As Grössing has shown, the dialogue confirms Peuerbach's position as Ladislaus's astrologer and his relations with members of this group. In addition to Nihili and others, they included Jurist Ulrich Sonnenberg (so called by Enea

of its indigenous members were Viennese masters who were now serving the court. The limited evidence for Peuerbach's activities in Wiener Neustadt is astrology-related. In 1456 he worked on the construction of a wall sundial, presumably in conjunction with the imperial residence; the following year, he also prepared an astrolabe for the emperor.[68] We know from a letter that he was again in Wiener Neustadt when a comet, now known as "Halley's," appeared in June 1456. Since Johann Nihili was not at court at this time, we may infer that it was with Peuerbach that the emperor consulted about this disturbing event. At the same time, Peuerbach is the presumed author of a German account of the comet written for the benefit of King Ladislaus.[69] This short judgment is notable for the author's reluctance to draw specific conclusions from the appearance of the comet. Although dutiful in reporting what past astrologers thought such phenomena foretold (including the "death of a great lord"), he cautiously avoided inferences of his own.[70] Peuerbach was once again with King Ladislaus in Melk in September 1457, three months after the appearance of another comet and some two months before Ladislaus's death under suspicious circumstances.[71]

Peuerbach's activities, as pieced together from these scattered data, show him moving between several courts while lecturing at the University in Vienna every other semester. When Johann Nihili and King Ladislaus both died in 1457,[72] Peuerbach's life became ostensibly less fragmented, with service limited to the imperial court and the university. Throughout these years, he was already collaborating in these activities with his student, friend, and associate Johannes Regiomontanus. It was Regiomontanus, for example, who at the request of Frederick III and Empress Eleonor of Portu-

Silvio Piccolomini in his *Memoirs*; called Sonnberger by Grössing), a master of arts in Vienna who served as *magister in actu regens* in 1431, later as bishop of Gurk and as Austrian chancellor. Finally, a Johann Hartung, who headed the imperial revenue service, was probably the same Johann Hartung who belonged to the Faculty of Law at Vienna, served as its dean in 1433, and authored an astrological text in German. He attended the Council of Basel as imperial envoy, and was in close contact with Enea Silvio Piccolomini. See Grössing, *Humanistische Naturwissenschaft* (n. 3), pp. 126–28; and Zinner, *Regiomontanus* (n. 29), p. 29.

[68] Grössing, *Humanistische Naturwissenschaft*, pp. 102–3.

[69] As the account is addressed to a king, Frederick III was not the addressee (he had been crowned emperor in 1452). Ladislaus Posthumus is the most plausible candidate: see Konradin Ferrari d'Ochieppo, "Weitere Dokumente zu Peurbachs Gutachten über den Kometen von 1456, nebst Bemerkungen über den Chronikbericht zum Sommerkometen 1457," *Österreichische Akademie der Wissenschaften*, Sitzungsberichte, math.-nat. Kl., Abt. II 169 (1961), Heften 5–10: 150, 163–66.

[70] "Dae wollt ich liber schweygen denn daz ich von dises Cometen Bedeutung schreyben soll" (ibid., p. 165).

[71] Zinner, *Regiomontanus* (n. 29), p. 30. Peuerbach and Regiomontanus together observed a lunar eclipse from Melk on 3 September 1457, but Zinner does not cite the evidence for their being with Ladislaus.

[72] On Nihili's presumed death date, see Grössing, *Humanistische Naturwissenschaft* (n. 3), p. 103.

gal cast the horoscope of their son Maximilian soon after the latter's birth in 1459.[73]

But these relationships dissolved rather suddenly in 1461. In that year Peuerbach died. Regiomontanus left Vienna and the imperial court, moving to Italy with the household of the visiting papal legate, Cardinal Bessarion. It is perhaps not a coincidence that the reigning pope was Pius II (1458–1464), none other than Enea Silvio Piccolomini. Regiomontanus made contacts with fellow astronomers and astrologers, lectured in Padua, and immersed himself in Bessarion's excellent library. Eventually, in the mid-1460s, he would return to Hungary, where he had retained close contacts with Johannes Vitéz, the archbishop of Esztergom, who had once been Ladislaus Posthumus's chancellor.[74] But so far as we know, he never returned to the imperial court.

The 1470s and 1480s, by way of epilogue, mark yet another turning point in the relations of Frederick III's court to both astrology and prophecy. These years, during which the prophet-astrologer Johannes Lichtenberger was associated with the emperor, mark a further shift away from not only the approach of medical astrology,[75] but also the circumspection of the mathematical astrology practiced by the masters of arts. Lichtenberger (d. 1503), whose academic credentials are still unknown, called himself an astrologer but thought of himself as a prophet, claiming to rely for his judgments on astrological knowledge, revelation, and long experience. Indeed, Dietrich Kurze has argued convincingly that Lichtenberger tried to use the scientific trappings of astrology to legitimize his prophecies. His merger of these two different traditions, which had until then interacted but sporadically, set a trend that flourished into the first quarter of the sixteenth century.[76]

---

[73] MS Vienna, ÖNB lat. 5179; see Grössing, *Humanistische Naturwissenschaft*, pp. 121–22; and Helmuth Grössing, "Der Humanist Regiomontanus und sein Verhältnis zu Georg von Peuerbach," in *Humanismus und Naturwissenschaften*, ed. Rudolph Schmitz and Fritz Krafft, Beiträge zur Humanismusforschung, Bd. 6 (Boppard: Harald Boldt Verlag, 1980), p. 79. See also Regiomontanus's "Prognosticon de anno 1455," in MS Vienna, ÖNB lat. 4756, fols. 70–75: Grössing, *Humanistische Naturwissenschaft* , p. 268 n. 22); Zinner, *Regiomontanus* (n. 29), pp. 50–51.

[74] See Piccolomini's letter to Vitéz from November 1453 in Wolkan, ed., *Briefwechsel*, (n. 54), Abt. III, 1: 356. Vitéz (d. 1471), a correspondent and patron of Regiomontanus (see Shank, "'Notes on al-Bitruji'" [n. 31], pp. 19–22), would play a key role in Hungarian politics under King Matthias Corvinus.

[75] Of the six *Leibärzte* identified in Frederick's court (mostly during his last years), only one—Johann von Linden—seems to have had explicit astrological connections. See Paul-Joachim Heinig, "Musik und Medizin am Hof Kaiser Friedrichs III. (1440–1493): Studien zum Personal der deutschen Herrscher im 15. Jahrhundert," *Zeitschrift für Historische Forschung* 16 (1989): 175.

[76] Dietrich Kurze, "Popular Astrology and Prophecy in the Fifteenth and Sixteenth Centuries: Johannes Lichtenberger," in *"Astrologi Hallucinati:" Stars and the End of the World in Luther's Time*, ed. Paola Zambelli (Berlin: De Gruyter, 1986), pp. 177–93, esp. 179, 184;

In the 1470s, Lichtenberger's prophecies continued to play directly into the messianic expectations surrounding Frederick III. Flattering though his pronouncements were, they already suggest that he and others were beginning to doubt that the indecisive man on the imperial throne was the expected one. Lichtenberger broached these awkward problems in his prophecies, initially by way of constructive encouragement. Within a decade, however, he had given up. In his lengthy, highly influential, and heavily illustrated and plagiarized *Pronosticatio* of 1488, he transferred his grand hopes to Frederick's son. Maximilian would indeed raise imperial mythology to new heights, only to be outdone in turn by both his son Charles V and his grandson Phillip II.[77]

CONCLUSIONS

The summary of evidence presented here is sufficiently spotty to limit sharply any generalizations. It is nevertheless safe to say that the Habsburg courts of the late fourteenth and fifteenth centuries relied consistently and extensively, but not exclusively, on academics and academically trained consultants for expertise—in theology, law, medicine, and also astrology. Indeed, this type of consultative relationship was, arguably, the primary

---

Paola Zambelli, "Forward" and "Introduction: Astrologers' Theory of History," in ibid., pp. vii, 5–6.

For James of Clusa's failed attempt to merge astrology and prophecy, see n. 34 above. Langenstein's opposition to judicial astrology was as harsh as his attacks on recent prophecy: see André Vauchez, "Les théologiens face aux prophéties à l'époque des papes d'Avignon et du Grand Schisme," *Mélanges de l'École Française de Rome—Moyen Âge 102* (1990): 581–86; Emil Donckel, "Studien über die Prophezeiung des Fr. Telesforus von Cosenza, O.F.M. (1365–1386)," *Archivum Franciscanum Historicum* 26 (1933): 29–104, esp. 96–99.

For the parallel French case, see Simon de Phares's involvement in prophecy and astrology (Boudet, "Simon de Phares" [n. 46], esp. pp. 629–31, 641–42) and Charles VIII as the counterpart to Frederick (Reeves, *Influence of Prophecy* [n. 45], passim). Barbara Bauer sees the professional development of the courtly astronomer/astrologer as starting from a Lichtenberger-type of prophetic astrologer in the late fifteenth century and shifting toward practitioners with greater scientific autonomy in the seventeenth. The implication of maturation here overlooks the equally sharp contrast between Lichtenberger and the Gmunden-Peuerbach-Regiomontanus era, which precedes him and yet resembles the Keplerian era highlighted by Bauer; see Barbara Bauer, "Die Rolle des Hofastrologen und Hofmathematicus als fürstlicher Berater," in *Höfischer Humanismus*, ed. August Buck, Mitteilung XVI der Kommission für Humanismusforschung (Weinheim, 1989), pp. 93–117, esp. 95–96. For outworkings of astrology and prophecy at the popular level in Italy, see Ottavia Niccoli, *Prophecy and People in Renaissance Italy*, trans. Lydia G. Cochrane (Princeton: Princeton University Press, 1990), chap. 6, esp. pp. 166–67.

[77] Lichtenberger's *Pronosticatio* was heavily plagiarized from that of the Urbino court astrologer Paul of Middelburg, and widely reprinted: Kurze, *Johannes Lichtenberger* (n. 44), pp. 13–14, 22, and passim; Kurze, "Popular Astrology and Prophecy" (n. 76), pp. 187–91; Reeves, *Influence of Prophecy* (n. 45), pp. 347–51; Tanner, *Last Descendant of Aeneas* (n. 43), pp. 123–26.

way in which the University of Vienna interacted with the surrounding courts.

Despite Henry of Langenstein's prominence as a reviver of the university, his skepticism about judicial astrology set the tone for neither the Habsburg courts nor the subsequent generation. Indeed, the eighty years between Langenstein and Regiomontanus witness an increasing engagement with astrology on the fringes of academic life, and a growing reliance at court on locally trained practitioners. At first, the most consistent evidence of interest in matters astrological is found among the Habsburgs' personal physicians, who were usually also members of the faculty of medicine, and often had spent time at Padua. The limited evidence points to the court rather than the university as the leading venue for their astrological practice.[78] From the 1430s to 1460, however, the most prominent practitioners had not gone beyond the arts degree; their primary writings and curricular interests were mathematical and astronomical, not medical. Three of these four were trained and taught at Vienna.

Throughout this entire period, a succession of Habsburg courts—the short-lived one of Ladislaus Posthumus (1452–1457), and the long-lived one of Frederick III (1440–1493)—provided not only an outlet, but also a stimulus for a kind of knowledge that had no official home in the Viennese curriculum. The judgments that survive for Peuerbach and Regiomontanus show that they used their knowledge with caution. During the Lichtenberger years, caution disappeared—with enthusiasm.

Did the changing astrological demands of the Habsburg courts in turn leave marks on the university? At present it is difficult to answer the question satisfactorily. The current state of the evidence suggests that the demand for astrological skills at court continued to have little noticeable effect on the *official* curriculum.[79] The more interesting and difficult question is, however, what effect it had on the *unofficial* teaching, and on what

[78] In addition to the small number of its members and the late beginning of its *Acta* (started by Galeazzo de Sancta Sophia in 1399), the medical faculty is the only faculty that Enea Silvio Piccolomini omitted from his description of the university—an oversight that seems to confirm its relative institutional marginality despite the high profile of its members; see Lhotsky, *Wiener Artistenfakultät* (n. 24), p. 136.

Cracow parallels the general pattern of late medieval astrology in Vienna—no official university status in the late fourteenth century, followed by the court's importation of foreign or foreign-trained medical astrologers. Cracow added a chair in astrology in the second half of the fifteenth century, and its scholars discussed astrology in commentaries on Aristotle's *Physics*; see the overview (not footnoted) by Mieczyslaw Markowski, "Die Astrologie an der Krakauer Universität in den Jahren 1450–1550," in *Magia, Astrologia e Religione nel Rinascimento—Convegno polacco-italiano* (Wroclaw/Warsaw: Polish Academy of Sciences, 1974), pp. 83–89; and especially idem, "Der Standpunkt der Gelehrten des späten Mittelalters und der Renaissance dem astrologischen Determinismus gegenüber," *Studia Mediewistyczne* 23 (1984): 11–44, esp. 29.

[79] There is at present no evidence for Habsburg stimulation of John of Gmunden's curricular innovations (see n. 32 above), which have yet to be explained.

might be called the *research* concerns of the academic consultants. Numerous clues point to a thriving intellectual life below the threshold of officialdom. The expertise in astrology and instrument-making that university masters like Gmunden, Peuerbach, and Regiomontanus were providing is but one example. Another is a well-known work: Peuerbach's *New theory [theoricae] of the planets*, which—curiously—he delivered as a set of lectures not at the university, but at the *Bürgerschule*, the "school for citizens" associated with St. Stephen's cathedral. Meanwhile, at the university Peuerbach was lecturing not on astronomy, but on the Latin poets—Juvenal, Horace, and Virgil.[80]

The research interests associated with astrological consulting at court obviously made demands on the time and activities of Peuerbach in particular, involving the construction of instruments, the computation of almanacs, the preparation of judgments, and the interpretation of comets. That his duties literally opened new doors is most dramatically illustrated by his petition for access to the books that the Faculty of Arts kept under lock and key. While astrological consulting used and sharpened the skills of its practitioners, the extent to which it also shaped their questions, and perhaps their answers, remains to be investigated, particularly for Peuerbach and Regiomontanus.[81] In other words, we face here the prospect of another intriguing shift "from social [or perhaps socio-political] into intellectual factors," as John Murdoch strikingly titled one of his seminal papers.[82] To answer this problem convincingly and in detail, we will need not only to read the texts that these university masters produced, but to reread them with an awareness of the imperial court nearby.

The figures of Peuerbach and Regiomontanus, with whom this case study culminates, mark the high-point of astronomical work in fifteenth-century Europe. Their joint efforts produced and diffused the most influential introduction to elementary astronomy of the following century (Peuerbach's *New theory of the planets*, first printed by Regiomontanus)

[80] On the apparent *institutional* marginality of the advanced mathematical tradition in Vienna (and a partial explanation based on differences in income from fees), see Michael H. Shank, "The Classical Scientific Tradition in Fifteenth-Century Vienna," in *Tradition, Transmission, Transformation: Proceedings of Two Conferences on Pre-Modern Science held at the University of Oklahoma*, ed. F. Jamil Ragep and Sally P. Ragep, with Steven Livesey (Leiden: Brill, 1996), pp. 115–36, esp. 120, 121 n. 14, 131–33 (an error on p. 118 of this article may be corrected by substituting "Heinricianum" for the two occurrences of "Privilegium maius").

[81] For tantalizing case studies of astrological concerns stimulating theoretical work, see Jerzy Dobrzycki and Richard L. Kremer, "Peurbach and Marâgha Astronomy? The Ephemerides of Johannes Angelus and Their Implications," *Journal for the History of Astronomy* 27 (1996): 187–237, esp. 191–93; and Robert S. Westman's forthcoming book on fifteenth-century Bologna prognostication. Richard Kremer and I have begun a major joint investigation of the cosmological assumptions that underlie the work of Peuerbach and Regiomontanus.

[82] Murdoch, "From Social into Intellectual Factors" (n. 4), pp. 271-89.

and the most astute critical introduction to Ptolemy's *Almagest*, namely the *Epitome of the Almagest* that Peuerbach began and that Regiomontanus completed in a critical manner that led directly to Copernicus's sun-centered universe.[83] For these reasons alone, the cultural context of their astronomical work is well worth pondering. A central element of that context was the proximity of the University of Vienna to imperial, royal, and ducal courts with astrological inclinations. Although this complex symbiosis is still poorly understood, the awareness of its existence can stimulate new questions and hypotheses, from the smaller-scale biographical to the larger-scale theoretical.[84]

Without a doubt, the image of the late-medieval university as a preserve of theoretical knowledge out of touch with the larger world must be modified. This image is not in every case impertinent, but it is misleading. Both the Viennese city council and the Habsburg courts suggest by their behavior that they sometimes found university masters, thanks to their specialized knowledge, to be useful. Conversely, the demands for academic consulting drew the masters into more practically oriented activities and issues than they would have faced in the lecture halls. Even at a preliminary stage of the investigation, the evidence already testifies to the transacademic character of late medieval learning.

[83] Noel Swerdlow, "The Derivation and First Draft of Copernicus's Planetary Theory: A Translation of the Commentariolus with Commentary," *Proceedings of the American Philosophical Society* 117 (1973): 423–512, esp. 471–78.

[84] For example, did the University of Vienna's proximity to Frederick III's court have anything to do with the mathematically inclined fourteen-year-old Regiomontanus's move in 1450 from Leipzig to Vienna, where the Leipzig alumnus Johann Nihili was imperial astrologer? Did the concrete astrological activities of the likes of Gmunden, Peuerbach, and Regiomontanus leave traces on their theoretical work?

DOMINGO DE SOTO'S "LAWS" OF MOTION: TEXT AND CONTEXT

WILLIAM A. WALLACE

Almost thirty years ago I first documented Domingo de Soto's use of the expression *uniformiter difformis* (with respect to time) and its application to falling motion—a passage in which he anticipated Galileo's concept of uniform acceleration in free fall, and this by over eighty years.[1] I planned then, as I put it, "to publish the principal portions of his Latin text with all significant variants together with an English translation and a commentary."[2] Circumstances beyond my control have prevented me from carrying out that intention in full, although I have worked at the enterprise from time to time over the intervening years.[3] The occasion of honoring John Murdoch now gives me what may be my last opportunity to fulfill this long-outstanding promissory note.

SOTO'S WORK IN CONTEXT

To review the salient facts: Soto explains Aristotle's rules for the comparison of motions (which is what I intend by the expression "'laws' of mo-

---

[1] William A. Wallace, "The Enigma of Domingo de Soto: *Uniformiter difformis* and Falling Bodies in Late Medieval Physics," *Isis* 59 (1968): 384–401. A preliminary version of that essay was presented in the Colloquium Series of the Department of the History of Science, Harvard University, on 11 May 1967.

[2] Ibid., p. 399 n. 58.

[3] For details, see William A. Wallace, "The Concept of Motion in the Sixteenth Century," *Proceedings of the American Catholic Philosophical Association* 41 (1967): 184–95; "The 'Calculatores' in Early Sixteenth-Century Physics," *British Journal for the History of Science* 4 (1969): 221–32; "Mechanics from Bradwardine to Galileo," *Journal of the History of Ideas* 32 (1971): 15–28; "Experimental Science and Mechanics in the Middle Ages," *Dictionary of the History of Ideas,* 2: 196–205; "Galileo Galilei and the *Doctores Parisienses,*" in *New Perspectives on Galileo,* ed. Robert E. Butts and Joseph C. Pitt (Dordrecht: Reidel, 1978), pp. 87–138; "The Early Jesuits and the Heritage of Domingo de Soto," *History and Technology* 4 (1987): 301–20; "Science and Philosophy at the Collegio Romano in the Time of Benedetti," in *Cultura, Scienze e Techniche nella Venezia del Cinquecento,* Atti del Convegno Internazionale di Studio "G. B. Benedetti e il suo tempo" (Venice: Istituto Veneto di Scienze, Lettere ed Arti, 1987), pp. 113–226; and "Late Sixteenth-Century Portuguese Manuscripts Relating to Galileo's Early Notebooks," *Revista Portuguesa de Filosofia* 51 (1995): 677–98. The first five of these essays, plus "The Enigma of Domingo de Soto," are reprinted in William A. Wallace, *Prelude to Galileo: Essays on Medieval and Sixteenth-Century Sources of Galileo's Thought,* Boston Studies in the Philosophy of Science, 62 (Dordrecht and Boston: Reidel, 1981); the next two, both published in 1987, in idem, *Galileo, the Jesuits and the Medieval Aristotle,* Collected Studies Series 346 (Aldershot, UK: Variorum, 1991).

tion" in my title), in his commentary on the *Physics*, Book 7, texts 35–37, first published at Salamanca in 1551–52.[4]

He then offers a critique and emendation of these "laws" in his questions on the *Physics*—particularly in his third and fourth questions on Book 7, which he precedes by a "Digression on Ratios" wherein he explains the meaning of *uniformiter difformis* and related expressions along with the mathematics of ratios and proportions required for their understanding. The questions were published in a companion volume to the commentary, at the same place and time.[5] An earlier edition of both commentary and "questionary" had appeared at Salamanca around 1545, but this was incomplete and did not extend to this portion of Aristotle's text. Seven subsequent editions of both commentary and questions were published down to 1613, the next five at Salamanca in the years 1555–56,[6] 1563, 1568–69, 1572, and 1582, the penultimate at Venice also in 1582, and the last at Douay in 1613. The late Vincente Beltrán de Heredia, who had worked for years in the Salamancan archives and prepared the definitive biography of Soto,[7] assured me that no manuscript copies of Soto's works on the *Physics* are extant. Thus the best one can hope for is a "variorum" edition of any relevant texts, showing the variants that might appear in the successive editions.

The burden of this essay is to present such an edition of the main portion of Soto's fourth question on the seventh book, which bears the title "Whether the velocity of motion depends with respect to its cause on the ratio of the ratios of the agents to their respective resistances?"[8] The third question dealt with the velocity of motion with respect to its effect, and it was in this question that Soto proposed his *uniformiter difformis* doctrine and explained how this could be used to calculate the distance of a body's fall through the application of the mean-speed theorem deriving from the English calculators. Since no significant variants were detected in the texts of this question cited in the "Enigma of Domingo de Soto" article, consid-

[4] The full title is *Reverendi Patris Dominici Soto Sogobiensis theologi ordinis Praedicatorum in inclyta Salamanticensi academia professoris ac Caesareae Maiestati à sacris confessionibus super octo libros Physicorum Aristotelis. Commentaria.* This is from the second edition (Salamanca 1555–56), prepared after Soto had returned from the Council of Trent, where he served as the representative of the Dominican Order at the Council and as imperial theologian there of Charles V, who had also named him his personal confessor.

[5] This title reads simply *Reverendi Patris Dominici Soto Segobiensis Theologi ordinis praedicatorum super octo libros Physicorum Aristotelis. Quaestiones.* The publication of the two volumes at the same place and time is true of all nine editions, which were produced as companion volumes of a set.

[6] An edition of 1557–58 also appeared, but this is the same as the 1555–56 edition with only a change in the title page.

[7] Vincente Beltrán de Heredia, *Domingo de Soto, O.P., Estudio biográfico documentado* (Salamanca: Biblioteca de Teólogos Españoles, 1960).

[8] "Utrum velocitas motus attendatur ex parte causae penes proportionem proportionum, quae sunt agentium ad suas ipsarum resistentias."

eration of the question is postponed to later in the essay and priority is given to the fourth question. The latter contains Soto's teaching on the use of "ratio of ratios" (*proportio proportionum*) in laws of motion as first explained by Thomas Bradwardine. There are difficulties in this text, both in the way Soto presents it and in the way he was understood by later thinkers. Thus a study of the text and its context may prove helpful for gauging the value of Soto's work in its own right, as well as the influence it may have exerted down to the time of Galileo.

VARIANTS IN THE EDITIONS

In what follows, the various editions are designated by the first nine capital letters of the Roman alphabet, as follows:

A. Salamanca: Juan de Junta, 1545? (incomplete)
B. Salamanca: A. de Portonaris, 1551–52
C. Salamanca: A. de Portonaris, 1555–56
D. Salamanca: A. de Portonaris, 1563
E. Salamanca: Domingo de Portonaris, 1569
F. Salamanca: Domingo de Portonaris, 1572
G. Salamanca: Ildefonso de Terranova y Neila, 1582
H. Venice: Apud Franciscum Zilettum, 1582
J. Douay: Ex typographia Petri Borromans, 1613

Edition *A* ends with the first page of the first question on the seventh book, containing only 16 of the 42 paragraphs that make up that question in later editions, with the last words being "et remissio est proprie motus alterationis." *B* is the first complete edition, with commentary and questions on all eight books. *C*, the last to appear in Soto's lifetime (he died at Salamanca on 15 November 1560), is presented as the second edition, recently revised and corrected by the author.[9] All of the editions up to and including *F* are set in Gothic type; the remainder are in Roman. *H* is dedicated by the publisher to a Dominican friar, "Dominico Bolano, Iacobi filio, Patricio Veneto." The Flemish Dominican Jacobus de Brouwer prepared the last edition, *J*, and, as will be seen in the critical apparatus, in the process he made many emendations in the text.

Throughout the nine printings there are many changes of paragraphing, punctuation, and spelling. The first step in preparing a "variorum" edition of the questions on the seventh book was to standardize the paragraphing, using *C* as the basic guide. This yielded the following composition: Q. 1, 42

[9] The Latin reads: "Secunda aeditio nuperrimè ab Authore recognita, multisque in locis aucta et à mendis quàm maximè fieri potuit repurgata."

par.; Q. 2, 39 par.; Digression on Ratios, 62 par.; Q. 3, 53 par.; and Q. 4, 41 par., for a grand total of 237 paragraphs. Since the middle "Digression" presented the most difficult typesetting problem, with an intractable number of variants, this was eliminated at the outset, bringing the total for consideration down to 175 paragraphs. Soundings were then made in these paragraphs to identify significant word variants that could be traced through the various editions.[10] These yielded 45 terms or phrases to be checked throughout all four questions relating to the seventh book.

The first use to which this information could be put would be tracing the degree of dependence of the various editions on those that preceded them. Since any changes after *C* were made by the publisher (except for *J*, whose editor is known), the presumption would be that the publisher simply revised the previous edition in preparing his text. With a few exceptions, this generally seems to have been borne out. The first complete edition, *B*, obviously was based on *A*. The incomplete edition, *A*, contained only a small portion of Book 7, but this included 6 of our 45 variants, and all of these read the same in *B* and *A*. Here and in what follows, if the same readings are found in a subsequent edition,—in this case, *B*'s dependence on *A*,—this is written *BA* = *n*, where *n* is the number of variants, here 6. Only in comparisons with *A* is 6 the total possible, and thus the agreement in this case is 6/6. In all other comparisons all 45 variants have been evaluated, and thus 45 is understood to be the denominator in estimating the degree of agreement between any two editions.

The second edition, *C*, being a revised edition, would not be expected to read with either *B* or *A*, and this turns out to be the case, for *CB* = 27 (that is, 27 agreements out of 45 possible, or 27/45) and *CA* = 2 (that is, 2 agreements out of 6 possible, or 2/6). Edition *D* was probably based on *C*, but it reads equally with *C* and *B*, since *DC* = 22 and *DB* = 22. *E*, on the other hand, seems clearly to be based on *D*, since *ED* = 28, whereas *EB* = 17 and *EC* = 9. *F* likewise seems to have been based on its predecessor, since *FE* = 33, as compared to *FD* = 27, *FB* = 16, and *FC* = 15.

The first edition not from the Portonaris family and the first in Roman type, *G*, presents more of a problem. Its agreements read as follows: *GD* = 28, *GE* = 25, *GB* = 25, *GC* = 22, and *GF* = 21; these figures indicate that *D* probably served as its basic exemplar, but others such as *B* and *E* could have been consulted. The Venice edition, *H*, the first not from Salamanca, reads basically with *C*, its probable exemplar, since *HC* = 40, whereas *HD* = 29, *HB* = 26, *HF* = 14, and *HE* = 13. Finally, the Douay edition, *J*, is a maverick that does not correlate well with any previous edition; its agree-

[10] In this task I was assisted by Paul G. Zomberg, who as a graduate student had worked with me on the editorial staff of the *New Catholic Encyclopedia* and subsequently earned a doctorate in English at the University of Toronto.

ments are *JD* = 28, *JG* = 25, *JF* = 24, *JC* = 24, *JB* = 23, *JH* = 22, and *JE* = 17, and it probably reflects the editor's originality more than his dependence on any particular previous edition.

SOTO'S COMMENTARY ON ARISTOTLE'S RULES

Aristotle's seven rules for the comparison of local motions are given in chapter 5 of Book 7 of the *Physics* (249b27–250b7).[11] Soto refers to these rules in his fourth question, having provided his commentary on them in what he identifies as texts 35 through 38 of that book, following Averroes's division. There is little that is remarkable in Soto's commentary, since the rules he finds are basically those found by modern commentators, such as W. D. Ross (cited in note 11). One noteworthy feature, however, is found in his explanation of Aristotle's fourth rule in text 35, where Soto uses capital letters to designate the numerical values to be applied to motive force, resistance, space, and time. In edition *C* such letters are signaled by periods on either side of the letter (e.g., .A., .B., etc.). In Soto's *Quaestiones* on the *Physics* a similar technique is used, except that lowercase letters are there used in edition *C* to designate variables (e.g., .a., .b., etc.), and also numerals (e.g., .2., .4., .8., etc.) where the latter are required. Printers apparently had difficulty preserving this convention in their punctuation, and this led to confusion, as will be seen below.

In what follows, relevant excerpts are provided from Soto's commentary as found in edition *C*. The Latin text is given first and then its translation into English. In the Latin text, Soto's convention of putting periods on both sides of the capital letters designating variables has been followed, but not in the translation. Nor has this been done in the last sentence of the translation, where Soto does not give numerals but simply writes out the numbers (*unum, duo, quatuor*) and does not himself employ numerals. So as to clarify the sense, throughout the translation notes have been inserted that put Soto's version of Aristotles "laws" in modern notation, as is common practise among those working on late medieval science.

Latin text:

> [Text 35] ... Duae ergo hinc colliguntur regulae. Prima per divisione mobilis et multiplicationem spatii, potentia et tempore manentibus aequalibus. Si aliqua potentia movet aliquam resistentiam per aliquod spatium in aliquo tempore, eadem, vel aequalis potentia in eodem vel aequali tempore movebit

[11] For the Greek text and a commentary identifying and explaining the seven rules, see W. D. Ross, *Aristotle's Physics: A Revised Text with Introduction and Commentary* (Oxford: Clarendon Press, 1936), pp. 683–87.

subduplam resistentiam per duplum spatium. Ratio est quod qualis est proportio medietatis resistentia [*sic*] ad totam, talis est spatii ad spatium duplum.

Secunda regula est per divisionem mobilis et temporis, potentia et spatio invariatis. Si aliqua potentia movet per aliquod spatium aliquam resistentiam, eadem vel aequalis potentia per aequale spatium movebit subduplam resistentiam in subduplo tempore. Probatur: quia quanta est proportio totius resistentiae ad medietatem, tanta est temporis ad tempus....

Tertia ergo est per divisionem spatii et temporis, potentia et mobili invariatis. Si aliqua potentia moveat aliquam resistentiam per aliquod spatium in aliquo tempore, eadem vel aequalis potentia movebit eandem resistentiam per dimidium spatium in dimidio tempore....

Quarta regula est econverso per divisionem potentiae et resistentiae, spatio et tempore invariatis. Si aliqua potentia movet aliquam resistentiam per aliquod spatium in aliquo tempore, subdupla potentia movebit subduplam resistentiam per idem spatium in eodem tempore. Ut si .A. potentia moveat .B. resistentiam per .C. spatium .D. tempore, medietas ipsius .A. quae sit .E. movebit similter medietatem ipsius .B. quae sit .F. Nam similiter se habent haec secundum analogiam, id est, proportionem. Nempe qualis est proportio potentiae ad subduplam, talis est resistentiae ad resistentiam....

In tex. 37. excludit duas alias regulas, quas falso quis posset elicere ex quarta. Cum enim dictum sit, quod si aliqua potentia moveat resistentiam per aliquod spatium in aliquo tempore, subdupla movebit subduplam etc., posset quis inferre quod eadem potentia movebit duplam resistentiam, aut per dimidium spatium in eodem tempore, aut per idem spatium in duplo tempore, ut additio fiat ad resistentiam et non ad potentiam. Quae tamen regula exceptionem tunc patitur, quando multiplicata resistentia, potentiam moventis rei aequat vel superat, tunc enim potentia nullatenus movebit resistentiam, nec per aliquod spatium, nec in aliquo tempore. Quoniam a proportione aequalitatis vel minoris inaequalitatis non fit actio. Si tamen multiplicatio resistentiae non aequaret nec excederet potentiam, tunc regula vera quidem esset. Et est numero quinta....

Perseverat ergo in exemplo quartae regulae, quod si .E. medietas ipsius .A. movet .F. medietatem ipsius .B. non est necessarium, ut ipsa virtus .E. moveat totum .B. duplum ad .F. per aliquod spatium, quantunque parvum, in aliquo tempore quantuncumque magno, eadem ratione non est certa secunda regula, subtrahendo a potentia, et non a resistentia. Ut si .A. virtus (inquit philosophus) movet resistentiam .B. per aliquod spatium in aliquo tempore, non est necesse, ut .E. medietas ipsius .A. moveat totam resistentiam .B. per aliquod minus spatium aut in aliquo maiori tempore: quia poterit potentia tantum minui, ut adaequetur resistentiae aut superetur ab ea: tunc non poterit eam movere. Alias (inquit) homo possit movere navim, imo quantancumque magnitudinem. Nam si decem homines (verbi gratia) movere possunt totam navem per aliquod spatium aliquo tempore, tunc, si teneret proportio, posset unus homo movere totam navem, aut per subdecuplum spatium in eodem tempore, aut in tempore decuplo per idem spatium. Si tamen potentia diminuta semper excedit resistentiam, vera est regula: est quae numero sexta.

Exempla harum regularum, quando verum habeant, quando vero minime, sic accipito. Si potentia ut quatuor movet resistentiam ut unum per aliquod spatium in aliquo certo tempore, eadem potentia movebit resistentiam ut duo per subduplum spatium in eodem tempore: sed tamen resistentiam ut quatuor nullatenus movebit: quoniam a proportione aequalitatis non fit actio. Pari ratione si potentia ut quatuor movet resistentiam ut unum per aliquod spatium in aliquo tempore, potentia ut duo movebit eandem resistentiam per subduplum spatium in eodem tempore: sed tamen potentia ut unum nullatenus movebit eam....

In tex. 38. subditur regula septima, a divisis ad coniuncta. Si duae sunt virtutes motivae, quarum seorsum singulae singulas resistentias scilicet pondera moveant, per aliquod spatium in aliquo tempore, virtus composita ex ambabus movebit aggregatum ex ambobus ponderibus per idem spatium in eodem tempore. Et ratio est, quia servatur semper proportio potentiae ad resistentias. Ut si singulae potentiarum sint ut duo, et singulae resistentiarum ut unum, ambae simul potentiae erunt ut quatuor, ambaeque resistentiae ut duo; atqui quatuor ad duo eadem est proportio, quae est duorum ad unum.

Translation:

[Text 35] ... From this, therefore, two rules are gathered. First, for the division of the thing moved and the multiplication of the distance, with the power and the time remaining equal: if any power moves any resistance through any distance in any time, the same or equal power will move half the resistance through double the distance in the same or equal time.[12] The reason is that whatever the ratio of half the resistance to the whole resistance, the same ratio will obtain between the distance and double the distance.

The second rule is for the division of the thing moved and the time, with the power and the distance remaining equal: if any power moves any resistance through any distance, the same or equal power will move half the resistance in half the time through an equal distance.[13] The proof: because whatever the ratio of the whole resistance to its half, the same ratio will obtain between the respective times.

[Text 36] ... The third rule is therefore for the division of the distance and the time, with the power and the distance remaining equal: if any power moves any resistance through any distance in any time, the same or equal power will move the same resistance through half the distance in half the time.[14]

The fourth rule is the converse, for the division of the power and the resistance with the distance and the time remaining equal: if any power moves any

[12] I.e., if A moves B over S in T, then T will move $B/2$ over $2S$ in T.
[13] I.e., if A moves B over S in T, then A will move $B/2$ over S in $T/2$.
[14] I.e., if A moves B over S in T, then A will move B over $S/2$ in $T/2$.

resistance over any distance in any time, half the power will move half the resistance over the same space in the same time....[15]

[Text 37] In text 37 he excludes two other rules, which one might falsely elicit from the fourth. For since it has been said that if any power moves a resistance through any distance in any time, one might infer that the same power will move double the resistance either through half the distance in the same time or through the same space in double the time, with an addition being made to the resistance and not to power. But then the rule admits of an exception, for when the resistance is multiplied such that it equals or exceeds the power, the power will not move the resistance over any distance or in any time, since no action results from a ratio of equality or lesser inequality. But if the multiplication of the resistance does not equal or exceed the power, the rule remains valid. And this is the fifth rule....[16]

He continues then with another case of the fourth rule. If E, which is half A, moves F, which is half B, it is not necessary that E will move B, the double of F, over any distance however small and in any time however great; the same would apply to the second rule, if one were to diminish the power and not the resistance. For example, if power A (says the Philosopher) moves resistance B over any space in any time, it is not necessary that E, which is half A, will move the entire resistance B over any less distance or in any more time; for if the power is diminished such that it is equal to the resistance or is less than the resistance, it will not be able to move the resistance.... But if the diminished power always exceeds the resistance, the rule is valid. And this is the sixth rule.[17]

In text 38 he adds the seventh rule, from separate to conjoined elements: if there are two motive powers each of which moves its respective resistance, say a weight, over any distance in any time, a power composed of both will move the aggregate of the two resistances or weights over the same distance in the same time.[18] And the reason is that the same ratio of power to resistances is always conserved. For example, if the individual powers are as 2, and the individual resistances as 1, both powers together will be as 4 and both resistances as 2; but 4 to 2 is the same ratio as 2 to 1.

SOTO'S *QUAESTIONES*: BOOK 7, QUESTION 4

Of the four questions that accompany Soto's commentary on the seventh book of Aristotle's *Physics*, only the last two are related directly to the so-called laws of motion. The first question inquires whether it is necessary

[15] I.e., if A moves B over S in T, then $A/2$ will move $B/2$ over S in T.

[16] I.e., if A moves A over S in T, then A will not move $2B$ over $S/2$ in T, nor will it move $2B$ over S in $2T$, in the event that $2B \geq$ A; otherwise it will do both.

[17] I.e., if A moves B over S in T, then $B/2$ will not move B over a shorter S in less T, unless $A/2 > B$, in which case the rule remains valid.

[18] I.e., if A moves B over S in T, and E moves F over S in T, then $A + E$ will move $B + F$ over S in T.

that mover and moved be in contact, and the second, whether any motion is always comparable to another. Following this, as already mentioned, comes the digression on ratios, and then the two questions dealing with the velocity of motions: the third, ascertaining it from its effect, and the fourth, from its cause. Here we supply the essentials of the last question, namely, whether or not this is determined by the ratio of the ratios (*proportio proportionum*) of the agent forces causing the motion to their respective resistances. As already indicated, the entire question is composed of forty-one paragraphs: the first seven of these pose various difficulties to be solved, in the style of a scholastic disputation; the eighth supplies the *sed contra*, which calls attention to Aristotle's seven rules as supplying a satisfactory answer; the ninth through the thirty-fifth give Soto's resolution of the question; and the last six, the thirty-sixth to the forty-first, resolve the difficulties noted at the outset. In what follows, the first seven and the last six paragraphs are omitted in order to save space, since they do not bear directly on Soto's resolution.

As previously, the Latin text that has been established is presented first and then, immediately following, its English translation. Notes are used in the Latin text to indicate variant readings, after the fashion of a critical apparatus. The term or expression for which there is a variant is given first, followed by a bracket, then the variant reading(s), and finally one or more capital letters designating the edition(s) listed in the second section above. Paragraphs are numbered throughout with the number of the paragraph as it is found in edition *C*. The same numbers are then used in the English translation, so as to permit easy access from the Latin to the English and vice versa. In the translation the notes serve a different purpose, providing a sort of commentary to the text in which different versions of the "laws" are put in modern notation, calculations are summarized, and other clarifications are provided. In the text of the translation, bracketed inserts provide the numerical equivalents of ratios given in Latin, so that the reader need not pause to look up terms such as "sesquialterate" and "double sesquitertian."

Latin text:

[8] In contrarium videntur agere septem regule Aristotelis, quas postremo capite huius libri statuendas curavit.

[9] Praenotare in primis[19] in quaestionis aditu, opus est quatuor esse (ut ait Aristoteles) consideranda in motu, scilicet id, quod movetur, et virtutem a qua, et tempus in quo, et spatium per quod movetur.

[19] in primis] imprimis *H*

[10] Porro autem virtus idem est que[20] virtus effectrix motus, sed id quod movetur est resistentia que a virtute motrice superanda est. Est enim regula hic supponenda, quod a proportione equalitatis aut minoris inequalitatis nulla fit actio, ut[21] lib. 2. dicere ceperamus, nempe, ubi resistentia sit equa aut maior quam virtus agentis, sed solum ubi proportio est maioris inequalitatis[22] agentis ad passum. Atque[23] tam virtus activa quam que resistit esse potest aut interne inherens mobili aut extrinsecus adiacens, ut dum grave cadit gravitas ipsa est interna virtus, et si quis illud expellat est extrinseca. Item medium ipsum per quod cadit, seu aqua sit seu aer, extrinseca est resistentia. In motibus itidem alterationis et augmentationis exempla etiam[24] sunt, sed que non sunt[25] presentis negotii. Preterea, virtus activa esse poterit aut spiritualis aut corporea, naturalis aut libera, fatigabilis et corruptibilis (ut in elementis et elementatis), atque infatigabilis et incorruptibilis, ut in corporibus celestibus. Titulus ergo questionis promiscue comprehendit omnes species, tam virtutis quam resistentie.

[11] Atqui varietas dignoscendi eiusmodi velocitatem motuum quatuor potest modis estimari. Primo, penes proportionem potentiarum motivarum. Secundo, penes proportionem inter resistentias. Tertio, penes proportionem inter excessus potentiarum super suas resistentias. Quarto, penes proportionem proportionum agentium super suas ipsarum resistentias.

[12] Exempli gratia, ut si sint .a. movens ut .8. et .b. movens ut .4. moventia pares resistentias ut .1, utrum .a. moveat iuxta primam regulam in duplo velocius quam .b, propterea quod activitas ut .8. est duplo maior quam[26] .4. An iuxta tertiam,[27] in proportione dupla sesquitertia, propterea quod excessus .8. ad .1. est .7. et .4. ad .1. est .3, et inter .7. et .3. est dupla sesquitertia. An vero secundum quartam, velocius moveat solum in sesquialtero, ex eo quod proportio octupla quae est inter .8. et .1. est sesquialtera ad quadruplum,[28] que est .4. ad .1, ut in proportionibus expositum est.

[13] Simili modo, si agens .a. virtutis ut .4. moveat resistentiam ut .2, et .b. eiusdem virtutis ut .4. moveat resistentiam ut .3, utrum .a. iuxta secundam regulam moveat in sesquialtero velocius quam .b, ideo quod resistentia ut .2. est in sesquialtero minor quam resistentia ut .3; an secundum tertiam, in duplo velocius, propterea quod excessus .4. ad .2. est duplus ad excessum .4. ad .3. An vero iuxta quartam, plusquam in duplo velocius, idcirco[29] quod proportio dupla .4. ad .2. est maior quam dupla ad proportionem sesquitertiam, que est .4. ad .3.

---

[20] que] quod *J*
[21] ut] et ut *H*
[22] inequalitatis] inequalitates *EF*
[23] atque] atqui *J*
[24] etiam] enim *F*, om. *J*
[25] que non sunt] non *J*
[26] quam] quam ut *J*
[27] iuxta tertiam] taxitertiam *EF*
[28] quadruplum] quadruplam *FJ*
[29] idcirco] id circo *BDEG*

[14] Porro autem, in prima et in secunda regula perquam[30] tenuis est probabilitas, sed inter tertiam et quartam relinquitur disceptatio, et tandem quinta est ex mente Aristotelis omnino amplectenda.

[15] Respondetur ergo ad quaestionem quatuor conclusionibus. Prima est adversus primam regulam. Velocitas motus non est attendenda penes proportionem virtutis unius agentis ad virtutem alterius. Conclusio in primis patet, quia ubi resistentie essent[31] inequales, nulla esset apparentia solam esse expectandam[32] proportionem inter virtutes motrices. Quis enim diceret quod si virtus ut .6. moveat resistentiam ut .1. et virtus ut .4. resistentiam ut .2, non moveat virtus ut .6. nisi in sesquialtero velocius (que est proportio, quam habet ad .4.), cum ad suam resistentiam habeat sextuplam .4. vero ad suam, non nisi duplam.

[16] Sed tamen, et ubi resistentia sit eadem conclusio nihilominus est notissima; alias, si virtus ut .4. moveret resistentiam ut .3. aliqua velocitate, virtus ut .2. subdupla ad .4. moveret eandem resistentiam subdupla velocitate, quod est impossibile, cum a proportione minoris inequalitatis nulla sit[33] actio. Qua de causa ait Aristoteles text. 37. quod non quotiescunque aliqua virtus movet aliquam resistentiam in aliquo tempore, subdupla movebit eandem in duplo[34] tempore.

[17] Miror tamen quosdam schole nostre qui aiunt regulam primam, contrariam huius conclusionis,[35] veram esse in universum, posita constantia potentie motive. Quod est dicere, fixa eadem resistentia, que sit minor utraque potentia. Existimant enim idem esse tunc metiri velocitates aut penes primam regulam, aut certe penes quartam, id quod est manifeste falsam. Nam si virtus ut .8. moveat resistentiam ut .1, et virtus ut[36] .4. moveat eandem resistentiam, velocitas motus prioris ad eam que est posterioris, secundum primam regulam habebit proportionem duplam, qualis est .8. ad .4, et secundum quartam habebit tantum sesquialteram, qualis est octuple proportionis ad quadruplam. Continet enim octupla quadruplam, et item duplam, que est medietas quadruple.

[18] Secunda conclusio contra secundam regulam. Neque velocitas motus attendenda est penes proportionem resistentiarum, que quidem conclusio eisdem penitus rationibus constat quibus secunda. Ubi nanque diverse sunt virtutes active, palam peccat secunda regula. Nulla enim est apparentia, quod ubi resistentia ut .2. movetur a potentia ut .6. et resistentia ut .3. a potentia ut .4, non moveatur prima resistentia velocius quam secunda, nisi in sesquialtero, qualis est proportio inter resistentias, sane cum prima moveatur a proportione tripla, et secunda nonnisi[37] a sesquitertia.

---

[30] perquam] per quam *CH*
[31] essent] essent sent *G*
[32] expectandam] spectandam *J*
[33] sit] fit *E*
[34] duplo] subduplo *J*
[35] huius conclusionis] huic conclusioni *J*
[36] ut] om. *EF*
[37] nonnisi] non, nisi *BCDEFG*

[19] Neque ubi virtus est eadem habet veram, alias si resistentia ut .2. moveretur a virtute ut .4. aliqua proportione, sequeretur quod resistentia ut .4, que est dupla, moveretur ab eadem virtute subdupla velocitate. Quod est falsum, eo quod a proportione equalitatis non sequitur actio.

[20] Neque si virtus utranque vinceret resistentiam, certa esset regula. Nam si resistentia ut .1. moveretur a virtute ut .8. et resistentia ut .4. ab eadem ipsa, tunc per secundam regulam velocitas prioris motus deberet excellere velocitatem posterioris quadrupla proportione, que est inter resistentias; et tamen secundum quartam, que est omnium norma, non excellit nisi proportione tripla, qualis est octuple proportionis (quam virtus habet super priorem resistentiam) ad proportionem duplam, quam habet ad posteriorem. Componitur enim octupla proportio ex tribus duplis.

[21] Tertia conclusio adversus tertiam regulam, in qua plusculum est difficultatis. Velocitas motus ex parte cause non attenditur penes proportionem arithmeticam, que est inter excessus virtutum[38] super suas singularum resistentias. Conclusio hec demonstrationem fortitur[39] ex quarta regula Philosophi tex. 36. Enimvero si regula[40] tertia, huic conclusioni contraria, veridica[41] esset, consequens fieret, ut si .a. virtus ut .8. moveret resistentiam ut .4. et .b. virtus ut .2. moveret resistentiam ut .1, velocitas primi motus esset velocitate secundi quadruplo maior, qualis est proportio excessus prime potentie super suam resistentiam ad excessum secunde super suam, scilicet .4. ad .1. Cuius tamen contrarium astruit dicta regula Aristotelis quarta que[42] talis est. Si aliqua virtus moveat aliquam resistentiam in aliquo tempore per aliquod spatium, subdupla virtus movebit subduplam resistentiam eodem tempore per idem spatium, et ita deinceps, dividendo virtutem et resistentiam. Quo fit, ut virtus ut .2. eadem velocitate moveat resistentiam ut .1. qua virtus ut .8. resistentiam ut .4. Et huius causa, illud est, quod utraque virtus movet suam resistentiam a proportione dupla.

[22] Hinc manifeste colligitur quarta conclusio, qua stabilitur quinta regula. Velocitas motus penes causa attenditur penes proportionem proportionum agentium super suas ipsorum resistentias. Que quidem proportio (ut tractatu de proportionibus prefati sumus) dicitur proportionalitas geometrica. Conclusio est, quam Paulus Venetus, Hentisberus, et fere ennatores Aristotelis consentienter affirmant. Nam etsi nusquam fuerit ab Aristotele sub his terminis constituta, colligi eam tamen existimant ex suis regulis septem.

[23] Primum ex quarta manifeste (ut modo dicebamus), eo enim quod qualis est proportio virtutis cuiuspiam ad eam quam superat resistentiam, talis est medietas virtutis ad medietatem resistentie, docet eadem velocitate movere subduplam virtutem, subduplam resistentiam, qua tota movebat totam. Namque ait eodem tempore per idem spatium eandem movere.

---

[38] virtutum] virtutem *DEFG*
[39] fortitur] sortitur *CF*
[40] Philosophi tex. 36. Enimvero si regula] om. *H*
[41] veridica] veri dica *BE*
[42] quarta que] quarta, que *BH*; et quarta que *J*

[24] Sequitur item eadem quarta conclusio dilucide ex septima regula. Nempe, quod si due virtutes seorsum moveant aliquo uno tempore duas resistentias, virtus composita ex ambabus virtutibus movebit resistentiam compositam ex ambabus resistentiis eodem tempore per idem spatium, quod est dicere eadem velocitate. Cuius profecto ratio est, quam asserit quarta conclusio. Nempe, quod talis est proportio ambarum simul virtutum ad ambas simul resistentias, qualis est alterutrius ad suam. Loquitur enim de virtutibus habentibus easdem proportiones ad suas resistentias, siquidem ait movere easdem eodem tempore. Etenim, si virtus ut .6. moveat resistentiam ut .4. et virtus ut .3. resistentiam ut .2, tunc virtus ut .9, que constat ex .6. et .3, movebit resistentiam ut .6, que constat ex .4. et .2, eadem velocitate. Quoniam omnes sunt proportiones sesquialtere. Et idem comperies in omni genere proportionis.

[25] At vero contra istam conclusionem, et pro examine regularum Aristotelis, arguitur primo authoritate eiusdem Aristotelis primo de celo, cap. de infinito, existimantis velocitatem motuum attendi penes excellentiam moventis ad mobile. Et idem videtur sentire Commentator, lib. 4. tex. 70. et lib. hoc .7. tex. 35.

[26] His tamen locis facile responsum dedimus,[43] Aristotelem solum affirmare, quod quanto excellentia erit maior, tanto motus erit celerior; nihil tamen loquitur de proportionibus, quod qualis fuerit proportio excessus ad excessum, talis sit velocitas[44] ad velocitatem. Et Commentator tex. hic .36. videtur se ipse[45] exponere, ut bene agnovit Paulus, scilicet, quod per excessum intelligebat proportionem. Et sic, comparando inter se duo agentia, comparat proportionem unius excessus super suam resistentiam proportioni alterius supra suam. Sed aliter urgentius sic arguitur.

[27] Etsi quarta nostra conclusio optime consonet[46] quarte regule Aristotelis, reliquis vero non similiter. Est nanque prima regula, quod si aliqua virtus moveat resistentiam in aliquo tempore per aliquod spatium, eadem virtus eodem tempore movebit subduplam resistentiam per duplum spatium, id est, dupla velocitate. Et quidem particulariter, ubi virtus movet resistentiam dupla proportione, consentit quarta conclusio. Etenim si virtus ut .8. moveat resistentiam ut .4. aliquo tempore per aliquod spatium, eadem virtus movebit resistentiam ut .2. in duplo velocius, quia proportio quam .8. habet ad .2. est dupla ad duplam quam habebat ad .4.

[28] Et per hoc item refutatur tertio iam impugnata regula de excessibus. Qui si essent hic aspiciendi, solum excederet velocitas in sesquialtero, qualis est proportio excessus quo .8. excedit .2. ad excessum quo excedit .4. At vero non subinde sequitur (salua quarta conclusione) quod eadem virtus eodem tempore moveat subquadruplam resistentiam per quadruplum spatium, id est, quadrupla velocitate, sed solum movebit in triplo velocius, qualis est proportio octuple, scilicet, .8. ad .1, ad duplam, puta .8. ad .4.

[43] responsum dedimus] responsumimus *B*
[44] velocitas] velocitatis *J*
[45] se ipse] seipsum *J*
[46] consonet] consonent *DEHJ*

[29] Qua ratione colligi videtur, quod vel quinta nostra conclusio est falsa, vel prima regula Aristotelis. Respondebitur forsan, quod regula Aristotelis solum loquitur de medietate resistentie, non tamen subinde procedit ad minores partes.

[30] Sed et contra hoc etiam nunc[47] pugnat aperte replica. Si virtus motiva ut .6. moveat in aliquo tempore resistentiam ut .4, eadem virtus non movebit medietatem (iuxta quartam conclusionem) eodem tempore per duplum precise spatium, quia non movebit solum duplo maiori velocitate; ergo quarta conclusio repugnat Aristoteli. Probatur antecedens: .6. ad .2. est proportio tripla, ad quatuor vero est sesquialtera; tripla autem ad sesquialteram est maior quam dupla. Constat enim ex duabus sesquialteris et una sesquitertia, ut tripla que est .12. ad .4.[48] constat ex sesquitertia, que est .12. ad .9, et sesquialtera que est .9. ad .6, et sesquialtera que est .6. ad .4;[49] ergo virtus ut .6.[50] plusquam duplo velocius movebit resistentiam ut .2.[51] quam resistentiam ut .4.

[31] Quinimo est regula generalis, quod quandocunque virtus excellit resistentiam minori proportione quam dupla, eadem virtus habebit ad medietatem resistentie maiorem proportionem quam duplam, et per consequens movebit eam plusquam duplo velocius. Et econverso quandocunque[52] virtus exuperat resistentiam maiori proportione quam dupla, excedit medietatem resistentie minori quam dupla. Ut si virtus ut .6. moveat resistentiam ut .2. aliqua velocitate, non movebit resistentiam ut .1. duplo maiori, sed citra. Quoniam .6. ad .1. habet proportionem sextuplam,[53] ad .2. vero triplam, et sextupla ad triplam est minus[54] quam dupla. Constat enim ex dupla et tripla, quare dupla non est eius medietas.

[32] Eodem omnino modo patescit, eandem conclusionem quartam non in universum consentire cum .2. regula Aristotelis, que venit in idem cum prima, videlicet, si aliqua virtus movet resistentiam in aliquo tempore per aliquod spatium, eadem virtus movebit subduplam resistentiam per idem spatium subduplo tempore, quod pariter est dicere, dupla velocitate. Quare si quartam conclusionem sustineas, nequibus verificare regulam nisi ubi virtutis ad resistentiam fuerit iuste proportio dupla.

[33] Tertia vero, et quinta, et sexta regule Aristotelis nihil habent hic peculiaris difficultatis.

[34] Ad hec igitur argumenta ferme Calculatores nihil aliud respondent, quam quod prima et secunda regula duntaxat habet verum, ubi virtutis ad resistentiam est proportio dupla, neque maior, neque minor. Nam tunc ad medietatem erit quadrupla, que est dupla ad duplam. Et certe ita necessarium est dicere.

---

[47] etiam nunc] etiamnum *B*
[48] .4.] quarta *F*
[49] .4;] quartam *F*
[50] .6.] sextam *F*
[51] .2.] secundam *F*
[52] quandocunque] quandoque *H*
[53] sextuplam] suxtuplam *DEG*
[54] minus] minor *J*

Nam in aliis proportionibus,[55] neque si velocitas attendatur penes proportionem proportionum, ut dicit quarta conclusio, et multo minus si attendatur penes proportionem inter excessus, possunt defensari eiusmodi regule.

[35] Attamen ne Aristoteles videatur tam mancus qui in solo casu particulari constituerit regulas. Et preterea, ut quarta nostra conclusio sit universalis, interpretande sunt dicte regule, quod nomine medietatis resistentie intelligit illam generaliter, ad quam virtus habet duplam proportionem ad proportionem quam habet ad totam resistentiam. Ut si virtus ut .6. moveat resistentiam ut .4. aliqua velocitate, eadem virtus movebit subduplam resistentiam, id est, illam super quam habebit proportionem duplam ad sesquialteram dupla velocitate, et per consequens, eodem tempore per duplum spatium, aut per idem spatium subduplo tempore. Proportio autem dupla ad sesquialteram est maior quam dupla, puta dupla sesquiquarta, qualis est .9. ad .4. Et pari ratione eadem virtus movebit tertiam partem resistentie (id est, illam, super quam virtus habet proportionem triplam ad proportionem quam habet[56] super totam resistentiam) velocitate triplo maiorem. Huiusmodi arte patere cuicunque potest via dignoscendi velocitates motus.

Translation:

[8] On the contrary there seem to stand the seven rules of Aristotle, which he saw fit to put in the last chapter of this Book.

[9] First, when approaching this question, one must consider the four elements involved in motion (as Aristotle says), namely, the thing moved, the power that moves it, the time during which it moves, and the space over which it moves.

[10] The power is the same as the power that effects the motion, while the thing moved is the resistance that is overcome by the motive power. Here a rule is to be presupposed, that there is no action from a ratio of equality or of lesser inequality, as we undertook to point out in Book 2, namely, when the resistance is equal to or greater than the power of the agent, but only when the ratio between agent and patient is one of greater inequality. Both the active power and the resistive force can be something either inhering internally in the moving object or coming to it from without, as when a heavy object falls its gravity is an internal power, and, if someone should throw it, this is extrinsic. Again, the medium through which it falls, whether this be water or air, is an extrinsic resistance. Examples are also available in the motions of alteration and of augmentation, but these do not pertain to the present inquiry. Again, the active power can be either spiritual or corporeal, natural or free, fatigable and corruptible (as in the elements and compounds) or indefatigable and incorruptible, as in the heavenly bodies. Thus the title of the question takes in all kinds of powers as well as resistances, indifferently.

---

[55] Nam in aliis proportionibus] om. *J*

[56] proportionem triplam ad proportionem, quam habet] om. *CDEFGHJ*

[11] There are four ways of looking at the different ways of judging the velocities of the motions of such objects. First, from the ratio between the motive powers;[57] secondly, from the ratio between the resistances;[58] thirdly, from the ratio between the excesses of the powers over the resistances;[59] fourthly, from the ratio of the ratios of the agents to their corresponding resistances.[60]

[12] For example, if movers *A* of 8 and *B* of 4 move equal resistances of 1, does *A* move with twice the velocity of *B* because its activity, being 8, is twice as great as 4, according to the first rule? Or, following the third rule, in a double sesquitertian ratio [i.e., 7/3], because the excess of 8 over 1 is 7 and that of 4 over 1 is 3, and between 7 and 3 there is a double sesquitertian? Or, following the fourth rule, would *A* move with a velocity that is greater only in a sesquialterate ratio [i.e., 3/2], because the octuple ratio that holds between 8 and 1 is sesquialterate with respect to the quadruple that holds between 4 and 1, as has been explained in the digression on ratios?[61]

[13] Similarly, if agent *A*, with a power of 4, moves a resistance of 2, and *B*, with the same power of 4, moves a resistance of 3, would *A*, according to the second rule, move with greater velocity than *B* in a sesquialterate ratio [i.e., 3/2], because a resistance of 2 is in a sesquialterate ratio less than a resistance of 3;[62] or, according to the third rule, with twice the velocity, because the excess of 4 over 2 is twice the excess of 4 over 3?[63] Or, according to the fourth rule, with more than double the velocity, from the fact that the ratio of 4 to 2 doubled is more than the sesquitertian ratio of 4 to 3 doubled?[64]

[14] Further, there seems to be but slight probability in the first and second rules, while the third and fourth leave room for debate; and, finally, according to the mind of Aristotle, a fifth[65] is to be completely accepted.

[15] The reply to the question may be made, therefore, under four conclusions. The first is against the first rule. The velocity of motion is not to be ascertained from the ratio of the power of one agent to the power of the other.

---

57 I.e., $V_2/V_1 = F_2/F_1$, presupposing that $R_2 = R_1$.

58 I.e., $V_2/V_1 = R_1/R_2$, presupposing that $F_2 = F_1$.

59 I.e., $V_2/V_1 = (F_2 - R_2)/(F_1 - R_1)$.

60 This statement permits of two formulations, depending on the sense of the expression "ratio of ratios" (*proportio proportionum*). If "ratio" is to be understood as a simple arithmetical ratio, then the meaning is $V_2/V_1 = (F_2/R_2)/(F_1/R_1)$—that is, the ratio of the respective ratios of the forces to the resistances. If "ratio" is to be understood in Bradwardine's sense as a geometrical or exponential ratio, then the meaning is more complex. Expressed in words it means that V_2/V_1 is the exponent to which F_1/R_1 must be raised to equal F_2/R_2. Stated algebraically, this becomes $(F_2/R_2) = (F_1/R_1)^{V_2/V_1}$ an expression that is termed "Bradwardine's function." From the example Soto gives at the end of par. [12], formulated in the next note, it is clear that he intends the second meaning here, not the first.

61 I.e., $8/1 = (4/1)^{3/2}$.

62 I.e., $3/(3/2) = 2$.

63 I.e., $(4 - 2) = 2(4 - 3)$.

64 I.e., with an exponent of two, since $(4/2)^2$ or 16/4 is greater than $(4/3)^2$ or 16/9.

65 Note the reference to a "fifth" conclusion, which is not explicitly indicated in Soto's fourth question, but which is mentioned again in pars. [22] and [29].

This conclusion is apparent, first of all, because in the case where the resistances are unequal there is no reason why one should take account only of the ratios between the motive powers. For, should one say that if a power of 6 were to move a resistance of 1 and a power of 4 a resistance of 2, a power of 6 would move with greater velocity only in a sesquialterate ratio [i.e., 3/2]—which is the ratio of 6 to 4—when 6 would have a sextuple ratio to its resistance [i.e., 6/1] and 4 only a double ratio to its [i.e., 4/2 = 2]?

[16] In the case where the resistances are the same, on the other hand, the conclusion is still obvious; otherwise, if a power of 4 were to move a resistance of 3 with a given velocity, a power of 2, the half of 4, would move the same resistance with half the velocity, and this is impossible, because there cannot be any action on the basis of a ratio of lesser inequality. For this reason Aristotle says in text 37 that it is not true that if a certain power can move a resistance in a given time, half the power will move the same in twice the time.

[17] I marvel, however, at certain members of our school, who say that the first rule, contrary to this conclusion, is true universally, granted the constancy of the motive power, that is to say, making the resistance the same but less than either power. For they think that determining the velocity by the first rule is the same as by the fourth, and this is obviously false. For, if a power of 8 moves a resistance of 1, and a power of 4 moves the same resistance, the velocity of the first motion to the second would be in double ratio, that of 8 to 4, according to the first rule,[66] whereas according to the fourth, it would be only in a sesquialterate ratio, that of the octuple ratio to the quadruple, for an octuple contains a quadruple and a double again, which is half the quadruple.[67]

[18] The second conclusion is against the second rule. For the velocity of motion is not to be ascertained from the ratios between the resistances, and this conclusion follows from the same reasons as the first. For the second rule is clearly inapplicable where there are different agent powers. Thus there is no reason why, if a resistance of 2 is moved by a power of 6 and a resistance of 3 by a power of 4, the first resistance is not moved with a greater velocity than the second, and in a sesquialterate ratio, which is the ratio between the resistances [i.e., 3/2]; and this because the first is moved with a triple ratio [i.e., (6/2) = (3/1)] and the second is not, but rather with a sesquitertian [i.e., 4/3].

[19] Nor is the rule true when the power is the same, for if a resistance of 2 were moved by a power of 4, it would follow from the ratio that a resistance of 4, which is double [i.e., twice two], would be moved by the same power with half the velocity. And this is false, by the fact that no action proceeds from a ratio of equality.

[20] Nor would the rule be correct if the power were sufficient to overcome both resistances. For, if a resistance of 1 were moved by a power of 8 and a

[66] I.e., $(8/1)/(4/1) = 8/4 = 2/1$.
[67] I.e., $(8/1) = (4/1)(2/1) = (4/1)^{3/2}$.

resistance of 4 by the same, then, according to the second rule, the velocity of the first motion should exceed the velocity of the second by the quadruple ratio that obtains between the resistances [i.e., 4/1]; yet, according to the fourth rule, which is the best norm, it would not exceed by more than a triple ratio, which is that of the octuple ratio (which the power has over the first resistance) to the double ratio it has over the second. For an octuple ratio is composed of three doubles.[68]

[21] The third conclusion is against the third rule, and in this there is somewhat more difficulty. The velocity of motion is not ascertained on the part of the cause from the arithmetical proportion that exists between the excesses of the powers over their respective resistances. This conclusion gets its demonstration from the fourth rule of the Philosopher, text 36. For if the third rule, which is contrary to this conclusion, were true, there would result as a consequence that if *A*, with a power of 8, were to move a resistance of 4, and *B*, with a power of 2, were to move a resistance of 1, the velocity of the first motion would be four times greater than the velocity of the second, for this is the ratio of the excess of the first power over its resistance [i.e., 8 – 4 = 4] to the excess of the second over its [i.e., 2 – 1 = 1], namely, 4 to 1. Contrary to this, however, one may oppose Aristotle's fourth rule, already mentioned, which is the following: if any power were to move any resistance in any period of time over any space, half the power would move half the resistance in the same time over the same space, and so on, dividing both the power and the resistance. From this it follows that a power of 2 would move a resistance of 1 with the same velocity as a power of 8 would move a resistance of 4 [since (2/1) = (8/4)]. And the reason for this is that each power moves its resistance in a double ratio [i.e., 2/1].

[22] From the foregoing the fourth conclusion is easily deduced, and with this the fifth[69] rule is established. The velocity of motion ascertained from its cause is determined by the ratio of the ratios of the agents to their resistances. And this ratio (as we have already said in the treatise on ratios) is said to be a geometrical ratio. The conclusion is one that Paul of Venice,[70] Heytesbury,[71] and practically all the commentators on Aristotle commonly affirm. For, even if Aristotle never did state this in these terms, nevertheless they feel that one can deduce it from his seven rules.

[23] This obviously follows, first, from [Aristotle's] fourth rule (as we were just saying), from the fact that the same ratio exists between any power and

---

[68] I.e., $(8/1) = (2/1)(2/1)(2/1) = (2/1)^3$.

[69] Note again the reference to a "fifth" rule, which has already been mentioned in par. [14] and will recur again in par. [29].

[70] See Paul of Venice, *Summa naturalium, Physica* (Venice, 1503), cap. 32, fol. 17vb: "Quarta conclusio. Velocitas in motu locali attenditur penes proportionem proportionum potentiarum moventium ad suas resistentias: ita quod ex equalibus proportionibus potentiarum moventium ad suas resistentias provenit equalis velocitas; ex inequalibus inequalis. scilicet. ad duplam dupla velocitas: et ad subdublam subdupla velocitas: ut si a. ad b. esset proportio quadrupla: et c. ad d. esset proportio dupla: dico quod a. moveret b. in duplo velocius quam c. d. eo quod proportio quadrupla est dupla duple."

[71] The reference is to William Heytesbury (fl. c. 1355), the Oxford calculator, but no indication is given of the work being referenced.

the resistance it overcomes as exists between half the power and half the resistance, and thus half the power will move half the resistance with the same velocity as the entire power moves the whole resistance. For he says that it moves the same resistance over the same space in the same time.

[24] The fourth conclusion also follows clearly from the seventh rule, namely, that if two powers were separately to move two resistances in any one time, the power composed of the two powers would move a resistance composed of both resistances in the same time over the same space, that is to say, with the same velocity. And the reason for this is the same as is asserted in the fourth conclusion, namely, that the same ratio exists between both powers taken together and both resistances taken together as exists between each taken separately. For to speak of powers as having the same ratios to their resistances is to say equivalently that they will move them in the same time. Thus, if a power of 6 moves a resistance of 4, and a power of 3 a resistance of 2, then a power of 9, which is composed of 6 and 3, will move a resistance of 6, which is composed of 4 and 2, with the same velocity. The reason is that all are sesquialterate ratios [i.e., $(9/6) = (6/4) = (3/2)$]. And the same will be found to be true in every other kind of ratio.

[25] But, against this conclusion, and to give closer examination to Aristotle's rules, there is the argument first from the authority of Aristotle himself, in the first book of *De caelo*, in the chapter on the infinite, where he proposes that the velocities of the motions should be ascertained from the superiority of the mover over the thing moved. And the same seems to be the opinion of the Commentator[72] in Book 4, text 70, and here in Book 7, text 35.

[26] But to these texts we can easily reply that Aristotle intended to state only that the more the superiority the greater the velocity; "more quickly" says nothing of ratios, nor does it say that the ratio of one excess to the other excess is that of one velocity to the other. Yet the Commentator himself, here in text 36, seems to say, as Paul [of Venice] properly recognized, that by excess he meant ratio.[73] And so, comparing two agents to each other, he compares the ratio of the excess of one over its resistance to the ratio of the other over its. But it seems that one can argue more cogently in a different manner as follows.

[27] Although our fourth conclusion agrees very well with Aristotle's fourth rule, the same cannot be said for the others. For the first rule is that, if some power moves a resistance over a given space in a particular time, the same power will move half the resistance over twice the space in the same time, that is, with double velocity. And, in this particular case, when the moving power and the resistance stand in the ratio of 2 to 1, the fourth conclusion is in agreement. For, if a power of 8 moves a resistance of 4 over a given space in a particular time, the same power will move a resistance of 2 with double the velocity, because the quadruple ratio that exists between 8 and 2 [i.e., $(8/2) = (4/1)$] is double the double ratio that exists between 8 and 4 [since $(8/2) = (8/4)^2$].

[72] That is, Averroes (1126–98), in his great commentary on the *Physics*.

[73] Paul of Venice, *Summa naturalium* (n. 70).

[28] In this way the third rule concerning excesses, which has already been disproved, is likewise refuted. For, if the excesses were here taken into account, the velocity would be greater only in a sesquialterate ratio, which is the ratio of the excess by which 8 is greater than 2 to the excess by which it is greater than 4.[74] But from this it does not immediately follow (independently of the fourth conclusion) that the same power would move a quarter the resistance four times the distance in the same time, i.e., with four times the velocity, but that it would move it only with three times the velocity, for this is the ratio of the octuple, namely, 8 to 1, with respect to the double, namely, 8 to 4.[75]

[29] From this argument it would seem to follow that either our fifth[76] conclusion, or the first rule of Aristotle, is false. Perhaps the reply to this is that the rule of Aristotle speaks only of half resistances, and that one cannot apply it directly to smaller parts.

[30] Now, however, another objection obviously militates even against this. If a motive power of 6 were to move a resistance of 4 in a certain time, the same power would not move half the resistance (according to the fourth conclusion) over exactly twice the distance in the same time, because the velocity would not be simply double; therefore the fourth conclusion is opposed to Aristotle. The antecedent is proved as follows: 6 to 2 is a triple ratio [i.e., 3/1], whereas 6 to 4 is sesquialterate [i.e., 3/2]; but the ratio of a triple ratio to a sesquialterate ratio is more than double.[77] For a triple ratio is composed of two sesquialterates and one sesquitertian,[78] just as the triple, 12 to 4, is composed of a sesquitertian, 12 to 9, and a sesquialterate, 9 to 6, and a sesquialterate, 6 to 4.[79] Therefore a power of 6 will move a resistance of 2 with more than twice the velocity it will move a resistance of 4.[80]

[31] There is the general rule, in fact, that, whenever a power exceeds a resistance in a ratio less than double, the same power will have a ratio to half the resistance greater than double the first ratio, and as a consequence it will move the resistance with more than twice the velocity. And, vice versa, whenever a power exceeds a resistance by a ratio greater than 2 to 1, it will exceed half the resistance by less than double the original ratio. Thus, if a power of 6 moves a resistance of 2 with a given velocity, it will move a resistance of 1, not with double the velocity, but with less. For 6 stands in sextuple ratio to 1, and in triple ratio to 2, and the ratio of the sextuple to the triple is less than double.[81] For the sextuple is composed of a double and a triple, and for this reason the double is not its half.[82]

---

[74] I.e., $(8-2)/(8-4) = (6/4) = (3/2)$

[75] Since $(8/1) = (8/4)(8/4)(8/4) = (8/4)^3$.

[76] Note yet again the reference to a "fifth" conclusion, previously mentioned in pars. [22] and [29].

[77] I.e., $(3/1)$ is greater than $(3/2)(3/2)$, which is equal to $(3/2)^2$, or 2 1/4.

[78] I.e., $(3/1) = (3/2)(3/2)(4/3)$.

[79] I.e., $(12/4) = (12/9)(9/6)(6/4)$.

[80] Since $(6/2) > (6/4)^2$, i.e., $3 > 36/16$, or 2 1/4.

[81] I.e., 6 or $(6/1)$ is less than $(3/1)(3/1)$, which is equal to $(3/1)^2$, or 9.

[82] I.e., $(6/1) = (2/1)(3/1)$ and $(6/1)^{1/2} \neq (2/1)$.

[32] And in exactly the same manner it becomes obvious that the fourth conclusion itself does not agree completely with the second rule of Aristotle, which agrees with the first, namely, if a given power moves a resistance over a determinate distance in a particular time, the same power will move half the resistance over the same distance in half the time, which is the same as to say with twice the velocity. For this reason, if one subscribes to the fourth conclusion, one cannot verify the rule, except in the case where the ratio of the power to the resistance is exactly double.

[33] The third, and the fifth, and the sixth rules of Aristotle here do not offer any special difficulty.

[34] To these arguments, therefore, nearly all the Calculators[83] say no more than that the first and second rules are true only when the ratio of the power to the resistance is double, and neither more nor less. For then the ratio to half will be quadruple, which is double with respect to double [i.e., $(4/1) = (2/1)^2$]. And certainly it is necessary to say this. For with other ratios, if the velocity is determined from the ratio of the ratios, as the fourth conclusion states, and even less if it is determined from the ratio between the excesses, these rules cannot be defended.

[35] Nevertheless, Aristotle should not be thought deficient for having formulated rules that are valid only in a special case. Moreover, since our fourth conclusion is universal, the foregoing rules are to be interpreted in such a way that by the expression "half the resistance" he means generally that to which the power has a ratio double the ratio it has to the entire resistance. Thus, if a power of 6 moves a resistance of 4 with a given velocity, the same power will move half the resistance (that is, that over which it will have a ratio double a sesquialterate ratio) with twice the velocity, and, as a consequence, it will move over twice the distance in the same time, or over the same space in half the time. A ratio double a sesquialterate ratio is more than a ratio of 2 to 1—it is a double sesquiquartan ratio, i.e., 9 to 4.[84] And, for the same reason, the same power will move one third part of the resistance (i.e., that over which the power has triple the ratio it has to the entire resistance) with a triple velocity.[85] And with a technique of this kind the way is clear for anyone to be able to ascertain the velocities of motions.

[83] This term is applied onomastically to Richard Swineshead or Suiseth (fl. c. 1340–55), who composed a *Liber calculationum* while at Merton College, Oxford; it is also used generally to refer to Bradwardine, Heytesbury, and others of that school who were adept at *calculationes* of the type being considered here.

[84] Since $(3/2)^2 = (3/2)(3/2) = 9/4$.

[85] The calculation Soto is proposing here, when his parenthetical expression is taken into account, would read $(3/2)^3 = (3/2)(3/2)(3/2)$, and this yields a ratio of 27/8. What this means is that an F/R ratio of 27 to 8 would be required to produce a velocity three times as fast as that produced by an F/R ratio of 3 to 2. So, if the force stays at 3, a resistance of 8/9 might be thought of as "one third part" of the resistance of 2, because the ratio of 3 to 8/9 is "three times" (i.e. the cube of) the ratio of 3 to 2. Alternatively, to take an example that involves only whole numbers, if a force of 8 moves a resistance of 4 with a given velocity, then a resistance of 1 will be called a resistance one-third of the resistance of 4 because the ratio 8/1 is "three times" (i.e. the cube of) the ratio 8/4. If so, the force of 8 will move the resistance of

In the Latin text, as has already been remarked, the variant readings are designated by note numbers 19 through 56, totaling 38 in all. (Only 10 of these variants, it should be noted, were included in the 45 soundings taken from all four questions on Book 7 to trace the provenance of the nine editions, but this is of no importance for what follows.) It turns out that few of the 38 variants are significant for determining the sense of Soto's exposition. The most significant is the last, note 56, which shows an omission of six words from the parenthetical expression in par. [35] that otherwise would clarify what Soto means by "one third part of the resistance," a point to be discussed later. Also important are the variants in notes 48 through 51 in par. [30], which show that edition *F* is flawed in its numerical examples, giving ordinal numbers in Latin for what were intended to be cardinal numerals. And it should be noted that the three occurrences of "fifth" (*quinta*) referring to a fifth "rule" or "law" in pars. [14], [22], and [29] show no variant readings, which indicates that no editor questioned the existence of such a law even though it is not stated explicitly in the text.

There are obviously difficulties in the foregoing text, the main one relating to the number of conclusions Soto is offering and precisely how his "fifth" rule or law, whatever it may be, is related to Bradwardine's "ratio of ratios" as this is invoked to explain the ratios of velocities produced by various powers or forces working against different resistances. One should consider, however, that the question in which the text is situated is really part of Soto's commentary on Aristotle's seventh book, in which he, like Bradwardine before him, was intent on improving Aristotle's seven rules and indeed assimilating them, as well as Averroes's teachings, within the new formulations. (Note that at the first occurrence of the "fifth" rule, par. [14], Soto does present this as being "according to the mind of Aristotle.") Going back and forth between Aristotle's seven rules and his own four or five understandably could have occasioned some confusion in Soto's exposition. However that may be, there can be no doubt that Soto understood Bradwardine's geometrical or exponential ratios correctly, as can be seen from the examples he provides in pars. [12], [13], [17], [20], [27], [28], [30], [31], [34], and [35].[86]

---

1 three times as fast as it moves the resistance of 4, since $8/1 = (8/4)(4/2)(2/1) = (2/1)^3$. Edith Sylla, who has proposed this interpretation to me, observes that this would be a clever way of "reading" Aristotle to agree with Bradwardine. And, as long as Aristotle never gives examples except those involving ratios of 2/1, no glaring counterinstances would rule out this "interpretation."

[86] The same cannot be said of the Augustinian Alonso de la Vera Cruz, who plagiarized much of Soto's fourth question on Book 7 and shows a misunderstanding of the ratios being presented there; see his *Physica speculatio* (Salamanca, 1562), lib. 7, spec. 5, pp. 141–43. Vera Cruz uses the same expression as Soto for his fourth conclusion on p. 143, which reads: "Velocitas motuum penes causam attenditur penes proportionem proportionum agentium super suas ipsorum resistentias, quae proportio geometrica proportionalitas dicitur." The example he then gives to instantiate this is the following: "Ut si virtus ut .6. moveat resisten-

Since the reception of Soto's ideas among later thinkers, and expecially the Jesuits, is of major interest, next to consider are two different settings where Soto exerted some influence, the first the famous Jesuit college in Rome, the Collegio Romano, and the second Jesuit institutions in Portugal, their *studia* at Evora and Coimbra. The second of these, in particular, may cast some light on the "fifth" rule as it was understood by those who had read Soto's text, if not by Soto himself.

SOTO AND THE COLLEGIO ROMANO

In previous writings I have traced the influence of various aspects of Soto's teachings on professors at the Collegio Romano, particularly those teaching between 1577 and 1591.[87] Here I focus on a remark made by the last of these, Ludovicus Rugerius, who taught the *Physics*, *De caelo*, etc., in the academic year 1590–91. When discussing how the motions of heavy bodies are related in terms of velocity at the beginning, the middle, and the end of their motions, Rugerius ties the answer to this question to Aristotle's rules for the comparison of motions in Book 7; he points out that several of those rules labor under severe difficulties, but for a fuller discussion of the ways in which they might be revised one should read Toletus and Domingo de Soto, among others.[88] The statement is of interest because by this time eight of the nine editions of Soto's text were already in print, with only the second of these, *B*, containing the full parenthetical expression in par. [35] to which reference has been made above. One can only wonder to which edition Rugerius was making reference, but it seems unlikely that edition *B* was what he had in mind.

The recommendation to read Toletus is not especially helpful in this regard. Toletus had been Soto's favorite student at Salamanca, so it is not remarkable that his combined commentary-questionary on the *Physics* should follow closely his teacher's exposition.[89] He too lists the four major

tiam ut quatuor, et virtus ut .3. resistantiam ut duo; quia utrobique est sexquialtera, virtus ut .9. quae constat ex .6. et .3. movebit resistantiam ut .6. quae constat ex .4. et .2. eadem velocitate, quia omnes istae sunt proportiones sexquialterae." That is, $6/4 = 3/2$, and $9/6 = (6+3)/(4+2) = 3/2$; none of these is a geometrical ratio, all being arithmetical and merely a restatement of Aristotle's seventh rule as given in his text 38.

[87] The main discussion is in William A. Wallace, *Galileo and His Sources: The Heritage of the Collegio Romano in Galileo's Science* (Princeton: Princeton University Press, 1984), but the subject is also touched on in several of the publications listed in n. 3 above.

[88] See ibid., p. 189. The Latin reads as follows: "Aliqua ex his theorematis graves patiuntur difficultates, quae tamen apud alios legi possunt. Diligenter Toletus, Sotus, et alii [legendi sunt], apud quos etiam potest videri quo modo Aristoteles hanc doctrinam a motu locali transferat etiam ad alios motus" (Ludovicus Rugerius, *In quatuor libros De caelo et mundo* [1591], MS Bamberg, SB 62–4, fol. 101r–v).

[89] Franciscus Toletus (1533–96) studied under Soto at Salamanca, later taught there himself, and, on becoming a Jesuit, was sent to the Collegio Romano, where he taught logic in

positions on the velocities of motion as they are found in Soto. Only his fourth conclusion, which is accompanied by the marginal notation "Quarto, proportio in velocitate motuum sumitur penes proportionalitatem proportionum virtutum ad suas resistentias," will be considered here. The text reads as follows:

> Quarto vero modo sumitur penes proportionalitatem virtutem [*sic*] ad suas resistentias, quod est dicere ut clare explicemus. Datis duobus motibus, considera in qua proportione virtutes excedant resistentias suas: iterum considera, quae proportio sit inter illas duas proportiones, nam in hac erit unus motus altero velocior. V.g. Sit virtus, ut octo, movens pondera quatuor: sit etiam virtus, ut octo, movens pondera, ut duo, prior virtus est in proportione dupla ad suam resistentiam: posterior vero in quadrupla, puta, octo ad duo. Quia ergo proportio quadrupla est dupla ad duplam, et iste motus erit duplo velocior. Hoc autem modo infallibiliter invenies proportionem velocitatis motuum: si tamen cognoscatur in qua proportione sit una ad aliam: et sic patet solutio argumenti. Non enim illo modo est sumenda proportio velocitatis motuum, puta, penes proportionem excessum [*sic*], sed modo, qui dictus est. Haec de libro septimo.[90]

This is very similar to the text of Paul of Venice, cited above in note 70, which discusses only the doubling and halving of the velocity, force, or resistance, and does not consider the more difficult cases of subalternate and triple ratios considered by Soto. The fact that 2+2, 2×2, and 2^2 all yield the same result, 4, does not permit one to discriminate between arithmetical operations and exponential or geometrical ratios, and thus to distinguish between the two possible interpretations of Bradwardine's "ratio of ratios" given in note 60 above. And, perforce, there is no mention of a "fifth" rule in either Paul of Venice or Toletus.

Another point to note is that around Rugerius's time at the Collegio Romano there seems to have been an interest in the experimental verification of laws such as Aristotle's. The Jesuit who preceded Rugerius and taught natural philosophy in the academic year 1589–90, Mutius Vitelleschi, remarks on the difficulty of deciding between alternative formulations by means of experiment, though he admits that the results he has seen seem to go contrary to Aristotle. In such a matter, he adds, the data of experience are of crucial importance, and yet there are many things that make him suspect all experiments of this kind. For one, it seems hardly possible that the shapes of the bodies that move, the resistances of the media, and all other things that influence their motions should be equal. For another, one cannot perceive a difference in velocity of this kind unless the motion takes

1559–60, natural philosophy in 1560–61, and metaphysics in 1561–62, after which he taught theology. He composed many textbooks for use in the order, and later was made a cardinal of the Roman Church.

[90] Franciscus Toletus, *In octo libros Physicorum Aristotelis* (Venice: Apud Iuntas, 1600), lib. 7, tex. 38, q. 5.

place over a very great distance, and over such a distance many things can happen that detract from the certitude of the experiment.[91] The attitude Vitelleschi expresses here may be a sign of a certain diffidence at that time among the Jesuits of the Collegio Romano toward both theoretical and experimental discussions of "laws" of falling bodies.

SOTO AND THE PORTUGUESE JESUITS

A more hopeful line of investigation is the reception of Soto's teaching at Jesuit institutions in Portugal, mainly those at Evora and Coimbra.[92] In fact, in a Coimbra manuscript dating from 1580 I came across a statement very similar to that of Rugerius with which I began the previous section. The context was a discussion of the two questions corresponding to Soto's third and fourth questions on Book 7 of Aristotle's *Physics*, those relating to velocities of motion as ascertained from effect and from cause, respectively. Here the author, pressed by limitations of time in his presentation, refers his readers to the fuller treatments to be found in Domingo de Soto's questions.[93] Following up this lead, I found in other manuscripts materials relating to Soto's fourth question, which I will now discuss—and also ma-

---

[91] See Wallace, *Galileo and His Sources* (n. 87), p. 185. The Latin of this passage reads as follows: "Equidem in hac re experientiam nullam omnino certam habeo, et quas habeo, illae potius obstant sententiae Aristotelis. Verum multa sunt qui suspectas mihi faciant omnes experientias de hac re, in qua tamen maxime standum esset experientiae. Primum enim vix videtur posse fieri ut figura corporum qui moventur, resistentia medii, et cetera omnia qui ad motum concurrunt sint omnino paria. Secundo, non potest percipi huiusmodi differentia in velocitate motus nisi motus fiat per magnam distantiam, ut constat etiam experientia. In magna autem distantia multa accidere possunt quae experientiae detrahant aliquid certitudinis. Quare cum haec ita se habeant nihil de hac re statuo, neque enim ad rei materiam necesse est in praesentia" (Mutius Vitelleschi, *In octo libros Physicorum et quatuor De caelo* [1589–90], MS Bamberg, SB 70, fol. 365v). Slightly before this passage Vitelleschi shows his acquaintance with both Bradwardine's treatise on ratios and Jean Taisnier's similar treatise (which was actually a plagiarization of Giovanbattista Benedetti's refutation of Aristotle's laws of free fall, a fact unknown to Vitelleschi), referring his readers to their works as follows: "Legendum est Thomas Bradwardinus in sua tractatione de proportione motuum, et Ioannes Thaisnerus in tractatione de eadem re, qui dicit id demonstrari" (ibid., fol. 365r).

[92] In January 1977, ten years after doing most of my work on the Collegio Romano materials, I spent a week in Portugal, mainly in Lisbon but with a side trip to Coimbra, tracing Jesuit materials relating to Galileo's early notebooks. I copied out by hand my transcription of many manuscripts composed between 1570 and 1587, and reported them in my 1995 article, "Late Sixteenth-Century Portuguese Manuscripts" (see n. 3 above). Unfortunately, my fear of writer's cramp limited what I could copy at the time, and there are more ellipses in my reporting than I would like, but at least it provides clues on which others might later work.

[93] The author is anonymous, but his Latin reads as follows: "Haec duae extremae questiones latius tractantur a Dominico de Soto; ibi possunt videri. Qua de causa praetermittuntur a nobis ac imprimis propter temporis brevitatem" (MS Coimbra BU Res. 2312, fol. 187r). For fuller information on this manuscript and others cited below, and their authors, see my article cited in the previous note.

terials relating to his third question, which I will treat in the following section.

The principal material relating to the velocities of motion as ascertained from their cause is found in two manuscripts, one from Coimbra dating from 1582 and the other from Lisbon dating from 1586.[94] Since these are in almost word-by-word agreement, I give here the reading of the first and supply the few variants of the second in notes. What is interesting about both is that, instead of having four conclusions, as does Soto, they supply five, and thus may throw some light on the "fifth" rule or law to which repeated reference has been made above.

In what follows the Latin text is given first, with paragraphs numbered for future reference, and then the English translation, with technical notes similar to those earlier provided for Soto's question.

Latin text:

> [1] Undenam tamquam ex causa accipienda sit velocitas motus, quaestio secunda....
>
> [2] Prima conclusio.[95] Velocitas motuum non attenditur penes proportionem quam habet una virtus activa ad aliam....
>
> [3] Quinque igitur causas fingere possumus e quarum aliquo[96] sumenda cognoscendaque sit motus[97] velocitas maior, aut minor, aut aequalis. Una earum est proportio virtutum agentium, altera est proportio resistentiarum, tertia est proportio excessuum virtutum agentium supra suas resistentias, quarta est proportio proportionum virtutum agentium ad suas resistentias, ultima est proportio excessuum virtutum agentium supra suas resistentias non simpliciter sed habita ratione quantitatis resistentiarum.
>
> [4] Secunda conclusio.[98] Velocitas motus[99] non attenditur penes proportionem resistentiarum praecise sumptarum....
>
> [5] Tertia conclusio.[100] Velocitas motuum non attenditur simpliciter penes proportionem quam habent excessus virtutum agentium supra suas resistentias....

[94] The first is MS Coimbra, BU Res. 2414, fol. 145v and following, attributed tentatively to Luis de Cerqueira, S.J., by Friedrich Stegmüller in his preliminary study of these manuscripts, and the second is MS Lisbon, BN FG 6283 (no foliation), containing the lectures of Antonio de Castelbranco, S.J. For details, see Friedrich Stegmüller, *Filosofia e teologia nas Universidades de Coimbra e Evora no Seculo XVI* (Coimbra: University of Coimbra, 1959).

[95] conclusio] assertio

[96] aliquo] altera

[97] motus] motuum

[98] conclusio] assertio

[99] motus] motuum

[100] conclusio] assertio

[6] Quarta assertio. Velocitas motuum ex parte causae non est attendenda penes proportionem proportionum virtutum agentium ad suas resistentias....

[7] Quinta assertio, ex iis duobus regulis, quarum prima est si virtutes motivae habuerint aequales proportiones ad suas resistentias tunc motus erunt aequalis velocitatis, ut si virtus ut sex moveat resistentiam ut quatuor, et virtus ut tria resistentiam ut duo, motus erunt aequalis velocitatis....

[8] Secunda regula est si virtutes motivae non habuerint aequales proportiones ad suas resistentias, tunc a virtute maioris proportionis sumatur pars quae habeat tantam proportionem ad suam resistentiam quantam habet virtus minoris proportionis ad suam. Deinde, qualis fuerit proportio excessus partis sumptae supra suam resistentiam ad excessum quem habet tota virtus supra eandem resistentiam, talis erit proportio velocitatis motus facti a virtute maioris proportionis ad velocitatem motus facti a virtute minoris proportionis. Verbi gratia, si virtus ut octo moveat resistentiam ut duo, et virtus ut sex moveat resistentiam ut quatuor, vellisque intelligere qualis sit proportio velocitatis unius motus ad velocitatem alterius facti[101] a virtute ut octo, quae habet maiorem proportionem summe virtutem[102] ut tria, quae habet[103] eandem proportionem ad resistentiam ut duo quam habet virtus ut sex ad resistentiam ut quatuor, cum autem excessus virtutis ut octo supra virtutem etiam[104] ut duo habeat proportionem sextuplam ad excessum quem virtus ut tria habet supra eandem resistentiam, dices virtutem ut octo movere[105] resistentiam ut duo sextuplo velocius quam virtus ut sex moveat resistentiam ut quatuor.

[9] Id quod ita suadebis ex dictis regulis, nam[106] virtus ut tria per primam regulam movet aequa celeritate resistentiam[107] ut duo, atque virtus ut sex resistentiam[108] ut quatuor; sed virtus ut octo movet resistentiam ut duo sextuplo velocius quam virtus ut tria; eo quod excessus virtutis ut octo supra resistentiam ut duo sit sextuplus ad excessum virtutis ut tria supra eandem resistentiam, ergo movet sextuplo velocius quam virtus ut sex movet resistentiam ut quatuor.

Translation:

[1] From what is the velocity of a motion to be ascertained as from a cause, second question....

[2] First conclusion. The velocity of motion is not to be ascertained from the ratio of one active force to another....

---

101 alterius facti] alterius
102 virtutem] virtute
103 habet] habeat
104 virtutem etiam] resistentiam
105 movere] movetur
106 nam] *om.*
107 resistentiam] resistentia
108 resistentiam] resistentia

[3] Five causes can thus be imagined from which one can take and know that the velocity of motions is greater, or less, or equal. One of these is the ratio of the agent forces,[109] another is the ratio of the resistances,[110] a third is the ratio of the excess of agent forces over their resistances,[111] a fourth is the ratio of the ratios of the agent forces to their resistances,[112] and the last is the ratio of the excess of the agent forces over their resistances, not absolutely but taking into account the quantity of the resistances.[113]

[4] Second conclusion. The velocity of motion is not ascertained from the ratio of the resistances taken alone....

[5] Third conclusion. The velocity of motions is not ascertained simply from the ratio that the excesses of the agent forces have over their resistances....

[6] Fourth assertion. The velocity of motions on the part of the cause is not to be ascertained from the ratios of the ratios of the agent forces to their resistances....

[7] Fifth assertion, from the following two rules: The first is that, if motive forces have equal ratios to their resistances, the motions will be of equal velocity. For example, if a force of six moves a resistance of four, and a force of three moves a resistance of two, the motions will be of equal velocity [i.e., $(6 - 4)/4 = (3 - 2)/2$].

[8] The second rule is that, if the motive forces do not have the same ratios to their resistances, then from the force of higher ratio let a part be taken that has the same ratio to its resistance as the force of lesser ratio has to its. Then, whatever the ratio may be that the excess of the part taken over its resistance to the excess that the entire force has over the same resistance, this will be the ratio of the velocity produced by the force of higher ratio to the velocity produced by the force of lesser ratio. For example, if a force of eight moves a resistance of two, and a force of six moves a resistance of four, and you wish to know the ratio of the velocity of one motion to the velocity of another, from the force of eight, which has a greater ratio, take a force equal to three, which has the same proportion to a resistance of two as six has to a resistance of four. Then, since the excess of a force of eight over a resistance also of two

[109] I.e., $V_2/V_1 = F_2/F_1$, basically Soto's formula given in n. 57.

[110] I.e., $V_2/V_1 = R_1/R_2$, Soto's formula in n. 58.

[111] I.e., $V_2/V_1 = (F_2 - R_2)/(F_1 - R_1)$, Soto's formula in n. 59.

[112] This fourth possibility presents more difficulty because of the different ways it may be written, depending on the meaning given to the first "ratio" in the expression "ratio of ratios," as explained in n. 60. If this is a simple arithmetical ratio, then the meaning is $V_2/V_1 = (F_2/R_2)/(F_1/R_1)$—that is, the ratio of the respective ratios of the forces to the resistances. If "ratio" is to be understood in Bradwardine's sense as a geometrical or exponential ratio, however, then the value of V_2/V_1 must be ascertained from the formula $F_2/R_2 = (F_1/R_1)^{V_2/V_1}$, that is, the exponent to which F_1/R_1 must be raised in order to equal F_2/R_2.

[113] I.e., $V_2/V_1 = [(F_2 - R_2)/R_2]/ [(F_1 - R_1)/R_1]$. Edward Grant notes that this formula represents one of the possibilities rejected by Bradwardine himself, although it was later adopted by Giovanni Marliani (d. 1483); see Edward Grant, ed., *A Source Book in Medieval Science* (Cambridge, Mass.: Harvard University Press, 1974), section entitled "Mathematical Representations of Motion," p. 296 n. 15.

would have a sixfold ratio to the excess that a force of three has over the same resistance, you may say that a force of eight moves a resistance of two six times faster than a force of six would move a resistance of four.[114]

[9] And you can argue for the same result from the said rules, for a force of three from the first rule moves a resistance of two with a speed equal to that with which a force of six moves a resistance of four; but a force of eight moves a resistance of two six times faster than a force of three; because the excess of a force of eight over a resistance of two is six times the excess of a force of three over the same resistance. Therefore, it will move [that resistance] six times faster than a force of six moves a resistance of four.[115]

Note in the above excerpts that the first four conclusions or assertions, pars. [2] and [4–6], restate the "laws" of motion approximately as they are found in Soto. The fifth rule is then explained in par. [7]. Here the author does not state this precisely as it appears in par. [3] but rather as a possibility that follows from two additional rules, which he explains in pars. [7–8] and then confirms in par. [9]. The rules seem promising in that they provide examples that can be put to numerical test. On my first reading, in view of the two different meanings that could be given to the first "ratio" in the expression "ratio of ratios," as explained in my notes 60 and 112 above, I tended to think that the author was attempting to carry out calculations found in Soto such as those I explain in my notes 61, 64, and 66 through 68. Now, however, having worked through his numerical examples, I have come to the conclusion that he did not, and, indeed, that when he used the expression "ratio of ratios" he could not have intended it in Bradwardine's sense. Thus he seems of little help in clarifying the referent of Soto's "fifth" law.

A Fifth Law or Not?

What, then, is to be said about Soto's "fifth" law? Did he have one or not? In my view his "fifth" law is basically his fourth, taking "ratio of ratios" in Bradwardine's sense, as he states in par. [22] of the excerpt from his fourth question on Book 7. But the presence of a fifth law in the two Portuguese manuscripts just discussed suggests a possible explanation of the two different numbers occurring in Soto's text. As Edith Sylla has informed me, in

[114] On the reading of the fifth possibility given in the previous note, let F_2 and R_2 take the values of 8 and 2 respectively, and F_1 and R_1 those of 6 and 4 respectively. Now reduce F_2 by 3, so that the formula will be instantiated as [(8 – 3) – 2]/2, which is equal to 6/4. Then, for the values $F_2 = 8$, $R_2 = 2$ and $F_1 = 3$, $R_1 = 2$, the ratio V_2/V_1 will be equal to [(8 – 2)/2]/[(3 – 2)/2], that is, 6, so that V_2 will be equal to $6V_1$.

[115] From the calculation at the end of par. [7], (3 - 2)/2 = (6 - 4)/4. Similarly, from the calculation at the end of the previous note, (8 - 2)/2 = 6[(3 - 2)/2]. Therefore (8 - 2)/2 = 6[(6 - 4)/4], as stated.

presenting the teaching on "ratios" it was not unusual for commentators to discuss five possible "laws" rather than the four entertained by Soto, with the additional "law" being the fifth just analyzed. It is not unlikely that Soto originally planned to include this in his list as a fourth possibility, leaving his fifth and last for his exposition of Bradwardine's version. Then, seeing that this would only confuse matters, he decided to drop the planned fourth possibility and present the Bradwardinian teaching in its place, making what was originally to be the fifth "law" actually his fourth. In cleaning up his manuscript, however, he was not successful in converting all his "fifth"s to "fourth"s, thus presenting us with the textual problem. There is no manuscript evidence to support this conjecture, of course. Yet the fact that none of the editors was concerned with the problem may indicate that the numbering of the "laws" was not a matter of great moment for them or their readership, and that the terms "fourth" and "fifth" were taken simply to refer to Soto's last "law."

Uniformiter Difformis Once Again

Turning now to the counterparts of Soto's third question on Book 7 in the Portuguese materials, one finds Luis de Cerqueira, whose manuscript has just been discussed (MS Coimbra, BU Res. 2414), treating this question explicitly. What is interesting in his account is a passage in which he makes reference to Soto's distinctive teaching—namely, on motion that is "uniformly difform" with respect to time. He does so in a context in which he first explains the term *proportio* and then addresses the question of how the velocity of a motion can be ascertained from its effect. This text is as follows:

> Commentarii in octo Physicorum libros. Annotationes in septimum librum....
>
> Quid et quotuplex sit proportio. Superiori capite docuit Aristoteles quinam motus cum quibus comparari possint. Modo vero tradit nonnulla documenta quibus intelligamus aequales vel inaequales motus quoad velocitatem, quantusque sit excessus velocitatis unius ad velocitatem alterius inter eos qui non sunt aequae celeritatis. Atque primum quidem in motu locali, tum quia motuum sit primus, tum quia in eo proportio haec magis appareat. Docet ergo in motu locali quatuor reperiri, scilicet, movens, mobile, spatium, atque tempus. Nomine autem moventis virtutem seu potentiam motivam, quaecumque ea sit, quae ad motum afferat adiumentum; nomine autem mobilis intelligit resistentiam, quaecumque ea sit, quae motum retardet....
>
> 'Utrum velocitas motus localis sit attendenda ab effectu penes quantitatem spatii, quaestio prima....

> Interdum enim mobile ita difformiter movetur ex parte temporis ut sumpta parte temporis in quo movetur semper in instanti medio tantum superet velocitatem quam habuit vel habebit sub uno instanti terminativo illius temporis quantum superatur a velocitate quam habuit vel habebit sub altero, hic autem motus dicitur uniformiter difformis ex parte temporis convenitque gravibus et levibus cum naturaliter movetur. Semper enim quo magis recedunt a termino a quo eo velocius moventur. Interdum vero mobile non ita uniformiter movetur diciturque talis motus difformiter difformis ex parte temporis, soletque reperiri in motibus progressivo animalium.

Translation:

> Commentaries on the eight books of the Physics. Annotations to the seventh book....
>
> What is a ratio and how many kinds are there? In the previous chapter Aristotle taught that certain motions can be compared with others. Now he treats several teachings that enable us to understand what motions are equal or unequal with regard to velocity, and, among those of unequal speed, how much the velocity of the one exceeds the velocity of the other. And he does so first for local motion, because this type of motion is primary, and because in it ratios are more apparent. He teaches therefore that four factors are found in local motion, namely, a mover, a mobile, space, and time. By a mover he means a motive force or power, whatever it may be, that would offer assistance to the motion. By a mobile he means a resistance, whatever it may be, that would retard the motion....
>
> Whether the velocity of a local motion is to be ascertained as from an effect according to the quantity of the space traversed, first question....
>
> Sometimes the mobile is moved so difformly with respect to time that, taken [any] part of time in which it moves, the velocity it has at the middle instant will exceed the velocity it had or will have at one terminal instant of that time by the same amount as it is exceeded by the velocity it had or will have at another terminal instant. Such a motion is said to be *uniformiter difformis* with respect to time, and it is found in heavy and light bodies when they move naturally, since the more they depart from their starting point the greater is the velocity with which they move. At other times the mobile is not moved uniformly in this way, and then the motion is said to be *difformiter difformis* with respect to time. This is usually found in the progressive motion of animals.

This text is quite unusual in that portions of it are quite similar to the anonymous commentary on the seventh book mentioned at the outset of the previous section, MS Coimbra, BU Res. 2312, whose author referred his readers to Soto's two questions on that book. Not only does that author make clear that the motion of heavy bodies is "uniformly difform with respect to time," but he goes on to explain precisely what this expression means. He writes:

> A motion that is *uniformiter difformis* with respect to time is one where the mobile is so moved that the first part of the motion achieved, namely, that in the first part of the time, surpasses the second [part] in the same ratio as the second surpasses the third. It surpasses, for example, [in this way]: if the stone descends over three squares it will descend *uniformiter difformiter*, for in the second square it will move as much faster than in the first square [as] in the third faster than in the second. And from these considerations, by way of opposition, we can gather what it is for a motion to be *difformiter difformis*, both with respect to the subject and with respect to time.[116]

This same text is found almost verbatim in Luis de Cerqueira's MS Coimbra, BU Res. 2414, of 1582, and MS Lisbon, BN FG 4921, which contains lectures on the *Physics* by the Trinitarian Fr. Marcus de Moura, composed in 1588. From the example given in this text there can be little doubt that all of these authors correctly understood the concept of uniform acceleration as this was later to be understood, and then experimentally verified, by Galileo.

A final question of interest is whether there is any connection between the ways these Portuguese authors conceived variations in velocity ascertained on the part of the cause with those ascertained on the part of the effect. In other words, how precisely did they conceptualize the physical causality involved in the natural motions of heavy and light bodies? Fortunately, this specific question is answered in a passage from commentaries on Aristotle's *De caelo* that is duplicated in manuscripts composed at both Evora and Coimbra. The Latin of the passage is now given from an anonymous exemplar composed in 1582, MS Lisbon, BN FG 4066 (at fol. 113v), which is similar to that contained in MS Coimbra, BU Res. 2414, attributed to Luis de Cerqueira, and likewise composed in 1582.

> Cur gravia et levia naturaliter mota celerius ad finem motus quam in principio ferantur, questio unica.
>
> Circa primam rationem Aristotelis quam sexto capito proposuit quaeritur hoc loco cur ea quae moventur naturaliter motu recto celerior in fine quam in principio motus moveantur, contra vero velocius in principio motus moventur quae violenter mutantur. Denique cur animalia in medio motus celerius moveantur quam in principio ac fine motus.
>
> Quod igitur attinet ad primum, praetermissis multorum responsionibus dicendum est causam huius rei esse quia quemadmodum vis quae in manu projicientis existit cum lapide coniuncta est movendo atque impellendo medio tali

[116] "Motus uniformiter difformis ex parte temporis ille est quando mobile ita movetur ut prima pars motus confecta, scilicet, in parte prima temporis eadem proportione vincat secundam quam secunda tertiam. Vincit, verbi gratia, si descendat lapis per tria quadrantes descendet uniformiter difformiter, nam in secundo quadrante tanto velocius movebitur quam in primo quadrato, in tertio velocius quam in secundo. Ex his per oppositum possumus colligere quisnam sit motus difformiter difformis, tam ex parte subiecti quam ex parte temporis" (fol. 90v).

> motu imprimit lapidi talem impulsum quo deinde a manu projicientis separatus movetur, sic etiam gravitas et levitas impellendo rem gravem aut levem ad suum locum naturalem per talem motum imprimit ei impulsum quendam quo interveniente talis motus rei gravis et levis celerior efficitur atque eo magis intenditur talis impetus quo gravia et levia ad sua loca naturalia magis accedunt, quod intelligendum est habita ratione eiusdem termini a quo. Nam si unus idemque lapis modo e medio turris descenderet, postea vero a summitate ipsius multo celerius in fine posterioris motus descenderet quam in fine prioris; quo enim longius est spacium quod percurritur eo maior est impetus a gravitate vel levitate per motum impressum cum continue per talem motum intendatur quousque res ad suum locum naturalem perveniat, quoniam vero impulsus hic efficit in re gravi vel levi similem motu ei qui a gravitate vel levitate proficiscitur. Appellatur idcirco ab Aristotele augmentum gravitatis aut levitatis, ab aliis dicitur gravitas aut levitas accidentaria, quoniam cessante motu amittitur.

Translation:

> Why heavy and light bodies when moved naturally are carried more swiftly at the end of the motion than in the beginning, a single question.
>
> Concerning the first argument that Aristotle proposed in the sixth chapter, the inquiry here is why things that are moved naturally in straight-line motion are moved more swiftly at the end of the motion than in the beginning, and, on the contrary, those that are moved violently are moved more swiftly in the beginning. Finally, why is it that animals move more swiftly in the middle of their motion than in the beginning or the end?
>
> With regard to the first, skipping over many replies, it should be said that the reason for this is that, just as the force that exists in the hand of the thrower when it is in contact with the stone and is moving it and impelling the medium impresses on the stone a certain impulse that moves it when separated from the hand of the thrower, so also gravity and levity, impelling a heavy or light body to its natural place, impresses by such motion a certain impulse through whose agency the motion of the heavy or light body is made swifter. And this impetus gets more intense as the heavy and light objects come closer to their natural places, which is to be understood in terms of the relation of each to the *terminus a quo*. For if one and the same stone were now to descend from the middle of a tower and later from its top, it would descend much more swiftly at the end of the later motion than at the end of the earlier. For the longer the space that is traversed, the greater is the impetus impressed by levity and gravity throughout the motion, since it is continually intensified until the body arrives at its natural place. And since this impetus effects in the heavy or light body a motion similar to that which arises in it from gravity and levity, Aristotle referred to it as an increase of gravity and levity; others, however, speak of it as an accidental gravity and levity, since it is lost as soon as the motion stops.

This passage is also contained in a manuscript attributed to Manuel a Lima, S.J., which comes from a course he taught at Evora in 1588, MS Lisbon,

BN FG 2533. It is notable that all three accounts invoke the analogy between *gravitas* and *impetus* pointed out by Soto in his third question on Book 8 of the *Physics*, where he states: "Just as the generator gives the heavy body a natural quality, which is the gravity by which it moves the body all the way to the center, so the projector imparts an impetus to the projectile by which it moves the projectile in a more eminent fashion."[117]

With this I conclude this study of Soto's "laws" of motion and how they were understood in the century in which they were written. Both Soto and the Jesuits who used his work were progressive Aristotelians who did not hesitate to read the concept of impetus into Aristotle's writings. This provided them with a consistent view of causality that enabled them to explain gravitational motion within a Thomistic context, seeing even that type of motion as coming under the influence of the First Unmoved Mover.[118] But as for the motive force that produces such motion, they were wide of the mark—as historians of medieval science would all agree. (The same, of course, could be said of Bradwardine in his more sophisticated attempt in the early fourteenth century.[119]) Thus it would be a mistake to see Soto's "laws" as in a direct line of progress that would lead ultimately to Newton's *Principia*. Soto's more important contribution, in my view, was his *uniformiter difformis* teaching. This continued to be developed by the Jesuits and, as I have argued elsewhere, could well have influenced Galileo in his formulation of the correct law of falling bodies.[120] But that is a different story, one that would take me well beyond the confines of "text and context."

---

[117] Soto, *Quaestiones* (n. 5), 1555-56: 100vb, "Sicut generans grave tribuit illi naturalem qualitatem, quae est gravitas, qua illud permovet usque ad centrum, sic et proiiciens impingat impetum proiecto, quo ipsum eminus moveat."

[118] On this point, see William A. Wallace, "Aquinas and Newton on the Causality of Nature and of God," in *Philosophy and the God of Abraham*, ed. R. J. Long (Toronto: Pontifical Institute of Mediaeval Studies, 1991), pp. 255–79.

[119] Whether, apart from Soto himself, any of the authors examined in this essay actually understood the expression "ratio of ratios" in Bradwardine's sense is still open to question. Alonso de la Vera Cruz surely did not, as I have explained in n. 86 above. Toletus's numerical example, in contradistinction to Vera Cruz's, checks out satisfactorily, but, as already noted, this is concerned only with the doubling and halving of values, and with these one cannot differentiate between arithmetical and geometrical ratios. The only other Jesuit example I have found is that in the two Portuguese manuscripts analyzed above, and the author of these did not subscribe to Bradwardine's teaching. Whether this would be true of other Jesuit readings is impossible to say on the evidence uncovered thus far. But the cryptic way in which Soto's fourth or fifth "law" is printed in editions *C* through *H* could well have militated against his views being correctly understood by anyone who read them.

[120] See Wallace, "Early Jesuits" (n. 3); and William A. Wallace, "Domingo de Soto and the Iberian Roots of Galileo's Science," in *Hispanic Philosophy in the Age of Discovery*, ed. Kevin White (Washington, D.C.: Catholic University of America Press, 1997), pp. 113-29..

ART, NATURE, AND EXPERIMENT AMONG SOME ARISTOTELIAN ALCHEMISTS

William R. Newman

A widely held view among one school of historians is that ancient and medieval science were largely devoid of experiment, and that this remained the case until the Scientific Revolution of the seventeenth century.[1] A number of scholars have gone further than this, not only stating that premodern science failed to employ systematic experimentation, but asserting that the reluctance to experiment derived from the concept that tinkering with nature would introduce an element of artificiality into one's research. According to this line of analysis, "experiment" can be taken to mean any intervention in natural processes that serves as the basis of a claim to knowledge about the natural world.[2] Shmuel Sambursky's well-known work *The Physical World of the Greeks* (1956) already makes the point quite forcefully that such intervention was largely absent from Greek science:

> The Ancient Greek believed fundamentally that the world should be *understood*, but that there is no need to *change* it. This remained the belief of subsequent generations up to the Renaissance. This passive attitude to the practical use of the forces of nature was reinforced by the complete ossification of the natural sciences in the Middle Ages in the condition to which Aristotle had brought them.... Systematic experimentation in the laboratory was first carried out in the seventeenth century. It was, to all intents and purposes, unknown to the ancient Greeks. Thus the riddle of the backwardness of terrestrial physics in comparison with astronomy in antiquity resolves itself into the problem of why there was almost no use of the laboratory experiment in

---

[1] I do not wish to deny, of course, that the quantity and quality of experimental endeavor increased dramatically in the seventeenth century, but to point out that the increase may not have been as marked as many scholars claim, once the evidence provided by alchemy has been taken into account. For the extreme viewpoint, see S. Sambursky, *The Physical World of the Greeks* (New York: MacMillan, 1956). For medievalists who subscribe to modified versions of Sambursky's position, see David Lindberg, *The Beginnings of Western Science* (Chicago: University of Chicago Press, 1992), pp. 52–53; and Edward Grant, *The Foundations of Modern Science in the Middle Ages* (Cambridge: Cambridge University Press, 1996), pp. 159–60.

[2] Needless to say, there are other ways of defining "experiment." My goal in this paper is not to determine the correctness of this or that definition, or even to give a comprehensive picture of experiment in the premodern world, but only to raise some questions about the validity of linking a putative disinterest in experiment to the issue of artificiality. For two attempts to deal with the meaning of "experiment" in an early modern context, see Ernan McMullin, "Medieval and Modern Science: Continuity or Discontinuity," *International Philosophical Quarterly* 4 (1965): 103–29; and Charles B. Schmitt, "Experience and Experiment: A Comparison of Zabarella's View with Galileo's in *De Motu*," *Studies in the Renaissance* 16 (1969): 80–138.

> Greek science. The main characteristic of an experiment is its *artificiality*, which distinguishes it from the observation of a process in its natural form.[3]

Although many medievalists might recoil at Sambursky's stark claim that science in the Middle Ages was a "complete ossification" of Aristotle's views, his notion that experiment was viewed as an artificial encroachment on the natural realm has found recent and influential adherents. Sarah Waterlow Broadie's *Nature, Change, and Agency in Aristotle's Physics* develops Sambursky's point much further, and places it within the context of Aristotle's distinction between artificial and natural substances. As Broadie points out, Aristotle makes this important disjunction in *Physics* II.1.-192b9–19, where he distinguishes natural products from artificial ones on the basis of the fact that the natural have an innate *principle of movement (or change)* [*echonta en heautois archên kinêseôs*], whereas the artificial have *no inherent trend toward change* [*oudemian hormên echei metabolês emphyton*].[4]

According to Broadie, Aristotle's claim that natural substances are only those that have "an inner principle of change or stasis" is fundamental to his philosophical system as a whole: indeed, the details of his cosmology and chemistry "can be traced back to this principle."[5] It is no surprise, then, that Broadie wants to ground what she views as Aristotle's lack of interest in experiment on the distinction between natural and artificial products:

> It is Aristotle's view that substance-typifying changes are as a rule successfully realized in the natural environment. Most of the conditions in which most individual natural substances find themselves are such as to permit their self-characterizing behaviour, or at least most of the time. It is therefore senseless to place a substance under artificial conditions for better observation.... [T]he artificial conditions are more likely than not to obstruct the typifying behaviour, and in that case we learn nothing at all about the substantial nature, since this could only have been revealed through changes other than those which are taking place. Experiment, in short, opens up no new access to the facts, and may succeed only in suppressing them.[6]

By drawing out the implications of Aristotle's distinction between the artificial and the natural, Broadie manages to supply a metaphysical basis to Sambursky's observation that the Greeks disliked experiment because it introduced artificiality into otherwise natural processes. Indeed, in Broadie's view, the avoidance of artificial intervention was a necessary consequence of the Aristotelian conception of natural science. Since Aristotle defined an object's "nature" as the sum of its regularly occurring properties,

[3] Sambursky, *Physical World of the Greeks* (n. 1), pp. 230–33.

[4] Aristotle, *The Physics*, trans. Philip H. Wicksteed and Francis M. Cornford (London: Heinemann, 1929), pp. 106–15.

[5] Sarah Waterlow [Broadie], *Nature, Change, and Agency in Aristotle's* Physics (Oxford: Clarendon Press, 1982), p. 1.

[6] Ibid., *Nature, Change, and Agency*, p. 34.

any attempt to isolate the object from its normal environment could only interfere with its nature. Since experiment relies on precisely such interference, Broadie asserts, it becomes ipso facto useless in Aristotelian natural science.

Surprisingly, Broadie's attractive-sounding claim does not take into account the extensive historical literature on experiment in antiquity. Relying only on Sambursky, she fails to consider the considerable evidence compiled by G. E. R. Lloyd and Heinrich von Staden— or, among the older authors, Albert Heidel and Ludwig Edelstein—that the Greeks, and even Aristotle, did indeed experiment.[7] It is not my purpose to argue here about the Aristotelian corpus as such, however. Instead, I wish to pass on to a discussion of the art/nature divide in medieval and early modern authors.

The opinion that Broadie represents has been independently adopted and expanded by historians of early modern science to include not only the ancient world, but the totality of natural science up until the seventeenth century. Such recent authors as Peter Dear and Antonio Pérez-Ramos explicitly argue along lines similar to Broadie that experiment was largely prohibited by the Aristotelian art/nature distinction until the end of the sixteenth century.[8] Dear claims, however, that the disciplinary category of mixed mathematics, which did not concern itself primarily with "natures" and causes in the terrestrial world, allowed Jesuit writers such as Franciscus Aguilonius and Christoph Scheiner to obviate the problem of artificial intervention. They could engage in "instrumentally contrived experiences"

---

[7] G. E. R. Lloyd, *Methods and Problems in Greek Science* (Cambridge: Cambridge University Press, 1991), pp. 70–99 (reprinted with new introduction from *Proceedings of the Cambridge Philological Society* n.s., 10 [1964]: 50–72); Heinrich von Staden, "Experiment and Experience in Hellenistic Medicine," *University of London Institute of Classical Studies Bulletin* 22 (1975): 178–99; Ludwig Edelstein, "Recent Trends in the Interpretation of Ancient Science," *Journal of the History of Ideas* 13 (1952): 573–604; Albert Heidel, *The Heroic Age of Science: The Conception, Ideals, and Methods of Science among the Ancient Greeks*, Carnegie Institution of Washington, Publication no. 442, (Baltimore: Williams and Wilkins, 1933).

[8] Peter Dear, *Discipline and Experience: The Mathematical Way in the Scientific Revolution* (Chicago: University of Chicago Press, 1995), p. 153: "The art/nature distinction impinged on the use of artificial contrivances in the making of natural knowledge—that is, it compromised the legitimacy of using in natural philosophy the sorts of procedures used by mathematicians"; and see p. 155: "The natural course of a process could [only] be subverted by man-made, artificial causes, becuase art replaced nature's purposes with human purposes. An aqueduct, for example, is not a natural watercourse; it reveals the intention of the human producer, which thwarts that of nature.... *The Aristotelian distinction between art and nature depended on seeing human purposes as separate from natural ones and hence irrelevant to the creation of a true natural philosophy* [my emphasis]." Antonio Pérez-Ramos, "Bacon's Forms and the Maker's Knowledge Tradition," in *The Cambridge Companion to Bacon*, ed., Markku Peltonen (Cambridge: Cambridge University Press, 1996), pp. 99–120; see p. 112: "What is the point, gnoseologically speaking, of making or constructing something in order to gain insight into Nature's mysteries if we posit from the very start that no productions of human technology can remotely equal or even approach the essence and subtlety of natural processes?"

only because their Aristotelianism was fused with an independent mathematical tradition stretching back to antiquity.[9] Pérez-Ramos, on the other hand, argues that Francis Bacon played a pivotal role in the emergence of experiment, by articulating the ideal of scientific truth as "maker's knowledge." Bacon's idea, as expressed by Perez-Ramos, is that the reproduction of a natural effect by implementation of a scientific theory is the touchstone of the theory's truth.[10] Needless to say, this Baconian precept presupposes a reconsideration of the art/nature distinction, since it makes "contrived experience" the very essence of truth-determination in natural science. And Bacon is famous, of course, for his "negation"—or rather reformulation—of the art/nature distinction. As he said in *De augmentis scientiarum*, "the artificial does not differ from the natural in form or essence, but only in the efficient."[11]

It is clear, then, that a broad scholarly conviction exists according to which the Aristotelian distinction between art and nature was inimical to the growth of experiment: only when the art/nature division was bridged in the Scientific Revolution could experiment play a central part in natural science. In the remainder of this paper I wish to raise some doubts about this common thesis by considering evidence drawn from the history of alchemy. The Latin alchemy of the Middle Ages was inveterately Aristotelian in its assumptions about the material world, and often adopted the rigorous disputational style of the scholastics.[12] This continued to be the case in important circles where alchemy was a focal point of discussion even during the sixteenth and seventeenth centuries, despite the success of such overtly anti-Aristotelian movements as Paracelsianism and Helmontianism.[13] As part of their engagement with Aristotle, alchemical authors were in the forefront of those addressing and reformulating the distinction between natural and artificial products.

In the following I shall first give an overview of the medieval alchemical debate about the artificial and the natural, and then focus in on an influential university figure of the early seventeenth century, Daniel Sennert (1572–1637). Sennert is known today primarily for his corpuscular theory of matter, which had a significant impact on figures in the "New Science"

[9] Dear, *Discipline and Experience*, p. 51. See also pp. 151–79.

[10] Pérez-Ramos, "Bacon's Forms" (n. 8), p. 111.

[11] Francis Bacon, *De augmentis scientiarum*, ed. James Spedding et al. (London: Longmans, 1870) 4: 294–95.

[12] For an example and discussion of this fact, see William R. Newman, *The "Summa perfectionis" of pseudo-Geber* (Leiden: Brill, 1991).

[13] For some discussion of alchemical disputation in the seventeenth century, see William R. Newman, *Gehennical Fire: The Lives of George Starkey, an American Alchemist in the Scientific Revolution* (Cambridge: Harvard University Press, 1994), esp. pp. 32–39 and 151–58.

such as Joachim Jungius and Robert Boyle.[14] But Sennert was also a strong defender of the scholastic theory of substantial forms, and is often identified as a peripatetic.[15] As I shall show, Sennert's Aristotelian allegiances did not prevent him from engaging in experiment—in fact he became an unusually vocal spokesman for "contrived experience" in the period directly before Bacon's famous experimental pronouncements of the 1620s.

Medieval Alchemy and the Art/Nature Debate

As I have argued elsewhere, from the thirteenth century onward European alchemy was a focal point of the art/nature debate.[16] This theme acquired the status of a locus classicus in alchemical writings, in reaction to the *De congelatione et conglutinatione* of the Persian philosopher Avicenna (980–1037), a work that was misleadingly attached to Aristotle's *Meteors* by Alfred of Sareshel at the beginning of the thirteenth century.[17] Under the involuntary pseudonym of Aristotle, Avicenna there debunked alchemy, by claiming generally that "art is inferior to nature, and cannot equal it, however much it strive."[18] It is important to realize that this attack on alchemy is thus embedded in a strongly worded ne plus ultra concerning technology as a whole. While Aristotle, at *Physics* II.8.199a16 was willing to concede that art sometimes "completes what nature cannot bring to a finish," Avicenna seemed in the *De congelatione* to deny any role for artificial

[14] Hans Kangro, *Joachim Jungius' Experimente und Gedanken zur Begründung der Chemie als Wissenschaft* (Wiesbaden: Steiner, 1968); Christoph Meinel, "Early Seventeenth-Century Atomism: Theory, Epistemology, and the Insufficiency of Experiment," *Isis* 79 (1988): 68–103; William Newman, "The Alchemical Sources of Robert Boyle's Corpuscular Philosophy," *Annals of Science* 53 (1997): 567–85.

[15] Kurt Lasswitz, *Geschichte der Atomistik* (Hamburg: Leopold Voss, 1890), 1: 436–54; Rembert Ramsauer, *Die Atomistik des Daniel Sennert* (Braunschweig: Vieweg, 1935); W. Subow, "Zur Geschichte des Kampfes zwischen dem Atomismus und dem Aristotelismus im 17. Jahrhundert (Minima naturalia und Mixtio)," in *Sowjetische Beitraege zur Geschichte der Naturwissenschaft*, ed. Gerhard Harig (Berlin: Deutscher Verlag der Wissenschaften, 1960), p. 161–191; Tullio Gregory, "Studi sull'atomismo del Seicento II," *Giornale Critico della Filosofia Italiana* 45 (1966): 44–63.

[16] William Newman, "Technology and Alchemical Debate in the Late Middle Ages," in *Isis* 80 (1989): 423–45. The centrality of alchemy in such discussion even in the seventeenth century is acknowledged by Mary Richard Reif, "Natural Philosophy in Some Early Seventeenth-Century Scholastic Textbooks" (Ph.D. diss., St. Louis University, 1962), p. 238: "One final question briefly touched upon by several authors concerns the possibility of producing a truly natural product by means of human skill. The question is usually posed in this way: "Can art effect certain works of nature?" The specific problem which they almost always have in mind is the transmutation of baser metals through the art of alchemy."

[17] Newman, "Technology" (n. 16), p. 427.

[18] Newman, *The "Summa perfectionis"* (n. 12), p. 49: "ars est debilior quam natura et non consequitur eam quamvis multum laboret."

processes in improving on natural processes.[19] Avicenna's attack, which includes the important phrase "sciant artifices alkimie species metallorum non posse transmutari," came to be known in Latin simply by the phrase *sciant artifices*. The *sciant artifices* was a challenge to alchemy that could not be ignored; as such, it was taken up by the army of subsequent alchemists, and rebutted in many an alchemical *Theorica*.

Of the many strategies that were available to alchemists who wished to argue against the *sciant artifices*, two are particularly relevant to our discussion. One, less common than the other, was to deny any necessary difference in essence between natural and artificial products. This position was adopted by an early *Book of Hermes* already circulating in the thirteenth century, though possibly older. The *Book of Hermes* had the following to say:

> human works are variously the same as natural ones, as we shall show in fire, air, water, earth, minerals, trees, and animals. For the fire of natural lightning and the fire thrown forth by a stone is the same fire. The natural ambient air and the artificial air produced by boiling are both air. The natural earth beneath our feet and the artificial earth produced by letting water sit are both earth. Green salt, vitriol, tutia, and sal ammoniac are both artificial and natural. But the artificial are *even better than the natural* [my emphasis], which anyone who knows about minerals does not contradict. The natural wild tree and the artificially grafted one are both trees. Natural bees and artificial bees generated from a decomposing bull are both bees. Nor does art make all these things; rather it helps nature to make them. Therefore the assistance of this art does not alter the natures of things. Hence the works of man can be both natural with regard to essence [*secundum essentiam*] and artificial with regard to mode of production [*secundum artificium*].[20]

The point of the *Book of Hermes*, obviously, is that human industry is capable of "making" such natural products as the four elements, plants, animals, and minerals. But the manufacture of such natural products by artificial means is possible only because nature can be induced by art to produce them. In other words, art is aiding nature, as Aristotle said at *Physics* II 8 199a 16, to perfect itself. But Hermes adds something unusual when he

---

[19] Aristotle, *Physica*, tr. R. P. Hardie and R. K. Gaye, in W. D. Ross, *The Works of Aristotle* (Oxford: Oxford University Press, 1966), II.8.199a16.

[20] Newman, *The "Summa perfectionis"* (n. 12), pp. 11–12. The text is sufficiently important to justify giving the Latin here (pp. 53–54): "Imo opera humana cum naturalibus multimode eadem sunt, ut in igne et aere, aqua, terra, mineris, arboribus, et bestiis ostendemus. Nam et ignis fulgoris naturalis et ignis de lapide eiectus uterque ignis est. Aer continens naturalis est et aer ex decoctione artificialis uterque aer. Terra substituens naturalis et terra ex reservatione aque artificialis utraque terra est. Sal vero viride et dragantum et thutia et sal armoniacus et naturalia et artificialia sunt. Immo et artificialia naturalibus potiora sunt, quod qui de mineriis sciunt non contradicunt. Arbor spontanea naturalis et arbor insita artificialis utraque arbor est. Apes naturales et apes ex tauro artificiales utreque apes sunt. Nec ars hec omnia facit, sed naturam facientem adiuvat. Auxilium igitur huius artis naturas rerum non permutat. Humana ergo opera et naturalia secundum essentiam et artificialia secundum artificium eadem esse possunt."

states that the same works can be natural with regard to essence and yet artificial with regard to mode of production (*naturalia secundum essentiam et artificialia secundum artificium*). This is very close, if not identical, to Francis Bacon's claim that artificial and natural products differ "not in form or essence, but only in the efficient." The distinction between natural and artificial production is reduced to one of agency, at least in certain instances.[21]

A more common approach to the art/nature debate among alchemists was to rely explicitly on *Physics* II.8.199a16, and simply to say that alchemy served as the *ministra naturae* while other, less-perfect arts did not. This is the approach taken, for example, by the early-fourteenth-century writer Petrus Bonus, where he argues that alchemy differs from building a house in that the latter art employs accidental "artificial forms," whereas genuine alchemy acts on the natural forms resident in bodies. In the case of alchemy, then, art "only administers the natural matter in artificial vessels," thus producing a natural product.[22] In building a house or a boat, on the other hand, art is not aiding nature but imposing an artificial form, which is necessarily superficial.

Now whether alchemists asserted that the difference between the artificial and the natural was merely a matter of agency or argued that alchemy, unlike "superficial" arts such as house-building, prodded nature to transform itself, it is clear that they were de facto employing a host of technological processes to operate upon natural ingredients while claiming that the products that they manufactured were entirely natural. Processes such as the subjection of metals to intense heat until they oxidize (calcination), separation of volatile minerals by means of sublimation and distillation, and the smelting of ores by means of reducing agents, were all routine features of the alchemical laboratory.[23] Yet by using the arguments adduced above, it was possible to view these radical techniques as leading to fully natural products.[24] It should not surprise us, then, to learn that alchemical authors also claimed that their artisanal techniques could be employed to derive knowledge from nature by means of experiment.

As I have argued elsewhere, one of the most popular alchemical texts of the Middle Ages, the *Summa perfectionis* attributed incorrectly to an Arabic

---

[21] I do not mean to suggest that the author of the *Book of Hermes* would have accepted that *all* artificial products can be viewed as natural. It is unlikely that Bacon himself would have accepted this view, however, for it would have forced him to the conclusion that unaided nature can produce such things as clocks and printing-presses. Like "Hermes," Bacon acknowledges that there *is* a distinction between natural and artificial products—namely, that their efficient causes differ.

[22] Petrus Bonus, *Margarita pretiosa*, in J. J. Manget, *Bibliotheca chemica curiosa* (Geneva: Chouet, G. de Tournes, Cramer, Perachon, Ritter, and S. de Tournes, 1702) 2: 59.

[23] Newman, *The "Summa perfectionis"* (n. 12), pp. 679–719.

[24] Ibid., pp. 1–56.

"Geber," makes use of a battery of metallurgical tests to determine the natures of the metals. These texts include, above all, cupellation and cementation, ancient assaying tests for determining the content of precious metal in an alloy. Cupellation worked by subjecting the fused alloy to oxidation and absorption of the oxides in a porous cupel containing added lead; cementation was employed to separate gold from silver by attacking the silver with corrosives, such as salt, in a heated crucible.[25] The *Summa* also prescribes firing or "ignition" to see if a metal incandesces before fusion, inspection of the fused metal after cooling to see if it has blackened, exposure to acid vapors with subsequent inspection of any efflorescence, extinction of the hot metal in salty or aluminous water or sulfur to produce color changes, burning with sulfur again to induce color changes or alterations of weight, the repetition of calcination and reduction for changes in color, weight, or volume, and the easy or difficult amalgamation with quicksilver.[26]

Although the *Summa* uses these tests in a perfectly ordinary fashion to determine whether a given sample of metal is precious or base, it extends their use to a determination of the "principles" of the metals. Like most medieval alchemists, "Geber" thinks that the metals are composed of sulfur and mercury. But he points out that the metals differ from one another because of a number of factors relating to their principles: they can contain varying amounts of mercury and sulfur, the mercury and sulfur can be mixed with varying amounts of impure "earth," a given principle can vary in color, the mercury or sulfur can be volatile or fixed, and the principles can have different melting points.[27] By subjecting tin to intense calcination, for example, one can determine the nature of its sulfur. The fact that tin repeatedly calcined gives off no sulfurous smell after the first calcination indicates that its volatile sulfurous principle has sublimed—its ability to withstand further violent heating is explained by the presence of a residual fixed sulfur.[28] Similarly, the fact that gold amalgamates more easily with mercury than do any of the other metals shows that it contains more mercury than they do.[29] Throughout his work, "Geber" employs such altered assaying tests as experimental tools for determining the fundamental natures of the metals. He is no more impeded by the art/nature division in this use of "contrived experience" than he is in his view that art can intervene in natural processes to yield an improved, yet natural, product.

[25] Robert Halleux, "Méthodes d'essai et d'affinage des alliages aurifères dans l'Antiquité et au Moyen Age," *Cahiers Ernest Babelon* 2 (1985): 39–77.

[26] Newman, *The "Summa perfectionis"* (n. 12), pp. 769–83.

[27] Geber believes that mercury, by being heated with salt, loses its humidity and becomes a solid. In modern terms the product of this would be "corrosive sublimate," mercury II chloride.

[28] Newman, *The "Summa perfectionis,"* p. 733.

[29] Ibid., pp. 721–26.

DANIEL SENNERT ON ART, NATURE, AND EXPERIMENT

Let us now pass to the early seventeenth century, in order to see how the rich heritage of late medieval alchemy fed into a Europe on the threshold of the scientific revolution. The subject of our examination will be Daniel Sennert, the well-known medical professor of the University of Wittenberg, whose *De chymicorum cum Aristotelicis et Galenicis consensu et dissensu* was published in 1619. As I have shown elsewhere, Sennert owed a direct and major debt to medieval alchemy. He had already adopted important elements of his influential corpuscular theory from "Geber's" *Summa* by 1611 at the latest.[30] Sennert was also a convinced believer in the reality of the philosophers' stone: the 1629 edition of his *De chymicorum* still maintains that metallic transmutation is an incontestable truth.[31] Let us therefore consider his attitude toward the art/nature distinction, and its implications for his use of experiment.

Sennert's *De chymicorum* of 1619 does in fact treat the distinction between artificial and natural products at some length. The pretext for doing so is supplied by opponents of chymistry, such as Thomas Erastus (1523–1583),[32] who deny that natural substances are really composed of the three Paracelsian principles, mercury, sulfur, and salt. After discussing a number of arguments proffered by Erastus and others, Sennert comes to the point that is of concern to us. Some authors, he states, deny that the "chymical resolutions" normally used to evince the existence of the three principles, such as the operations of sublimation and distillation, can ever reveal the true principles of things, because they are "not natural, but artificial."[33] Such opponents continue by saying that whatever nature has mixed, art cannot dissolve, so that the *tria prima* must be artifacts rather than genuine components of things. To this Sennert responds as follows:

> But if you consider the proximal agent of mixture, it must be denied that chymical resolutions are not natural, even if an artisan participates in his own fashion. For they are brought about by fire and heat, by means of a natural cause.[34]

---

[30] William Newman, "Alchemical Sources" (n. 14), pp. 573–76.

[31] Meinel, "Early Seventeenth-Century Atomism" (n. 14), pp. 95–96. As Meinel points out, some posthumous editions of *De chymicorum* reveal more reserve on this point. The whole matter needs to be examined more thoroughly, however, since the editors may have taken liberties with Sennert's work after his death.

[32] Sennert is responding to Thomas Erastus, *Disputationes de medicina nova Paracelsi* (Basel: Petrus Perna, 1572–73).

[33] Sennert, *De chymicorum*, (Wittenberg: 1619), p. 287.

[34] Ibid., p. 288: "Verumtamenvero, si proximum agens respicias, negandum, quod resolutiones Chymicae non sunt naturales; etiamsi artifex suo modo concurrat. Fiunt enim ab igne & calore, causa naturali."

Here Sennert develops the important point that one could always argue that the immediate causes of a given transformation were natural, even if they were initiated by an artificer. Thus when a plant is burnt up to obtain its salt, the operation is entirely natural, being due to the activity of fire. When spirit of wine is distilled in order to make a cordial, it is the same situation as occurs in the human body when vapors rise from the liver and are drawn into the brain, or when something is evaporated by the heat of the sun. Since the immediate agent, heat, is natural, "the things which are produced in this fashion are natural, not artificial."[35]

Sennert then addresses a related issue, that chymical resolutions could be considered "violent," and hence opposed to nature. He is willing to admit that chymical operations can be violent in the sense that they do not conform to the ordinary course of nature, for it is not part of the ordinary life-cycle of a plant to be calcined into a salt. But it does not follow from this, Sennert continues, that chymical operations are "merely artificial," and their products therefore artifacts, for who would deny the fire produced from the burning wood of the plant to be natural? Indeed,

> these actions are artificial insofar as the matter to be treated is exposed to a definite degree of fire in an artificial vessel. But the action of the heat on the matter is obviously natural. Thus the operation that is effected is natural with respect to the efficient and the matter.[36]

Sennert then draws on traditional alchemical arguments to assert that if the mere presence of "human industry and labor" implied artificiality, then grafted trees and chicks raised with the aid of incubation would necessarily be artificial. His argument reveals the ease with which one could obviate the strictures imposed by the art/nature distinction: one had only to focus on the proximal efficient cause in order to claim a natural pedigree for the product of his art.

The same glib facility is employed when Sennert addresses the Aristotelian claim that only natural products have an innate principle of change. Quoting Aristotle's famous dictum from *Physics*.II.1—*hoson estin apo technês, oudemian hormên echei metabolês emphyton*—he breezily responds that "the things that are made by means of chymistry all have this innate principle, and the heat is only an external agent that leads this *emphyton hormên* into act."[37] Clearly Sennert did not feel imprisoned by the Aristotelian claim that natural and artificial products differ. In order to avoid the conclusion that chymical products are mere artifacts, he had only

---

[35] Ibid., p. 288.

[36] Ibid., p. 289: "Artificiales sunt istae actiones, quatenus in artificialibus vasis materia elaboranda certo ignis gradui objicitur. At actio Caloris in materiam plane naturalis est. Et proinde opus, quod producitur, respectu efficientis & materiae naturale est."

[37] Ibid., p. 290: "At quae a Chymia fiunt, omnia istum internum impetum habent; calorque saltem externum agens est, qui illam emphyton hormên ad actum deducit."

to claim that they too displayed the signal mark of the natural product—an innate principle of change. And who indeed could dispute his claim? If any ordinary mineral possesses such a principle, then what empirical evidence could one adduce to show that a mineral "produced" by chymistry does not?

Having exhausted this topic in such perfunctory fashion, Sennert then passes to a new type of argument. The three principles are not only resolved by art, but actually appear without the aid of chymistry. Animal excrements obviously contain sulfur, as one can tell from their stench. The resinous sap of trees is clearly sulfurous. Salt, on the other hand, is "naturally rendered from the exhausted piles of cinders and calces around niter factories."[38] Similarly, a copious salt is naturally deposited on the side of wine-barrels, which coalesces into tartar. Nor should spirit of wine—whose inflammability is due to its high sulfur content—be considered artificial.[39] It is obviously present in the wine that inebriates drunkards, and is sometimes even separated naturally by the intense cold of winter. All of these arguments expose the extremely tenuous boundary that delimited the artificial from the natural, and underscore the ease with which one could pass from one to the other.

After having rejected the argument that the chymical principles are mere artifacts, Sennert responds to another objection—that there are no genuine principles more proximal than the four elements. This discussion will lead him from a defense of the natural status of chymical products to an explicit justification of experiment, expressed in surprisingly general terms. He argues, first, that upholders of this argument should go and smell a rotting corpse—they will soon abandon their view that decomposition arrives directly at the elements without some intervening product. If this fails to convince them, they should examine the sublimed product on the walls of an alembic: it obviously a "mixt," and not a pure element. The same is true of the cinder produced by miners in the process of refining ores: since the cinders are only partly soluble, they must be a mixture of salt and earth.

Having already disposed of the potential accusation that he is describing "merely artificial" processes, Sennert is now able to pass from a discussion of art and nature to a defense of "contrived experience" as such:

> One must not, therefore, make such superficial and incidental judgments about the works of nature. Rather they must be inspected a bit more deeply. And what is not presented to us casually must be sought from art and labor [*industria*]. In judging the unrestricted [*externis*] resolutions of nature, many impediments arise: the resolved parts do not present themselves for examina-

---

[38] Ibid., p. 291: "Sal naturaliter redditur e cinerum & calcium exhaustis cumulis, circa officinas nitrarias."

[39] Ibid. For the sulfurous character of spirit of wine, see p. 297: "Sulphur enim suum, ob quod inflammatur, habet Spiritus vini, pinguis tamen non est."

> tion, so that it might be known whether the elements are pure or of another type; instead they are spread out in vapors and dissipate. But there is greater certitude in the works of art, where nothing is lost, but all the materials are treated in sealed vessels, and the resolved parts are collected within, and the heterogeneities are separated from the homogeneities, so that a correct judgment may be made of all.[40]

Sennert's argument here is a remarkably clear justification for the isolation of an experimental subject from its natural environment. In order to determine whether there are intermediate principles between the elements and composed bodies, it is necessary to perform controlled resolutions in laboratory apparatus; otherwise, some of the fumes and vapors produced during sublimation and combustion will pass off into the atmosphere, making a genuine test impossible. It would be hard to overstate the degree to which Sennert's argument conflicts with the picture of Aristotelian science given by Dear and Pérez-Ramos, not to mention Broadie and Sambursky. Despite the fact that he accepts Aristotle's definition of the artificial as that which lacks an innate principle of change, Sennert openly turns to artificial instruments and techniques in order to isolate the natural components whose genuineness is in question. He is quite happy to intervene in natural processes in order to arrive at a knowledge of nature.

CONCLUSION

I began this paper by recapitulating the common view that Greek science, in particular Aristotelian science, was indifferent—even inimical—to experiment. I then described how Sarah Broadie, among others, has argued that this aversion to experiment was a necessary consequence, in Aristotle's philosophy, of his view that natural substances have an intrinsic principle of change. Still others, such as Peter Dear and Antonio Pérez-Ramos, have extended this principle to cover the peripatetic philosophy of the schools up until their reform either by Jesuit mixed-mathematicians or by followers of Francis Bacon in the seventeenth century.

Yet none of these scholars has taken into sufficient account the rich experimental literature provided by medieval and early modern alchemy. Their mistake is a serious one, for alchemy provides a powerful counterinstance to their speculations. Despite the fact that alchemy was permeated by

[40] Ibid., p. 292: "Non ergo ita obiter & superficialiter de naturae operibus judicandum. Sed paulo penitius ea introspicienda sunt. Et quod nobis fortuito non objicitur, Arte & industria inquirendum. Nimirum in naturae externis resolutionibus judicandis, multa occurrunt impedimenta: Neque partes resolutae ad examen sese sistunt, ut cognosci possint, An elementa pura sint, vel alterius generis, sed in auras diffunduntur & aufugiunt. Major autem certitudo est in operibus artis, ubi nihil perit, sed vasis clausis omnia administrantur, partesque resolutae inclusae colliguntur & heterogenea ab homogeneis discernuntur, ut rectum de omnibus possit fieri judicium."

an Aristotelian vision of the cosmos, alchemical writers managed to interpret Aristotle's (and Avicenna's) comments about art and nature in such a way that "contrived experience" presented no problem for them. To the contrary, we find "Geber" systematically employing experimental tests in the High Middle Ages to arrive at the genuine natures of the metals: and in Daniel Sennert's work, we see a "peripatetic" making a remarkably clear justification for the isolation of an experimental subject from its natural environment.

These examples, to put it baldly, throw the enterprise of assuming an aprioristic Aristotelian injunction against experiment into serious disarray. Although there can be no doubt that the seventeenth century witnessed a massive increase in the volume and quality of experimentation in the natural sciences, it seems clear now that we cannot assume that this was due to a sudden abandonment of the Aristotelian precept that artificial and natural things are in certain respects different.

INDEX CODICUM

GENERAL INDEX

In the following index, well-known medieval thinkers are usually listed by their second names, as Bacon, Bradwardine, Buridan, etc., except when the name includes a preposition, as Arnald of Villanova. The names of kings and emperors are anglicized, but other names may be in Latin or the vernacular. Arabic names are, for the most part, given the spelling they have in the articles—limitations of our word processing program prevented the consistent use of diacritical marks. With a few exceptions—for instance for the manuscripts listed in the separate Index Codicum—we have not indexed the footnotes. For concepts that occur very frequently, we have not listed every occurrence.

www.ingramcontent.com/pod-product-compliance
Lightning Source LLC
LaVergne TN
LVHW010631300926
849162LV00001B/122
* 9 7 8 9 0 0 4 1 0 8 2 3 3 *